ADVANCES IN PID CONTROL

Edited by **Valery D. Yurkevich**

Advances in PID Control
Edited by Valery D. Yurkevich

Published by InTech
Janeza Trdine 9, 51000 Rijeka, Croatia

Publishing Process Manager Silvia Vlase

Technical Editor Teodora Smiljanic

Cover Designer InTech Design Team

Additional hard copies can be obtained from orders@intechopen.com

Advances in PID Control, Edited by Valery D. Yurkevich
p. cm.
ISBN 978-953-307-267-8

For EU product safety concerns:
IN TECH d.o.o., Prolaz Marije Krucifikse Kozulić 3, 51000 Rijeka, Croatia,
info@intechopen.com or visit our website at intechopen.com.

Meet the editor

Valery D. Yurkevich received the Dipl.Eng. (1974), Ph.D. (1986) and D.Sc. (1997) from Novosibirsk State Technical University, where he is currently a professor in the Automation Department and teaches courses on control principles for undergraduates, and courses on control methods under conditions of incomplete information at the graduate level. His research interests are in nonlinear control systems, digital control systems, flight control, distributed parameter control systems, robotics, switching controllers for power converters, singular perturbations in control. He partakes in collaborative international research programs and was a visiting professor at the Silesian Technical University (Poland), University of Twente (The Netherlands), Concordia University (Canada), University of Ulsan (Korea), Harbin University of Science and Technology (China), and National University of Singapore. He has produced about 200 papers and international conference presentations and holds four patents. He is member of IEEE Control Systems Society and has wide-ranging experience as a reviewer of international journals and conferences.

Contents

Preface

Since the foundation and up to the current state-of-the-art in control engineering, the problems of PID control steadily attract great attention of numerous researchers and remain inexhaustible source of new ideas for process of control system design and industrial applications. PID control effectiveness is usually caused by the nature of dynamical processes, conditioned that the majority of the industrial dynamical processes are well described by simple dynamic model of the first or second order. The efficacy of PID controllers vastly falls in case of complicated dynamics, nonlinearities, and varying parameters of the plant. This gives a pulse to further researches in the field of PID control. Consequently, the problems of advanced PID control system design methodologies, rules of adaptive PID control, self-tuning procedures, and particularly robustness and transient performance for nonlinear systems, still remain as the areas of the lively interests for many scientists and researchers at the present time. The recent research results presented in this book provide new ideas for improved performance of PID control applications.

The brief outline of the book "Advances in PID Control" is as follows.

In Chapter 1 the predictive control methods for non-minimum phase systems are considered. In particular the classical approach is discussed where Smith predictor and internal model control structures are used to derive the predictive PID control constants. Then a modern approach to predictive PID control is treated and a generalized predictive control algorithm is considered where the model predictive controller is reduced to the same structure as a PID controller for second-order systems.

In Chapter 2 an adaptive PID control system design approach based on the almost strictly positive real (ASPR) property for linear continuous-time systems is presented. It has been shown that the presented approach guarantees the asymptotic stability of the resulting PID control system. In order to overcome the difficulties caused by absence of ASPR conditions, a robust parallel feedforward compensator (PFC) design method is proposed, which render the resulting augmented system with the PFC in parallel ASPR system. As an example, the proposed method is applied to an unsaturated highly accelerated stress test system.

In Chapter 3 the authors discuss sufficient conditions for global asymptotic stability of a class of nonlinear PID type controllers for rigid robot manipulators. By using a passivity approach, the asymptotic stability analysis based on the energy shaping methodology is presented for the systems composed by the feedback interconnection of a state strictly passive system with a passive system. Simulation results are included in the chapter and demonstrate that the proposed class of nonlinear PID type controllers for rigid robot manipulators have good precision and also possess better robustness. The performance of the proposed nonlinear PID type controllers has been verified on a two degree of freedom direct drive robot arm.

In Chapter 4 some class of nonlinear second order systems is considered. The proposed controller is a version of the classical PID controller, where an extra feedback signal and integral term are added. The authors show based on simulation results for simple pendulum and 2 DOF planar robot, that the proposed PI2D controller yields better performance and convergence properties than the classical PID controller. The stability analysis is provided via Lyapunov function method and conditions for gain tuning are presented, which guarantee asymptotic convergence of the closed loop system.

Chapter 5 is devoted to the comparison between the conventional PI controller tuned according to Zhuang-Atherton rules with other PI-like controllers such as PI controller with variable integral component, an adaptive PI controller, and a fuzzy adaptive PI controller. The comparison and conclusions concern the control performance are made by authors based on simulation results including simulations for a 3 DOF model of a low-speed marine vessel.

In Chapter 6 an extension to the traditional PID controller for mechanical system has been presented that incorporates an adaptive gain. The asymptotic stability of the closed-loop system is analyzed based on Lyapunov function method. The tuning rules for controller gains are derived.

In Chapter 7 an approach to continuous as well as digital PI/PID control system design via singular perturbation technique is discussed that allows to guarantee the desired output transient performances in the presence of nonlinear plant parameter variations and unknown external disturbances. The tuning rules for controller parameters are derived. Numerical examples with simulation results are included in the chapter to demonstrate the efficacy of the proposed approach.

In Chapter 8 a new PID control method is proposed that includes a nonlinear compensator. The algorithm of the nonlinear compensator is based on sliding mode control with chattering compensation. The effect of the proposed control method is evaluated for single-axis slide systems experimentally.

In Chapter 9 robust and intelligent multi-objective approaches are discussed for tuning of PID controllers to improve the performance of the closed-loop systems where the introduced tuning strategies are based on mixed H2/H-infty, multi-objective genetic algorithm, fuzzy logic, and particle swarm optimization techniques.

In Chapter 10 the digitally controlled switch mode power supply is investigated based on frequency domain approach where, in order to provide the desired frequency characteristic, pole-zero-cancellation technique is used. The proposed control technique is examined by using buck converter as a simple example.

In Chapter 11 the room temperature and humidity control systems with the conventional PID control using fixed reset and the modified control using adjustable resets which compensate for thermal loads upset are examined. The simulation results for one-day operation are presented.

In Chapter 12 a tele-tuning architecture is described which is based on the interconnection of the industrial plant, the server, and client. Identification tests were performed to validate the proposed architecture by means of simulation of the first-order-plus-dead-time systems using local and remote identification in a corporate network.

In Chapter 13 a PID application in real-time locating system is described. It has been shown that the proposed P-control and PID control algorithms require less calculation and show robust performance in compare with the conventional direct calculation method. The presented results can be used in embedded locating systems, home networking systems and robotics positioning systems.

Chapter 14 is devoted to PID control implementation using field programmable gate array technology. Experimental results for the second order system with P, PI, PD, and PID controllers are presented.

This book is intended for researchers and engineers interested in PID control systems. Graduate and undergraduate students in the area of control engineering can find in the book new ideas for further research on PID control techniques. The editor would like to thank all the authors for their contributions in the book. Finally, gratitude should be expressed also to the team at InTech for the initiative and help in publishing this book.

Prof. Valery D. Yurkevich
Novosibirsk State Technical University,
Russia

Part 1

Advanced PID Control Techniques

1

Predictive PID Control of Non-Minimum Phase Systems

Kenny Uren and George van Schoor
North-West University, Potchefstroom Campus
South Africa

1. Introduction

Control engineers have been aware of non-minimum phase systems showing either undershoot or time-delay characteristics for some considerable time (Linoya & Altpeter, 1962; Mita & Yoshida, 1981; Vidyasagar, 1986; Waller & Nygardas, 1975). A number of researchers that addressed this problem from a predictive control point of view mainly followed one of two approaches: a classical (non-optimal) predictive approach or a modern optimisation based predictive approach (Johnson & Moradi, 2005). The common characteristic of all these approaches is that they are model-based. Predictive control allows the controller to predict future changes in the output signal and to use this prediction to generate a desirable control variable. The classical predictive controllers that are most widely considered include the Smith predictor structure and the internal model control (IMC) structure (Katebi & Moradi, 2001; Morari & Zafiriou, 1989; Tan et al., 2001). Modern predictive controllers consider generalised predictive control (GPC) or model-based predictive control (MPC) structures (Johnson & Moradi, 2005; Miller et al., 1999; Moradi et al., 2001; Sato, 2010).

The performance of a PID controller degrades for plants exhibiting non-minimum phase characteristics. In order for a PID controller to deal with non-minimum phase behaviour, some kind of predictive control is required (Hägglund, 1992). Normally the derivative component of the PID controller can be considered as a predictive mechanism, however this kind of prediction is not appropriate when addressing non-minimum phase systems. In such a case the PI control part is retained and the prediction is performed by an internal simulation of plant inside the controller.

This chapter starts with a quick review of the system-theoretic concept of a pole and zero and then draws the relationship to non-minimum phase behaviour. The relationship between the undershoot response and time-delay response will be discussed using Padé approximations. Classical and modern predictive PID control approaches are considered with accompanying examples. The main contribution of the chapter is to illustrate the context and categories of predictive PID control strategies applied to non-minimum phase systems by:

- Considering the history of predictive PID control;
- The use of models in predictive control design;
- Exploring recent advances in predictive PID control where GPC (Generalised Predictive Control) algorithms play a prominent role;

- Appreciating the control improvements achieved using predictive strategies.

2. The influence of poles and zeros on system dynamics

When considering the compensation of systems it is of great importance to first understand the system-theoretic concept of a system pole and zero in the realm of system dynamics and control theory. Consider a continuous-time single-input, single-output (SISO) system

$$\dot{\mathbf{X}}(t) = \mathbf{A}\mathbf{X}(t) + \mathbf{B}u(t), \tag{1}$$

$$y(t) = \mathbf{C}\mathbf{X}(t) + Du(t), \tag{2}$$

where $u(t)$ and $y(t)$ are the scalar-valued input and output respectively. The column vector $\mathbf{X}(t)$ is called the state of the system and comprises n elements for an nth-order system. The $n \times n$ matrix $\mathbf{A}$ is called the system matrix and represents the dynamics of the system. The $n \times 1$ column vector $\mathbf{B}$ represents the effect of the actuator and the $1 \times n$ row vector $\mathbf{C}$ represents the response of the sensor. D is a scalar value called the direct transmission term. If $D = 0$, it is assumed that the input $u(t)$ cannot affect the output $y(t)$ directly.
If $\mathbf{X}(0) = \mathbf{0}$ and $D = 0$ (in the case where the output is not directly influenced by the input), then the system transfer function $G(s)$ is given by

$$G(s) = \frac{Y(s)}{U(s)} = \mathbf{C}(s\mathbf{I} - \mathbf{A})^{-1}\mathbf{B}. \tag{3}$$

The poles and zeros can be determined by writing $G(s)$ as

$$G(s) = \frac{N(s)}{D(s)}, \tag{4}$$

where the numerator polynomial is

$$N(s) \triangleq \det \begin{bmatrix} s\mathbf{I} - \mathbf{A} & -\mathbf{B} \\ \mathbf{C} & 0 \end{bmatrix}, \tag{5}$$

and the denominator polynomial is

$$D(s) \triangleq \det(s\mathbf{I} - \mathbf{A}). \tag{6}$$

Then the roots of $N(s)$ and $D(s)$ are defined as the *zeros* and *poles* of $G(s)$ respectively (Franklin et al., 2010; Hag & Bernstein, 2007). This holds only in the case where $N(s)$ and $D(s)$ do not have common roots.
The poles of $G(s)$ can be used to determine damping and natural frequencies of the system, as well as determining if the system is stable or unstable. As can be seen from Eq. (6) the poles depend only on the system matrix $\mathbf{A}$, but the zeros depend on matrices $\mathbf{A}$, $\mathbf{B}$ and $\mathbf{C}$. This leads to the question as to how the zeros influence the dynamic response of a system?
Consider a normalised transfer function of a system with two complex poles and one zero (Franklin et al., 2010):

$$T(s) = \frac{(s/a\zeta\omega_n) + 1}{s^2/\omega_n^2 + 2\zeta(s/\omega_n) + 1}. \tag{7}$$

The zero is therefore located at $s = -a\zeta\omega_n$. By replacing the s/ω_n with s results in a frequency normalising effect and also a time normalising effect in the corresponding step response. Therefore the normalised version of Eq.(7) can be rewritten as

$$T_n(s) = \frac{s/a\zeta + 1}{s^2 + 2\zeta s + 1}. \tag{8}$$

The normalised transfer function can be written as the sum of two terms

$$T_n(s) = T_1(s) + T_2(s), \tag{9}$$

$$= \frac{1}{s^2 + 2\zeta s + 1} + \frac{1}{a\zeta}\frac{s}{s^2 + 2\zeta s + 1}, \tag{10}$$

where $T_1(s)$ can be viewed as the original term with no added zeros, and $T_2(s)$ is introduced by the zero. Since the Laplace transform of a derivative dy/dt is $sY(s)$, the step response of $T_n(s)$ can be written as

$$y_n(t) = y_1(t) + y_2(t) = y_1(t) + \frac{1}{a\zeta}\dot{y}_1(t) \tag{11}$$

where y_1 and y_2 are the step responses of $T_1(s)$ and $T_2(s)$ respectively. The step responses for the case when $a > 0$ (introduction of a left half plane zero, $a = 1.1, \zeta = 0.5$) are plotted in Fig. 1(a). The derivative term y_2 introduced by the zero lifts up the total response of $T_n(s)$ to produce increased overshoot. The step responses for the case when $a < 0$ (introduction of a right half plane zero, $a = -1.1, \zeta = 0.5$) are plotted in Fig. 1(b). In this case the right half plane zero, also called a non-minimum phase zero causes the response of $T_n(s)$ to produce an initial undershoot.

In general a substantial amount of literature discusses the dynamic effects of poles, but less is available on the dynamic effects of zeros.

3. A closer look at non-minimum phase zeros

Before a formal definition of non-minimum phase zeros can be given, some definitions and assumptions are given. In this chapter only proper transfer functions will be considered. Eq. (4) may be expanded so that

$$G(s) = \frac{N(s)}{D(s)} = \frac{b_m s^m + b_{m-1}s^{m-1} + \cdots + b_1 s + b_0}{s^n + a_{n-1}s^{n-1} + \cdots + a_1 s + a_0}. \tag{12}$$

$G(s)$ is *strictly proper* if the order of the polynomial $D(s)$ is greater than that of $N(s)$ (i.e. $n > m$) and *exactly proper* if $n = m$ (Kuo & Golnaraghi, 2010). If $G(s)$ is *asymptotically stable*, that is, when the roots of $D(s)$ are all in the left half plane, each zero has a specific effect on the system for specific inputs. The roots of $N(s)$ (the zeros) can either be real or complex.

In general, a zero near a pole reduces the effect of that term in the total response. This can be shown by assuming that the poles, p_i, are real or complex but distinct and $G(s)$ can be written as a partial fraction expansion

$$G(s) = \frac{C_1}{s - p_1} + \frac{C_2}{s - p_2} + \cdots + \frac{C_n}{s - p_n}. \tag{13}$$

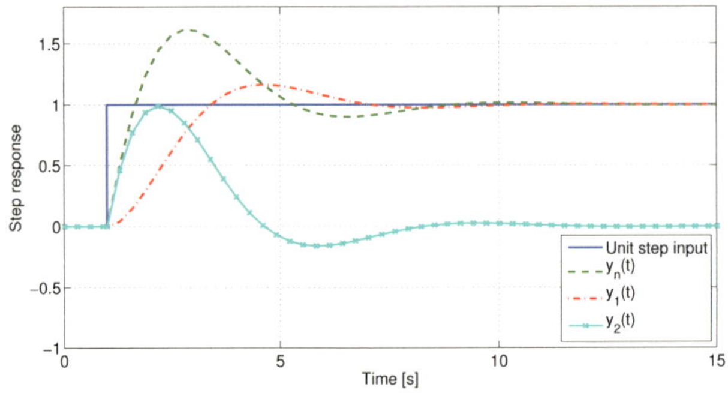

(a) Effect of a left half plane zero

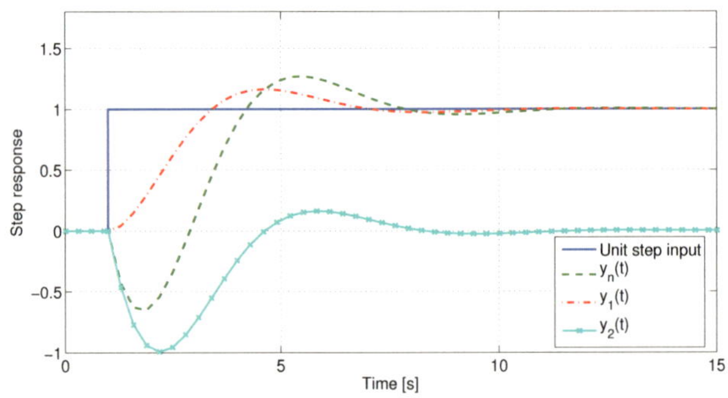

(b) Effect of a right half plane zero

Fig. 1. Step response of $T_n(s)$

When considering Eq. (13), and the equation for the coefficient C_1 given by

$$C_1 = (s - p_1)G(s)|_{s=p_1}, \tag{14}$$

it can be seen that in the case where $G(s)$ has an left half plane zero near the pole at $s = p_1$, the value of C_1 will decrease. This means that the coefficient C_1, which determines the contribution of the specific term in the response will be small. From this observation it can also be said that in general, each zero in the left half plane blocks a specific input signal (Hag & Bernstein, 2007). The question is what happens in the case of a right half plane zero?

(Hag & Bernstein, 2007) illustrated this by looking at the response of a transfer function to an unbounded input signal such as $u(t) = e^t$. Fig. 2 shows the responses of two transfer functions, $G_1(s) = 2(s+1)/(s+1)(s+2)$ and $G_2(s) = 2(s-1)/(s+1)(s+2)$. It can be seen that what distinguishes a right half plane zero is the fact that it blocked the unbounded signal. With a better understanding of the character of right half plane zeros, a formal definition of a non-minimum phase system will be given. Interesting enough, a non-minimum phase system is defined as a system having either a zero or *a pole* in the right-half s-plane (Kuo & Golnaraghi, 2010). (Morari & Zafiriou, 1989) defined a non-minimum phase system as having a transfer function that contains zeros in the right half plane or time delays or both.

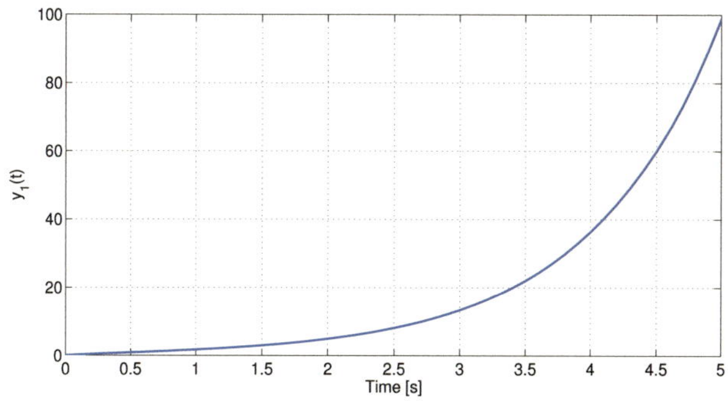

(a) Unbounded response of $G_1(s)$

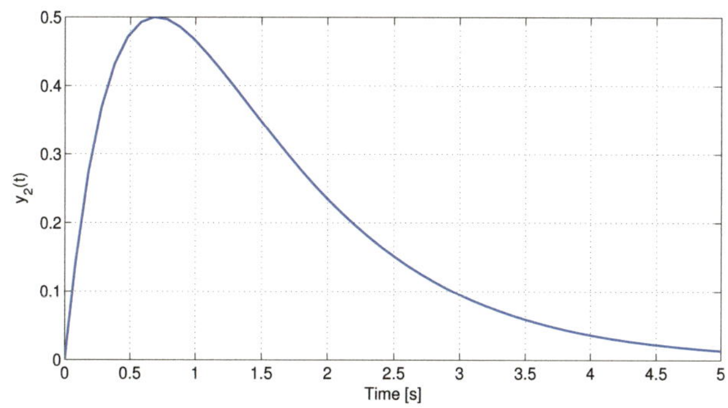

(b) Bounded response of $G_2(s)$

Fig. 2. Responses due to an unbounded input signal $u(t) = e^t$

In this chapter the focus is on non-minimum phase systems showing either *inverse response* (undershoot) or *time-delays*. Undershoot refers to an initial response in the opposite direction from the steady state. According to (Bernardo & Leon de la Barra, 1994) continuous systems having an odd number of real open right half plane zeros are characterised by an initial inverse response to a step input.
Systems having a time-delay characteristic is a special case of non-minimum phase systems (Waller & Nygardas, 1975). The Padé approximation is often used to approximate a time delay by a rational transfer function. Consider a first-order system with time-delay given by

$$G(s) = \frac{K}{1 + s\tau} e^{-sT}. \tag{15}$$

K represents the gain constant, τ the time constant, and T the time-delay of the system. The Padé approximation for the term e^{-sT} is given by

$$e^{-sT} \cong \frac{N_r(sT)}{D_r(sT)} \tag{16}$$

where

$$N_r(sT) = \sum_{k=0}^{r} \frac{(2r-k)!}{k!(r-k)!}(-sT)^k \tag{17}$$

$$D_r(sT) = \sum_{k=0}^{r} \frac{(2r-k)!}{k!(r-k)!}(sT)^k \tag{18}$$

and r is the order of the approximation (Silva et al., 2005).
Consider the function $G(s) = 2e^{-s}/(s+1)(s+2)$. The time-delay term can be approximated by a first order Padé approximation given by

$$e^{-s} \cong \frac{2T-s}{2T+s} = \frac{2-s}{2+s}, \tag{19}$$

and therefore the rational approximated version of $G(s)$ is given by

$$G_r(s) = \frac{2(-s+2)}{(s+1)(s+2)(s+2)}. \tag{20}$$

Fig. 3 plots the step responses of G and G_r and it is interesting to note that the response of G_r exhibits an initial inverse response. This also demonstrates a link between time delay-systems and inverse response systems.

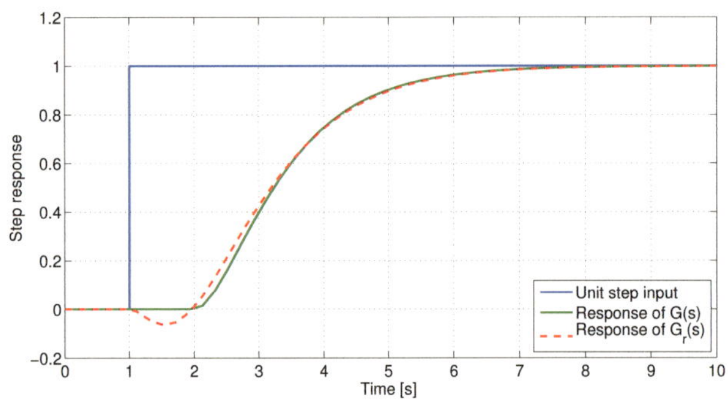

Fig. 3. First order Padé approximation of a time-delay system

4. Practical systems exhibiting non-minimum phase characteristics

Control engineers are often confronted with practical systems exhibiting an inverse response or a time-delay. An inverse response is found in boiler level control systems. An inverse response occurs when the flow rate of the feed water is increased by a step change, and the total volume of boiling water decreases for a short period (Astrom & Hägglund, 1995; Linoya & Altpeter, 1962). The same effect occurs in some aircraft regarding the step response of the elevator deflection to pitch angle (Franklin et al., 2010). In some high temperature gas-cooled reactors (HTGRs) such as the pebble bed modular reactor (PBMR) the power output also shows undershoot phenomena when helium is injected at the low-pressure side of the power conversion unit (PCU) (Uren et al., 2010).

5. Predictive PID controllers

5.1 Classical predictive approaches

Considering industrial applications, the simplicity and effectiveness of a control structure are vital attributes. This consideration can be more important than trying to implement more complex control structures. Therefore the basic structure of PID control is still considered, but with some enhancing adjustments. Like the Zielger-Nichols PID tuning rules, the Smith predictor concept have been around since the late 1950s (Bahill, 1983; Meyer et al., 1976; Smith, 1957; 1958). The internal model control (IMC) method which will also be discussed can be viewed as an extension of the Smith predictor (Astrom & Hägglund, 1995; Rivera et al., 1986)

5.1.1 Smith predictor structure

The undesirable effect of time-delays in feedback control loops are well known. One of the most popular ways to control systems with time delays effectively is by using a Smith predictor. The goal of the Smith predictor is mainly to eliminate the time delay from the characteristic equation and consequently allowing a larger controller gain to be used. The control structure of the Smith predictor is shown in Fig. 4. Let the plant be represented by

$$G_p(s) = G(s)e^{-Ts}, \tag{21}$$

where $T > 0$ is the time-delay. Along with the controller $G_c(s)$ an internal loop is added that simulates the plant dynamics. The plant model is given by

$$G_m(s) = \hat{G}(s)e^{-\hat{T}s}. \tag{22}$$

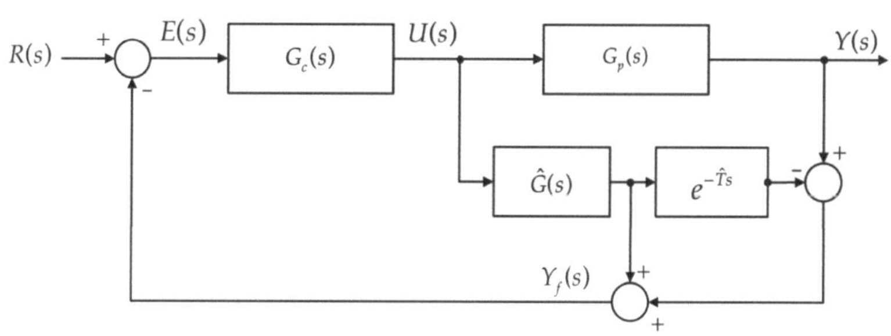

Fig. 4. Smith predictor control structure

Considering Fig. 4 the feedback signal, $Y_f(s)$ is given by

$$Y_f(s) = \hat{G}(s)U(s) + (G(s)e^{-Ts} - \hat{G}(s)e^{-\hat{T}s})U(s) \tag{23}$$

If a "perfect" model of the plant is considered then

$$G(s) = \hat{G}(s), \tag{24}$$
$$T = \hat{T}. \tag{25}$$

This means that the feedback is only dependent on the model of the plant, that is

$$Y_f(s) = \hat{G}(s)U(s). \tag{26}$$

The relationship between the control variable and the system output is

$$U(s) = \frac{1}{G(s)e^{-Ts}} Y(s), \tag{27}$$

and since $G(s) = \hat{G}(s)$, Eq.(26) becomes

$$Y_f(s) = \hat{G}(s) \frac{1}{G(s)e^{-Ts}} Y(s) = e^{Ts} Y(s). \tag{28}$$

This shows that the internal loop containing the plant model feeds back a signal that is a prediction of the output, since e^{Ts} represents a prediction $y(t+T)$ in the time domain. The closed loop transfer function of the system can be determined by using

$$Y(s) = G(s)e^{-Ts}U(s), \tag{29}$$

$$U(s) = G_c(s)(R(s) - Y_f(s)), \tag{30}$$

and Eq. (26) to obtain

$$\frac{Y(s)}{R(s)} = \frac{G(s)e^{-Ts}G_c(s)}{1 + G(s)G_c(s)}. \tag{31}$$

According to (Dorf & Bishop, 2011) the sensitivity expression in this case can be defined as

$$S(s) = \frac{1}{1 + G(s)G_c(s)}. \tag{32}$$

As can be seen, the controller can now be designed without considering the effect of the time delay. (Hägglund, 1992; 1996) combined the properties of the Smith predictor with a PI controller to control a first order plant with a time delay. The transfer function of the plant is given by

$$G_p(s) = \frac{Ke^{-Ts}}{\tau s + 1}, \tag{33}$$

where $K > 0$ is the plant gain, τ the time constant and T the time-delay of the plant. The PI controller is given by

$$G_c(s) = K_p\left(1 + \frac{1}{\tau_i s}\right), \tag{34}$$

where the K_p is the proportional gain, and τ_i is the integral time constant. The control structure is given in Fig. 5

The time delay can be approximated by a first order Padé approximation with the time delay $\hat{T} > 0$. This control structure results in five parameters that need tuning $(K_p, \tau_i, \hat{K}, \hat{\tau}, \hat{T})$.

Example

Consider the following first order plant with a time-delay of two seconds

$$G_p(s) = G(s)G_d(s) = \frac{2}{2s+1}e^{-2s}, \tag{35}$$

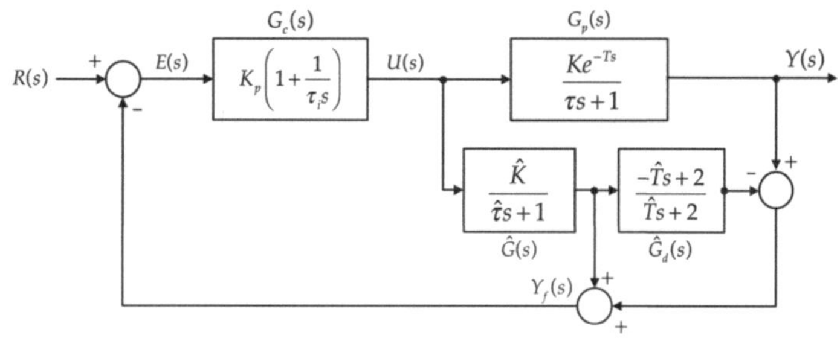

Fig. 5. PI with Smith predictor control structure

where $G_d(s)$ represents the time-delay dynamics. Let the model of the plant be given by

$$G_m(s) = \hat{G}(s)\hat{G}_d(s) = \frac{2}{2s+1}\frac{(-2s+2)}{(2s+2)}, \tag{36}$$

where $\hat{G}_d(s)$ represents the Padé approximation of the time-delay. The PI control constants are set to $K_p = 1$ and $\tau_i = 1.67$, resulting in the following PI controller

$$G_c(s) = (1 + \frac{0.6}{s}). \tag{37}$$

A predictive PID controller $C(s)$ as shown in Fig. 6 needs to be derived based on the predictive properties of the Smith predictor. PID controllers are sometimes augmented with a filter $F(s)$ to improve stability and dynamic response. By comparing the system transfer functions of the

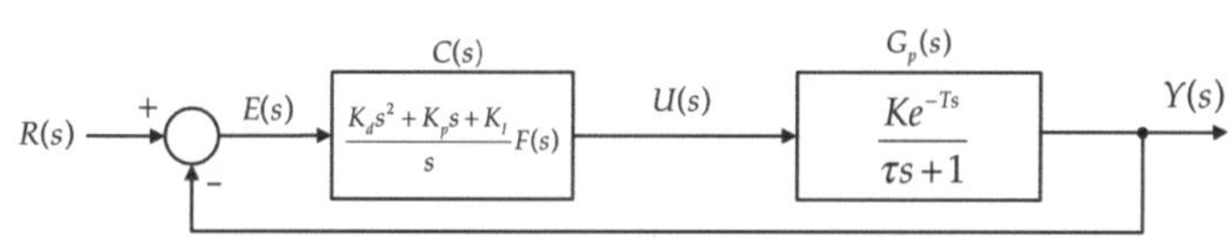

Fig. 6. PID controller based on Smith predictor characteristics

PI with Smith predictor control structure in Fig. 5 and the PID control structure in Fig. 6 a PID controller can be derived based on the Smith predictor qualities:

$$T_{Smith}(s) = T_{PID}(s), \tag{38}$$

$$\frac{\hat{G}(s)\hat{G}_d(s)G_c(s)}{1+\hat{G}_d(s)G_c(s)} = \frac{C(s)\hat{G}(s)\hat{G}_d}{1+C(s)\hat{G}(s)\hat{G}_d}, \tag{39}$$

$$C(s) = \frac{G_c(s)}{1+\hat{G}(s)G_c(s) - \hat{G}(s)G_c(s)\hat{G}_d(s)} \tag{40}$$

$C(s)$ can therefore be considered as a predictive PID controller. Substituting the numerical values leads to

$$C(s) = \frac{4s^4 + 14.4s^3 + 16.2s^2 + 7.4s + 1.2}{4s^4 + 20s^3 + 17.8s^2 + 4.4s}. \tag{41}$$

Applying model reduction techniques $C(s)$ reduces to a PID control structure which is a second order transfer function

$$C(s) = \frac{1.002s^2 + 2.601s + 1.098}{s(s + 4.025)}, \tag{42}$$

where $K_d = 1.002$, $K_p = 2.601$, $K_i = 1.098$ and $F(s) = 1/(s + 4.025)$. Fig. 7 shows the time response of the system output along with the control variable. It can be seen that the control signal acts immediately and not after the occurrence of the time-delay, demonstrating the predictive properties of the PID controller. Fig. 8 shows the time response of the

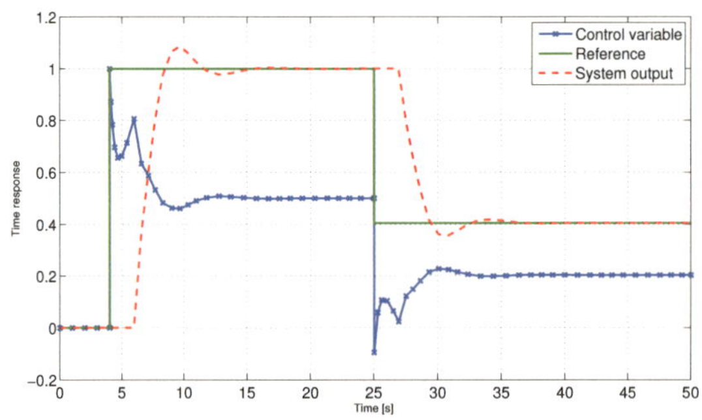

Fig. 7. Time response of system with predictive PID controller $C(s)$ based on Smith predictor

system for larger time-delays. It can be seen that the control performance deteriorates as the time-delay increases. This is due to the limited approximation capabilities of the first order Padé approximation.

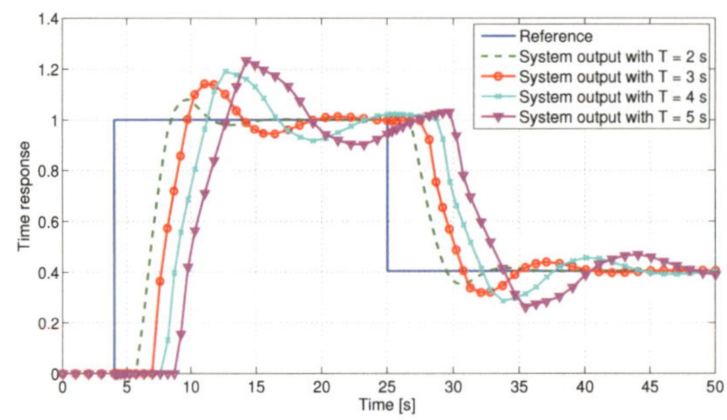

Fig. 8. Time responses of control system based on Smith predictor for different time-delays

5.1.2 Internal model control

The internal model control (IMC) design method starts with the assumption that a model of the system is available that allows the prediction of the system output response due to a output of the controller. In this discussion it is also assumed that the model is a "perfect" representation of the plant. The basic structure of IMC is given in Fig. 9 (Brosilow & Joseph, 2002; Garcia & Morari, 1982). The transfer functions of the plant, the IMC controller and plant model is given by $G_p(s, \varepsilon)$, $G_{IMC}(s)$ and $G_m(s)$ respectively. In the case when the model is not a perfect representation of the actual plant the tuning parameter ε is used to compensate for modelling errors.

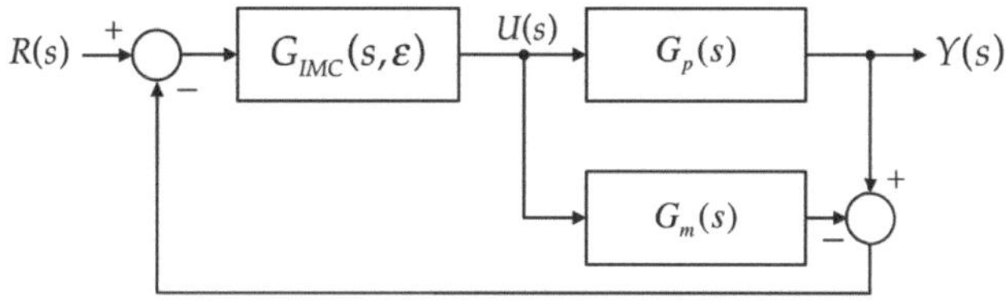

Fig. 9. Internal model control structure

The structure of Fig. 9 can be rearranged into a classical PID structure as shown in Fig. 10. This allows the PID controller to have predictive properties derived from the IMC design.

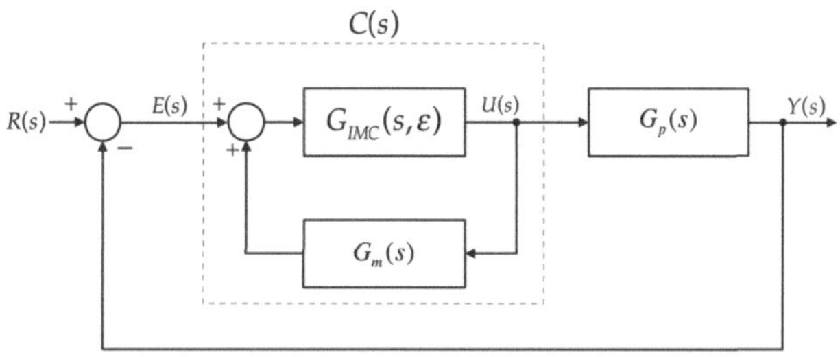

Fig. 10. Classical feedback representation of the IMC structure

The transfer function of the classical controller $C(s)$ is given by

$$C(s) = \frac{U(s)}{E(s)} = \frac{G_{IMC}(s,\varepsilon)}{1 - G_m(s)G_{IMC}(s,\varepsilon)}, \tag{43}$$

and the transfer function of the system is given by

$$T(s) = \frac{Y(s)}{R(s)} = \frac{G_p(s)C(s)}{1 + G_p(s)C(s)}. \tag{44}$$

A "perfect" controller $C(s)$ would drive the output $Y(s)$ of the system to track the reference input $Y(s)$ instantaneously, that is

$$Y(s) = R(s), \tag{45}$$

and this requires that

$$G_{IMC}(s,\varepsilon)G_p(s) = 1, \tag{46}$$

$$G_m(s) = G_p(s). \tag{47}$$

To have a "perfect" controller, a "perfect" model is needed. Unfortunately it is not possible to model the dynamics of the plant perfectly. However, depending on the controller design method, the controller can come close to show the inverse response of the plant model. Usually the design method incorporates a tuning parameter to accommodate modelling errors.

The plant considered is a non-minimum phase system of the following form

$$G_p(s) = \frac{N(s)}{D(s)}e^{-Ts} = \frac{N_-(s)N_+(s)}{D(s)}e^{-Ts}, \tag{48}$$

where $N_-(s)$ represents a polynomial containing only left half plane zeros, and $N_+(s)$ a polynomial containing only right half plane zeros. The IMC controller of the plant in Eq.(48) is given by

$$G_{IMC}(s,\varepsilon) = \frac{D(s)}{N_-(s)N_+(-s)(\varepsilon s+1)^r}, \tag{49}$$

where the zeros of $N_+(-s)$ are all in the left half plane and are the mirror images of the zeros of $N_+(s)$. The filter constant ε is a tuning parameter that can be used to avoid noise amplification and to accommodate modelling errors; and r is the relative order of $N(s)/D(s)$ (Brosilow & Joseph, 2002).

Example

Consider the following non-minimum phase system

$$G_p(s) = \frac{2(-2s+2)}{(2s+1)(2s+2)}. \tag{50}$$

The IMC controller can be derived by using Eq.(49), but in order to ensure zero offset for step inputs $G_p(s)$ is adapted as follows

$$G_p(s) = \frac{2(-2s+2)}{2(2s+1)(2s+2)}. \tag{51}$$

Then

$$G_{IMC}(s) = \frac{(2s+1)(s+1)}{(s+1)(\varepsilon s+1)^r}, \tag{52}$$

and let $\varepsilon = 1$ and $r = 1$ then

$$G_{IMC}(s) = \frac{(2s+1)(s+1)}{(s+1)(s+1)}. \tag{53}$$

The classical controller for this case is given by

$$C(s) = \frac{G_{IMC}(s)}{1 - G_p(s)G_{IMC}(s)} = \frac{1}{2}\frac{(2s+1)(s+1)}{s^2+3s} = \frac{s^2+1.5s+0.5}{s(s+3)}. \tag{54}$$

The form of $C(s)$ corresponds to the form of a PID controller (Dorf & Bishop, 2011):

$$C_{PID}(s) = \frac{K_d(s^2+as+b)}{s} \tag{55}$$

where $a = K_p/K_d$ and $b = K_i/K_d$. The IMC-based controller, Eq.(54), is therefore a PID controller augmented with a filter $F(s) = 1/(\varepsilon s+1)^r$ and is called and IMC-PID controller (Lee et al., 2008). Fig.11 shows the time response of the system output along with the control variable.

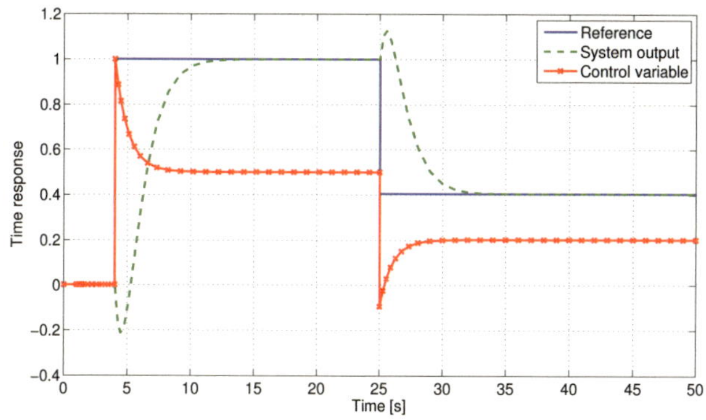

Fig. 11. Time response of control system based on IMC

5.2 Modern predictive approaches

One of the most successful developments in modern control engineering is the area of model predictive control (MPC). It is an optimal control structure utilising a receding horizon principle. This method have found wide-spread application in process industries and research in the field is very active (Wang, 2009). In MPC the control law is computed via optimisation of a quadratic cost function and a plant model is used to predict the future output response to possible future control trajectories. These predictions are computed for a finite time horizons, but only the first value of the optimal control trajectory is used at each sample instant.

Following a model predictive approach for the design of PID controllers is a challenging task. Two routes can be followed namely a *restricted model approach* or a *control signal matching approach* (Johnson & Moradi, 2005; Tan et al., 2000; 2002). In this section the restricted model approach will be considered. This approach formulates the control problem in terms the generalised predictive control (GPC) algorithm. The model used by the controller is restricted to second order such that the predictive control law that emerges has a PID structure. The following control algorithm is discussed in discrete-time since it offers a more natural setting for the derivation of predictive control techniques. It also simplifies the description of the design process and has a strong relevance to industrial applications when presented in discrete-time (Wang, 2009).

5.2.1 The GPC-based algorithm

Augmented state space model

The main idea is to derive an MPC control law equivalent to the second order control law of a PID controller. This can be done by developing an MPC control law, but considering a second-order general plant (Tan et al., 2000; 2002). Consider a single-input, single-output model of a plant described by:

$$\mathbf{X}_m(k+1) = \mathbf{A}_m\mathbf{X}_m(k) + \mathbf{B}_m u(k), \tag{56}$$

$$y(k) = \mathbf{C}_m\mathbf{X}_m(k), \tag{57}$$

where $u(k)$ is the input variable and $y(k)$ is the output variable; and $\mathbf{X}_m$ is the state variable vector of dimension $n = 2$, since a second order plant is considered. Note that the plant model has $u(k)$ as its input. This needs to be altered since a predictive controller needs to be designed. A common first step is to augment the model with an integrator (Wang, 2009). By taking the difference operation on both sides of Eq.(56) the following is obtained

$$\mathbf{X}_m(k+1) - \mathbf{X}_m(k) = \mathbf{A}_m(\mathbf{X}_m(k) - \mathbf{X}_m(k-1)) + \mathbf{B}(u(k) - u(k-1)). \tag{58}$$

The difference of the state variables and output is given by

$$\Delta\mathbf{X}_m(k+1) = \mathbf{X}_m(k+1) - \mathbf{X}_m(k), \tag{59}$$

$$\Delta\mathbf{X}_m(k) = \mathbf{X}_m(k) - \mathbf{X}_m(k-1), \tag{60}$$

$$\Delta u(k) = u(k) - u(k-1). \tag{61}$$

The integrating effect is obtained by connecting $\Delta\mathbf{X}_m(k)$ to the output $y(k)$. To do so the new augmented state vector is chosen to be

$$\mathbf{X}(k) = \left[\Delta\mathbf{X}_m(k)^T \; y(k)\right]^T. \tag{62}$$

where the superscript ^{T} indicates the matrix transpose. The state equation can then be written as

$$\Delta\mathbf{X}_m(k+1) = \mathbf{A}_m\Delta\mathbf{X}_m(k) + \mathbf{B}_m\Delta u(k), \tag{63}$$

and the output equation becomes

$$y(k+1) - y(k) = \mathbf{C}_m(\mathbf{X}_m(k+1) - \mathbf{X}_m(k)) = \mathbf{C}_m\Delta\mathbf{X}_m(k+1) \tag{64}$$

$$= \mathbf{C}_m\mathbf{A}_m\Delta\mathbf{X}_m(k) + \mathbf{C}_m\mathbf{B}_m\Delta u(k). \tag{65}$$

Eqs. (63) and (64) can be written in state space form where

$$\begin{bmatrix} \Delta\mathbf{X}_m(k+1) \\ y(k+1) \end{bmatrix} = \begin{bmatrix} \mathbf{A}_m & \mathbf{O}_m^T \\ \mathbf{C}_m\mathbf{A}_m & 1 \end{bmatrix} \begin{bmatrix} \Delta\mathbf{X}_m(k) \\ y(k) \end{bmatrix} + \begin{bmatrix} \mathbf{B}_m \\ \mathbf{C}_m\mathbf{B}_m \end{bmatrix} \Delta u(k), \tag{66}$$

$$y(k) = \begin{bmatrix} \mathbf{O}_m & 1 \end{bmatrix} \begin{bmatrix} \Delta\mathbf{X}_m(k) \\ y(k) \end{bmatrix}, \tag{67}$$

where $\mathbf{O}_m = \begin{bmatrix} 0\,0 \cdots 0 \end{bmatrix}$ is a $1 \times n$ vector, and $n = 2$ in the predictive PID case. This augmented model will be used in the GPC-based predictive PID control design.

Prediction

The next step in the predictive PID control design is to predict the second order plant output with the future control variable as the adjustable parameter. This prediction is done within one optimisation window. Let $k > 0$ be the sampling instant. Then the future control trajectory is denoted by

$$\Delta u(k), \Delta u(k+1), \cdots, \Delta u(k+N_c-1), \tag{68}$$

where N_c is called the control horizon. The future state variables are denoted by

$$\mathbf{X}(k+1|k), \mathbf{X}(k+2|k), \cdots, \mathbf{X}(k+m|k), \cdots, \mathbf{X}(k+N_p|k), \tag{69}$$

where N_p is the length of the optimisation window and $\mathbf{X}(k+m|k)$ is the predicted state variables at $k+m$ with given current plant information $\mathbf{X}(k)$ and $N_c \leq N_p$.

The future states of the plant are calculated by using the plant state space model:

$$
\begin{aligned}
\mathbf{X}(k+1|k) &= \mathbf{A}_m\mathbf{X}(k) + \mathbf{B}_m\Delta u(k),\\
\mathbf{X}(k+2|k) &= \mathbf{A}_m\mathbf{X}(k+1|k) + \mathbf{B}_m\Delta u(k+1),\\
&= \mathbf{A}_m^2\mathbf{X}(k) + \mathbf{A}_m\mathbf{B}_m\Delta u(k) + \mathbf{B}_m\Delta u(k+1),\\
&\vdots\\
\mathbf{X}(k+N_p|k) &= \mathbf{A}_m^{N_p}\mathbf{X}(k) + \mathbf{A}_m^{N_p-1}\mathbf{B}_m\Delta u(k) + \mathbf{A}_m^{N_p-2}\mathbf{B}_m\Delta u(k+1)\\
&+ \cdots + \mathbf{A}_m^{N_p-N_c}\mathbf{B}_m\Delta u(k+N_c-1).
\end{aligned}
$$

The predicted output variables are as follows:

$$
\begin{aligned}
y(k+1|k) &= \mathbf{C}_m\mathbf{A}_m\mathbf{X}(k) + \mathbf{C}_m\mathbf{B}_m\Delta u(k),\\
y(k+2|k) &= \mathbf{C}_m\mathbf{A}_m^2\mathbf{X}(k) + \mathbf{C}_m\mathbf{A}_m\mathbf{B}_m\Delta u(k) + \mathbf{C}_m\mathbf{B}_m\Delta u(k+1),\\
y(k+3|k) &= \mathbf{C}_m\mathbf{A}_m^3\mathbf{X}(k) + \mathbf{C}_m\mathbf{A}_m^2\mathbf{B}_m\Delta u(k) + \mathbf{C}_m\mathbf{A}_m\mathbf{B}_m\Delta u(k+1)\\
&+ \mathbf{C}_m\mathbf{B}_m\Delta u(k+2),\\
&\vdots\\
y(k+N_p|k) &= \mathbf{C}_m\mathbf{A}_m^{N_p}\mathbf{X}(k) + \mathbf{C}_m\mathbf{A}_m^{N_p-1}\mathbf{B}_m\Delta u(k) + \mathbf{C}_m\mathbf{A}_m^{N_p-2}\mathbf{B}_m\Delta u(k+1)\\
&+ \cdots + \mathbf{C}_m\mathbf{A}_m^{N_p-N_c}\mathbf{B}_m\Delta u(k+N_c-1).
\end{aligned}
$$

The equations above can now be ordered in matrix form as

$$
\mathbf{Y} = \mathbf{F}\mathbf{X}(k) + \Phi\Delta\mathbf{U}, \tag{70}
$$

where

$$
\mathbf{Y} = \left[y(k+1|k)\ y(k+2|k)\ y(k+3|k)\ \ldots\ y(k+N_p|k)\right]^T, \tag{71}
$$

$$
\Delta\mathbf{U} = \left[\Delta u(k)\ \Delta u(k+1)\ \Delta u(k+3)\ \ldots\ \Delta u(k+N_c-1)\right]^T, \tag{72}
$$

and

$$
\mathbf{F} = \begin{bmatrix} \mathbf{C}_m\mathbf{A}_m \\ \mathbf{C}_m\mathbf{A}_m^2 \\ \mathbf{C}_m\mathbf{A}_m^3 \\ \vdots \\ \mathbf{C}_m\mathbf{A}_m^{N_p} \end{bmatrix}, \tag{73}
$$

$$
\Phi = \begin{bmatrix}
\mathbf{C}_m\mathbf{B}_m & 0 & 0 & \ldots & 0\\
\mathbf{C}_m\mathbf{A}_m\mathbf{B}_m & \mathbf{C}_m\mathbf{B}_m & 0 & \ldots & 0\\
\mathbf{C}_m\mathbf{A}_m^2\mathbf{B}_m & \mathbf{C}_m\mathbf{A}_m\mathbf{B}_m & \mathbf{C}_m\mathbf{B}_m & \ldots & 0\\
\vdots & & & & \\
\mathbf{C}_m\mathbf{A}_m^{N_p-1}\mathbf{B}_m & \mathbf{C}_m\mathbf{A}_m^{N_p-2}\mathbf{B}_m & \mathbf{C}_m\mathbf{A}_m^{N_p-3}\mathbf{B}_m & \ldots & \mathbf{C}_m\mathbf{A}_m^{N_p-N_c}\mathbf{B}_m
\end{bmatrix}. \tag{74}
$$

Optimisation and control design

Let $r(k)$ be the set-point signal at sample time k. The idea behind the predictive PID control methodology is to drive the predicted output signal as close as possible to the set-point signal. It is assumed that the set-point signal remains constant during the optimisation window, N_p. Consider the following quadratic cost function which is very similar to the one obtained by (Tan et al., 2002)

$$J = (\mathbf{r} - \mathbf{y})^T(\mathbf{r} - \mathbf{y}) + \Delta\mathbf{U}^T\overline{\mathbf{R}}\Delta\mathbf{U}, \tag{75}$$

where the set-point information is given by

$$\mathbf{r}^T = \begin{bmatrix} 1 & 1 & \ldots & 1 \end{bmatrix} \times r(k), \tag{76}$$

and the dimension of $\mathbf{r}$ is $N_p \times 1$. The cost function, Eq.(75) comprises two parts, the first part focus on minimising the errors between the reference and the output; the second part focus on minimising the control effort. $\overline{\mathbf{R}}$ is a diagonal weight matrix given by

$$\overline{\mathbf{R}} = r_w \times \mathbf{I} \tag{77}$$

where $\mathbf{I}$ is an $N_c \times N_c$ identity matrix and the weight $r_w \geq 0$ is used to tune the closed-loop response. The optimisation problem is defined such that an optimal $\Delta\mathbf{U}$ can be found that minimises the cost function J. Substituting Eq.(70) into Eq.(75), J is expressed as

$$J = (\mathbf{r} - \mathbf{FX}(k))^T(\mathbf{r} - \mathbf{FX}(k)) - 2\Delta\mathbf{U}^T\Phi^T(\mathbf{r} - \mathbf{FX}(k)) + \Delta\mathbf{U}^T(\Phi^T\Phi + \overline{\mathbf{R}})\Delta\mathbf{U}. \tag{78}$$

The solution that minimises the cost function J can be obtained by solving

$$\frac{\partial J}{\partial \Delta\mathbf{U}} = 2\Phi^T(\mathbf{r} - \mathbf{FX}(k)) + 2(\Phi^T\Phi + \overline{\mathbf{R}})\Delta\mathbf{U} = \mathbf{0}. \tag{79}$$

Therefore, the optimal control law is given as

$$\Delta\mathbf{U} = (\Phi^T\Phi + \overline{\mathbf{R}})^{-1}\Phi^T(\mathbf{r} - \mathbf{FX}(k)) \tag{80}$$

or

$$\Delta\mathbf{U} = (\Phi^T\Phi + \overline{\mathbf{R}})^{-1}\Phi^T\mathbf{e}(k) \tag{81}$$

where $\mathbf{e}(k)$ represents the errors at sample k.

Emerging predictive control with PID structure

The discrete configuration of a PID controller has the following form (Huang et al., 2002; Phillips & Nagle, 1995):

$$u(k) = K_p e(k) + K_i \sum_{n=1}^{k} e(n) + K_d(e(k) - e(k-1)), \tag{82}$$

or

$$u(z) = \frac{q_0 + q_1 z^{-1} + q_2 z^{-2}}{1 - z^{-1}} e(z), \tag{83}$$

where K_p, K_i and K_d are the proportional, integral and derivative gains, respectively, and

$$q_0 = K_p + K_i + K_d, \tag{84}$$
$$q_1 = -K_p - 2K_d, \tag{85}$$
$$q_2 = K_d. \tag{86}$$

By taking the difference on both sides of Eq.(82), the velocity form of the PID control law is obtained:

$$\Delta u(k) = K_p[e(k) - e(k-1)] + K_i e(k) + K_d[e(k) - 2e(k-1) + e(k-2)]. \tag{87}$$

This equation can be written in matrix form as (Katebi & Moradi, 2001):

$$\Delta \mathbf{U}(k) = \mathbf{K}\mathbf{e}(k) = \mathbf{K}[\mathbf{r}(k) - \mathbf{y}(k)] \tag{88}$$

where

$$\mathbf{K} = \begin{bmatrix} K_p & K_i & K_d \end{bmatrix} \begin{bmatrix} 0 & -1 & 1 \\ 0 & 0 & 1 \\ 1 & -2 & 1 \end{bmatrix}, \tag{89}$$

and

$$\mathbf{y}(k) = \begin{bmatrix} y(k-2) & y(k-1) & y(k) \end{bmatrix}^T \tag{90}$$
$$\mathbf{e}(k) = \begin{bmatrix} e(k-2) & e(k-1) & e(k) \end{bmatrix}^T \tag{91}$$
$$\mathbf{r}(k) = \begin{bmatrix} r(k-2) & r(k-1) & r(k) \end{bmatrix}^T. \tag{92}$$

By equating Eq.(81) to Eq.(88)the following is obtained

$$\Delta \mathbf{U}(k) = (\Phi^T\Phi + \overline{\mathbf{R}})^{-1}\Phi^T \mathbf{e}(k) = \mathbf{K}^T \mathbf{e}(k) \tag{93}$$

and therefore the predictive PID controller constants are given by

$$\mathbf{K}^T = (\Phi^T\Phi + \overline{\mathbf{R}})^{-1}\Phi^T, \tag{94}$$

or

$$\begin{bmatrix} K_d & (-2K_d - Kp) & (K_d + K_i + K_p) \end{bmatrix}^T = (\Phi^T\Phi + \overline{\mathbf{R}})^{-1}\Phi^T. \tag{95}$$

Example

Consider the following discrete-time state space model of a non-minimum phase system

$$\dot{\mathbf{X}}(k) = \begin{bmatrix} -0.0217 & -0.3141 \\ 0.3141 & 0.7636 \end{bmatrix} \mathbf{X}(k) + \begin{bmatrix} 0.3141 \\ 0.2364 \end{bmatrix} u(k), \tag{96}$$
$$y(k) = \begin{bmatrix} -1 & 2 \end{bmatrix} \mathbf{X}(k). \tag{97}$$

The first step is to create the augmented model for the MPC design, and choose the values of the prediction and control horizon. In this example the control horizon is selected to be $N_c = 3$ and the prediction horizon is $N_p = 20$. Also the sampling period in this case is chosen as 1 second and a 100 samples is considered. Then the predicted output is given by Eq. 70 where

$$\mathbf{F} = \begin{bmatrix} 0.6500 & 1.8413 & 1.0000 \\ 1.2143 & 3.0432 & 1.0000 \\ 1.5796 & 3.7836 & 1.0000 \\ \vdots & \vdots & \vdots \\ 2.1515 & 4.9290 & 1.0000 \\ 2.1516 & 4.9292 & 1.0000 \\ 2.1517 & 4.9294 & 1.0000 \end{bmatrix}, \Phi = \begin{bmatrix} 0.1587 & 0 & 0 \\ 0.7982 & 0.1587 & 0 \\ 1.2595 & 0.7982 & 0.1587 \\ \vdots & \vdots & \vdots \\ 1.9996 & 1.9993 & 1.9989 \\ 1.9998 & 1.9996 & 1.9993 \\ 1.9998 & 1.9998 & 1.9996 \end{bmatrix}, \tag{98}$$

are matrices having 20 rows and 3 columns. By choosing a weight $r_w = 0.9$ the optimal control law (Eq. (81)) is given by

$$\Delta\mathbf{U} = \begin{bmatrix} 0.0628 & 0.2602 & 0.2108 & \cdots & -0.0144 & -0.0144 & -0.0145 \\ -0.0554 & -0.1681 & 0.0617 & \cdots & 0.0035 & 0.0035 & 0.0035 \\ -0.0085 & -0.0976 & -0.2766 & \cdots & 0.0452 & 0.0453 & 0.0453 \end{bmatrix} \mathbf{e}(k), \tag{99}$$

where the matrix multiplied with the error vector has 3 rows and 20 columns.
Fig. 12 shows the closed loop response of the system output along with the control variable. It can be seen that the control variable acts immediately and not after the occurrence of the

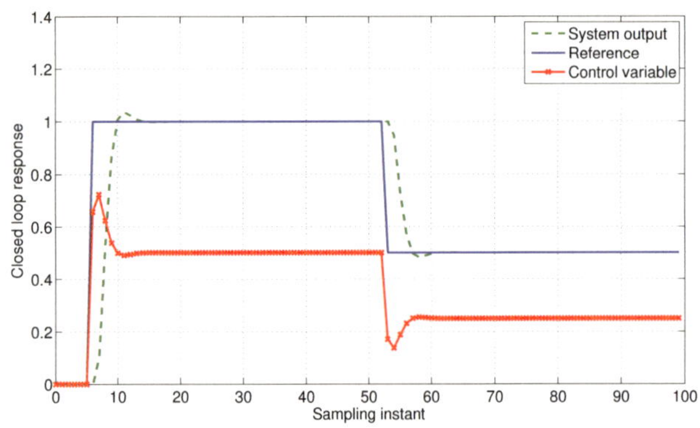

Fig. 12. Closed loop response of a system with an MPC controller having a PID control structure

time-delay. This shows that the MPC controller with a PID structure demonstrates predictive properties. An improvement in the control performance can be seen compared to the previous classical predictive controllers. This is due to the fact that the control law is computed via the optimisation of a quadratic cost function.

6. Conclusions

In this chapter both classical and modern predictive control methods for non-minimum phase systems were considered. Two popular methods considered in the classical approach were the Smith predictor and internal model control (IMC). These two methods utilise a plant model to predict the future output of the plant. This results in a control law that acts immediately on the reference input avoiding instability and sluggish control. In the classical approach the

Smith predictor and IMC structures were used to derive the predictive PID control constants. The predictive PID controller can effectively deal with the non-minimum phase effect.
A modern approach to predictive PID control features a different methodology. A generalised predictive control algorithm was considered. In this approach the model predictive controller is reduced to the same structure as a PID controller for second-order systems (Eq. (87)). In this case the equivalent PID constants changes at every sample since an optimisation routine using a cost function (Eq. (78)) is followed at each sample. The controller structure can further be adapted to be used as a design method to derive optimal values of PID gains (Eq. (95)). The novelty of this method lies in the fact that time-delays are incorporated without any need for approximation.

7. References

Astrom, K. & Hägglund, T. (1995). *PID Controllers:Theory, Design, and Tuning*, 2nd edn, Instrument Society of America.

Bahill, A. (1983). A simple adaptive smith-predictor for controlling time-delay systems: A tutorial, *Control Systems Magazine, IEEE* 3(2): 16 – 22.

Bernardo, A. & Leon de la Barra, S. (1994). On undershoot in siso systems, *Automatic Control, IEEE Transactions on* 39(3): 578 –581.

Brosilow, C. & Joseph, B. (2002). *Techniques of model-based control*, Prentice-Hall international series in the physical and chemical engineering sciences, Prentice Hall.

Dorf, R. & Bishop, R. (2011). *Modern control systems*, 12th edn, Pearson Prentice Hall.

Franklin, G. F., Powell, D. J. & Emami-Naeini, A. (2010). *Feedback Control of Dynamic Systems*, 6th edn, Prentice Hall PTR, Upper Saddle River, NJ, USA.

Garcia, C. E. & Morari, M. (1982). Internal model control. a unifying review and some new results, *Industrial & Engineering Chemistry Process Design and Development* 21(2): 308–323.

Hag, J. & Bernstein, D. (2007). Nonminimum-phase zeros - much to do about nothing - classical control - revisited part ii, *Control Systems Magazine, IEEE* 27(3): 45 –57.

Hägglund, T. (1992). A predictive PI controller for processes with long dead times, *Control Systems Magazine, IEEE* 12(1): 57 –60.

Hägglund, T. (1996). An industrial dead-time compensating PI controller, *Control Engineering Practice* 4(6): 749 – 756.

Huang, S., Tan, K. & Lee, T. (2002). *Applied predictive control*, Advances in industrial control, Springer.

Johnson, M. & Moradi, M. (2005). *PID Control: New Identification and Design Methods*, Springer.

Katebi, M. & Moradi, M. (2001). Predictive pid controllers, *Control Theory and Applications, IEE Proceedings* - 148(6): 478 –487.

Kuo, B. C. & Golnaraghi, F. (2010). *Automatic Control Systems*, 9th edn, John Wiley & Sons, Inc., New York, NY, USA.

Lee, M., Shamsuzzoha, M. & Vu, T. N. L. (2008). Imc-pid approach: An effective way to get an analytical design of robust pid controller, *Control, Automation and Systems, 2008. ICCAS 2008. International Conference on*, pp. 2861 –2866.

Linoya, K. & Altpeter, R. J. (1962). Inverse response in process control, *Industrial & Engineering Chemistry* 54(7): 39–43.

Meyer, C., Seborg, D. E. & Wood, R. K. (1976). A comparison of the smith predictor and conventional feedback control, *Chemical Engineering Science* 31(9): 775 – 778.

Miller, R. M., Shah, S. L., Wood, R. K. & Kwok, E. K. (1999). Predictive pid, *ISA Transactions* 38(1): 11 – 23.

Mita, T. & Yoshida, H. (1981). Undershooting phenomenon and its control in linear multivariable servomechanisms, *Automatic Control, IEEE Transactions on* 26(2): 402 – 407.

Moradi, M., Katebi, M. & Johnson, M. (2001). Predictive pid control: a new algorithm, *Industrial Electronics Society, 2001. IECON '01. The 27th Annual Conference of the IEEE*, Vol. 1, pp. 764 –769 vol.1.

Morari, M. & Zafiriou, E. (1989). *Robust process control*, Prentice Hall.

Phillips, C. & Nagle, H. (1995). *Digital control system analysis and design*, Prentice Hall.

Rivera, D. E., Morari, M. & Skogestad, S. (1986). Internal model control: Pid controller design, *Industrial & Engineering Chemistry Process Design and Development* 25(1): 252–265.

Sato, T. (2010). Design of a gpc-based pid controller for controlling a weigh feeder, *Control Engineering Practice* 18(2): 105 – 113. Special Issue of the 3rd International Symposium on Advanced Control of Industrial Processes.

Silva, G. J., Datta, A. & P., B. S. (2005). *PID controllers for time-delay systems*, Birkhauser Boston.

Smith, O. (1957). Close control of loops with dead time, *Chemical Engineering Progress* 53: 217–219.

Smith, O. (1958). *Feedback Control Systems*, McGraw-Hill, New York.

Tan, K. K., Huang, S. N. & Lee, T. H. (2000). Development of a gpc-based pid controller for unstable systems with deadtime, *ISA Transactions* 39(1): 57 – 70.

Tan, K. K., Lee, T. H., Huang, S. N. & Leu, F. M. (2002). Pid control design based on a gpc approach, *Industrial & Engineering Chemistry Research* 41(8): 2013–2022.

Tan, K. K., Lee, T. H. & Leu, F. M. (2001). Predictive pi versus smith control for dead-time compensation, *ISA Transactions* 40(1): 17 – 29.

Uren, K., Van Schoor, G. & Van Niekerk, C. (2010). Optimal power control of a three-shaft brayton cycle based power conversion unit, *South African institute of electrical engineers* 101: 60–67.

Vidyasagar, M. (1986). On undershoot and nonminimum phase zeros, *Automatic Control, IEEE Transactions on* 31(5): 440 – 440.

Waller, K. V. T. & Nygardas, C. G. (1975). On inverse repsonse in process control, *Industrial & Engineering Chemistry Fundamentals* 14(3): 221–223.

Wang, L. (2009). *Model predictive control system design and implementation using MATLAB*, Advances in Industrial Control, Springer.

2

Adaptive PID Control System Design Based on ASPR Property of Systems

Ikuro Mizumoto[1] and Zenta Iwai[2]
[1]*Department of Mechanical Systems Engineering, Kumamoto University*
[2]*Kumamoto Prefectural College of Technology*
Japan

1. Introduction

PID control is one of the most common control schemes applied to many industrial processes and mechanical systems. Because, the PID can be tuned according to the experience of operators and can applied to uncertain system without a certain system's model. However in cases where there are some changes of system properties, it has been pointed out the difficulties of maintaining the desired control performance and stability during operation, and in some cases, it might be difficult to tune the PID parameters so as to satisfy the desired performance. Furthermore, the control plays a very important role in the improvement of production quality, accuracy and in reducing production costs. As a result a great deal of attention has been focused on automatic or self tuning of PID controllers (Astrom & Hagglund, 1995), and in recent decades several kinds of auto-tuning PIDs including self-tuning schemes and adaptive control strategies have been proposed (Chang et al., 2003; Iwai et al., 2006; Kono et al., 2007; Ren et al., 2008; Tamura & Ohmori, 2007; Yamamoto & Shah, 2004; Yu et al., 2007). Unfortunately, most PID auto-tuning methods did not pay sufficient attention to the stability of the resulting PID control system and the tuned PID parameters did not guarantee the stability of the control system after any change of the systems.

In this Chapter, an adaptive PID control system design strategy based on the almost strictly positive real (ASPR) property for linear continuous-time systems will be presented. The adaptive PID scheme based on the ASPR property of the system can guarantee the asymptotic stability of the resulting PID control system and since the method presented in this chapter utilizes the characteristics of the ASPR-ness of the controlled system, the stability of the resulting adaptive control system can be guaranteed with certainty. The stability analysis will also be shown for ASPR systems. However, since most practical systems do not satisfy ASPR conditions, difficulties will appear in the practical application of the ASPR based adaptive PID control. In order to solve this problem, a robust parallel feedforward compensator (PFC) design method, which render the resulting augmented system with the PFC in parallel ASPR, will be provided.

The proposed adaptive PID control system can guarantee the stability, and by adjusting PID parameters adaptively, the method maintains a better control performance even if there are some changes of the system properties. In order to confirm the usefulness an effectiveness of the proposed adaptive PID design scheme for real world processes, the proposed method is applied to an unsaturated highly accelerated stress test system.

2. Problem statement

Consider a SISO continuous-time system with a relative degree of γ.

$$\begin{aligned}\dot{\boldsymbol{x}}(t) &= A\boldsymbol{x}(t) + \boldsymbol{b}u_f(t) + C_{d1}\boldsymbol{w}_d(t) \\ y(t) &= \boldsymbol{c}^T\boldsymbol{x}(t) + \boldsymbol{d}_1^T\boldsymbol{w}_d(t)\end{aligned} \tag{1}$$

where $\boldsymbol{x}(t) \in R^n$ is a state vector, $u(t)$ and $y(t) \in R$ are the input and the output of the system, respectively. $\boldsymbol{w}_d(t) \in R^{m\times 1}$ is a disturbance. The system (1) is not required to be stable and/or minimum-phase.

Suppose that the disturbance $\boldsymbol{w}_d(t)$ and a reference signal $r(k)$ which the system output $y(t)$ is required to follow are generated by the following known exosystem:

$$\begin{aligned}\dot{\boldsymbol{w}}_d(t) &= A_m\boldsymbol{w}_d(t) \\ r(t) &= \boldsymbol{c}_m^T\boldsymbol{w}_d(t)\end{aligned} \tag{2}$$

with a characteristic polynomial,

$$det(\lambda I - A_m) = \lambda^m + \alpha_{m-1}\lambda^{m-1} + \cdots + \alpha_1\lambda + \alpha_0 \tag{3}$$

We assume that the exosystem is stable or neutrally stable. That is, all its eigenvalues are located on the left half-plane and/or the imaginary axis.

The objective is to design an adaptive PID controller so as to have the output $y(t)$ track the reference signal $r(t)$.

Remark 1: The exosystem is divided into two parts for the disturbance model and the reference signal. The part of reference signal is available so that $r(t)$ is known, but the part of disturbance is just a model of the disturbance and practical signal of the disturbance is not available, only the characteristic polynomial is known.

2.1 Transformed system

For the system (1) with a relative degree of γ, there exists a nonsingular variable transformation:

$$\begin{bmatrix} \boldsymbol{z}(t) \\ \boldsymbol{\eta}(t) \end{bmatrix} = \boldsymbol{\Phi}\boldsymbol{x}(t) \tag{4}$$

such that the system (1) can be transformed into the form (Isidori, 1995):

$$\dot{\boldsymbol{z}}(t) = A_z\boldsymbol{z}(t) + \boldsymbol{b}_z u_f(t) + C_z\boldsymbol{\eta}(t) + D_{d1}\boldsymbol{w}_d(t) \tag{5}$$

$$\dot{\boldsymbol{\eta}}(t) = Q_\eta\boldsymbol{\eta}(t) + \boldsymbol{c}_\eta z_1(t) + F_{d1}\boldsymbol{w}_d(t) \tag{6}$$

$$y(t) = [1 0 \cdot 0]\begin{bmatrix} \boldsymbol{z}(t) \\ \boldsymbol{\eta}(t) \end{bmatrix} + \boldsymbol{d}_1^T\boldsymbol{w}_d(t) \tag{7}$$

where

$$A_z = \begin{bmatrix} \boldsymbol{0} & \boldsymbol{I}_{\gamma-1\times\gamma-1} \\ -a_0 & \cdots - a_{\gamma-1} \end{bmatrix} \qquad \boldsymbol{b}_z = \begin{bmatrix} 0 \cdots 0 \ b_z \end{bmatrix}^T \qquad C_z = \begin{bmatrix} \boldsymbol{0} \\ \boldsymbol{c}_z^T \end{bmatrix}$$

$$\boldsymbol{c}_\eta = \begin{bmatrix} \boldsymbol{0} \\ 1 \end{bmatrix} \quad b_z = \boldsymbol{c}^T A^{\gamma-1}\boldsymbol{b} \quad D_{d1} \in R^{\gamma\times m} \quad F_{d1} \in R^{n-\gamma\times m} \tag{8}$$

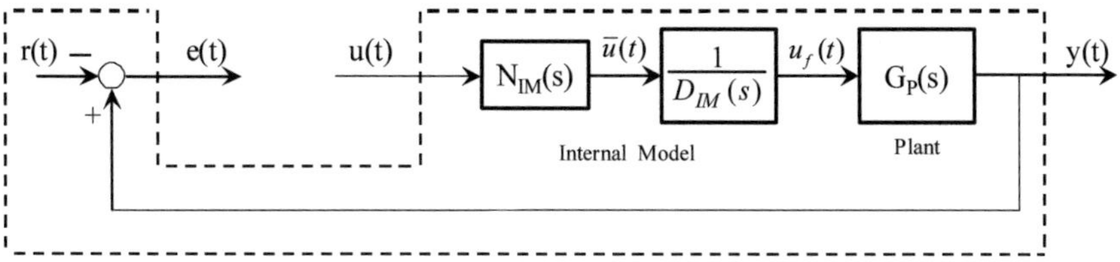

Fig. 1. Error System

2.2 Error system with an internal model filter

In order to alleviate the affect from disturbances and a PFC to be introduced later, we first consider the following internal model filter of the form:

$$u_f(t) = G_{IM}(s)\,[u(t)] = \frac{N_{IM}(s)}{D_{IM}(s)}\,[u(t)] \tag{9}$$

where

$$D_{IM}(s) = det(sI - A_m) \tag{10}$$

and $N_{IM}(s)$ is any stable polynomial of order m of the form:

$$N_{IM}(s) = s^m + \beta_{m-1}s^{m-1} + \cdots + \beta_1 s + \beta_0 \tag{11}$$

Defining the output following error by $e(t) = y(t) - r(t)$, consider the error system from $u(t)$ to $e(t)$ as shown Fig. 1.
Define new variables $\mathbf{X}_1(t) \in R^{\gamma\times 1}$, $\mathbf{X}_2(t) \in R^{n-\gamma\times 1}$ as follows:

$$\mathbf{X}_1(t) = \overset{(m)}{\mathbf{z}}(t) + \alpha_{m-1}\overset{(m-1)}{\mathbf{z}}(t) + \cdots + \alpha_1\dot{\mathbf{z}}(t) + \alpha_0\mathbf{z}(t) \tag{12}$$

$$\mathbf{X}_2(t) = \overset{(m)}{\boldsymbol{\eta}}(t) + \alpha_{m-1}\overset{(m-1)}{\boldsymbol{\eta}}(t) + \cdots + \alpha_1\dot{\boldsymbol{\eta}}(t) + \alpha_0\boldsymbol{\eta}(t) \tag{13}$$

It follows from (2), (3), (5), (6) that

$$\dot{\mathbf{X}}_1(t) = A_z\mathbf{X}_1(t) + \mathbf{b}_z\bar{u}(t) + C_z\mathbf{X}_2(t) \tag{14}$$

$$\dot{\mathbf{X}}_2(t) = Q_\eta\mathbf{X}_2(t) + C_\eta\mathbf{X}_1(t) \tag{15}$$

where

$$\bar{u}(t) = \overset{(m)}{u_f}(t) + \alpha_{m-1}\overset{(m-1)}{u_f}(t) + \cdots + \alpha_1\dot{u}_f(t) + \alpha_0 u_f(t) \tag{16}$$

(17)

and

$$C_\eta = \begin{bmatrix} \mathbf{0}\,\mathbf{0} \\ \mathbf{1}\,\mathbf{0} \end{bmatrix} \tag{18}$$

Next defining

$$\mathbf{E}(t) = \left[e(t)\ \dot{e}(t)\cdots\overset{(m-1)}{e}(t)\right]^T \tag{19}$$

we have from (2), (3) and (7) that

$$\dot{\boldsymbol{E}}(t) = A_E \boldsymbol{E}(t) + C_E \boldsymbol{X}_1(t) \tag{20}$$

where

$$A_E = \begin{bmatrix} \mathbf{0} & \boldsymbol{I}_{m-1 \times m-1} \\ -\alpha_0 & \cdots - \alpha_{m-1} \end{bmatrix} \in R^{m \times m}, \; C_E = \begin{bmatrix} \mathbf{0} \\ 1\,0 \cdots 0 \end{bmatrix} \in R^{m \times \gamma} \tag{21}$$

Thus, the error system can be represented by the following form without the term of disturbances:

$$\begin{aligned} \dot{\boldsymbol{E}}(t) &= A_E \boldsymbol{E}(t) + C_E \boldsymbol{X}_1(t) \\ \dot{\boldsymbol{X}}_1(t) &= A_z \boldsymbol{X}_1(t) + \boldsymbol{b}_z \bar{u}(t) + C_z \boldsymbol{X}_2(t) \\ \dot{\boldsymbol{X}}_2(t) &= Q_\eta \boldsymbol{X}_2(t) + C_\eta \boldsymbol{X}_1(t) \\ e(t) &= [1\; \mathbf{0}]\, \boldsymbol{E}(t) \end{aligned} \tag{22}$$

This system with an input $\bar{u}(t)$ and the output $e(t)$ is of the order $n+m$ with a relative degree of $\gamma + m$. Since the error system has the relative degree of $\gamma + m$, the error system can be transformed into the following canonical form by an appropriate non-singular variable transformation.

$$\begin{aligned} \dot{\boldsymbol{z}}_{ze}(t) &= A_{ze} \boldsymbol{z}_{ze}(t) + \boldsymbol{b}_{ze} \bar{u}(t) + C_{ze} \boldsymbol{\eta}_{ze}(t) \\ \dot{\boldsymbol{\eta}}_{ze}(t) &= Q_{ze} \eta_{ze}(t) + \boldsymbol{c}_{\eta e} z_{e1}(t) \\ e(t) &= z_{e1}(t) \end{aligned} \tag{23}$$

with

$$\boldsymbol{z}_{ze}(t) = \begin{bmatrix} z_{e1}(t) \\ \vdots \\ z_{e\gamma+m}(t) \end{bmatrix} \qquad \boldsymbol{\eta}_{ze}(t) = \begin{bmatrix} z_{e\gamma+m+1}(t) \vdots \\ z_{e\; n+m}(t) \end{bmatrix}$$

$$A_{ze} = \begin{bmatrix} \mathbf{0} & \boldsymbol{I}_{(\gamma+m-1)\times(\gamma+m-1)} \\ -\theta_0 & \cdots \; -\theta_{\gamma+m-1} \end{bmatrix} \qquad \boldsymbol{b}_{ze} = \begin{bmatrix} 0 \\ \vdots \\ 0 \\ b_{ze} \end{bmatrix}$$

$$C_{ze} = \begin{bmatrix} \mathbf{0} \\ \boldsymbol{c}_{ze}^T \end{bmatrix} \qquad \boldsymbol{c}_{\eta e} = \begin{bmatrix} \mathbf{0} \\ 1 \end{bmatrix} \tag{24}$$

and θ_i, b_{ze} and $\boldsymbol{c}_{ze}$ are appropriate constants and vector. Further it follows from (9) that

$$\begin{aligned} & u_f^{(m)}(t) + \alpha_{m-1} u_f^{(m-1)}(t) + \cdots + \alpha_1 \dot{u}_f(t) + \alpha_0 u_f(t) \\ &= u^{(m)}(t) + \beta_{m-1} u^{(m-1)}(t) + \cdots + \beta_1 \dot{u}(t) + \beta_0 u(t) \\ &= \bar{u}(t) \end{aligned} \tag{25}$$

Defining

$$\bar{z}_{IM}(t) = \left[u(t) \, , \, \dot{u}(t) \, , \, \cdots \, , \, u^{(m-1)}(t) \right]^T \tag{26}$$

we have the following system representation from $\bar{u}(t)$ to $u(t)$.

$$\begin{aligned} \dot{\bar{z}}_{IM}(t) &= A_{IM}\bar{z}_{IM}(t) + \boldsymbol{b}_{IM}\bar{u}(t) \\ u(t) &= \bar{c}_{IM}^T \bar{z}_{IM}(t) \end{aligned} \tag{27}$$

where

$$A_{IM} = \begin{bmatrix} \mathbf{0} & \boldsymbol{I}_{m-1\times m-1} \\ -\beta_0 & \cdots - \beta_{m-1} \end{bmatrix} \, , \, \boldsymbol{b}_{IM} = \begin{bmatrix} \mathbf{0} \\ 1 \end{bmatrix} \, , \, \bar{c}_{IM}^T = [1 0 \cdots 0] \tag{28}$$

Consider the following variable transformation using the state variable $\bar{z}_{IM}(t)$ in (27).

$$\xi_k(t) = -b_{ze} u^{(k-\gamma-1)}(t) + e^{(k-1)}(t) + \sum_{i=1}^{k-1} C_{\xi i} e^{(k-i-1)}(t) \qquad (\gamma + 1 \le k \le \gamma + m) \tag{29}$$

where

$$C_{\xi i} = \begin{cases} -\Delta_i + \theta_{\gamma+m-i} & (1 \le i \le \gamma + m - 1) \\ \theta_0 - \sum_{j=1}^{m} \beta_{j-1} C_{\xi \gamma+j-1} & (i = \gamma + m) \end{cases} \tag{30}$$

$$\Delta_i = \begin{cases} \beta_{m-1} & (i = 1) \\ \beta_{m-i} + \sum_{j=1}^{m-1} \beta_{m-i+j} C_{\xi j} & (2 \le i \le m) \\ \sum_{j=1}^{m} \beta_{j-1} C_{\xi j-m+i-1} & (m + 1 \le i \le \gamma + m - 1) \end{cases} \tag{31}$$

Then it is easily to confirm that the error system (22) or (23) with $\bar{u}(t)$ as the input can be transformed into the following form with $u(t)$ as the input.

$$\begin{aligned} \dot{\boldsymbol{e}}(t) &= A_e \boldsymbol{e}(t) + \boldsymbol{b}_e u(t) + C_e \boldsymbol{\xi}(t) \\ \dot{\boldsymbol{\xi}}(t) &= A_{IM} \boldsymbol{\xi}(t) + B_{\xi} \boldsymbol{e}(t) + C_{\xi} \boldsymbol{\eta}_{ze}(t) \\ \dot{\boldsymbol{\eta}}_{ze}(t) &= Q_{ze} \boldsymbol{\eta}_{ze}(t) + \boldsymbol{c}_{\eta e} e(t) \\ e(t) &= [1 0] \, \boldsymbol{e}(t) \end{aligned} \tag{32}$$

where

$$\boldsymbol{e}(t) = \left[e(t), \dot{e}(t), \cdots, e^{(\gamma-1)}(t) \right]^T, \ \boldsymbol{\xi}(t) = \left[\xi_{\gamma+1}(t), \cdots, \xi_{\gamma+m}(t) \right]^T \tag{33}$$

and

$$\begin{aligned} A_e &= \begin{bmatrix} \mathbf{0} & \boldsymbol{I}_{\gamma-1\times\gamma-1} \\ -C_{\xi\gamma} & \cdots - C_{\xi 1} \end{bmatrix} , \ C_e = \begin{bmatrix} \mathbf{0} & \mathbf{0} \\ 1 & \mathbf{0} \end{bmatrix} \\ \boldsymbol{b}_e &= [0 \cdots 0 b_{ze}]^T \, , \, b_{ze} = \boldsymbol{c}_1^T A_1^{\gamma+m-1} B_1 \in R \end{aligned} \tag{34}$$

$$A_{IM} = \begin{bmatrix} \mathbf{0} & \boldsymbol{I}_{m-1\times m-1} \\ -\beta_0 & \cdots - \beta_{m-1} \end{bmatrix}, \; B_{\xi} = \begin{bmatrix} -C_{\xi\gamma+1} & \mathbf{0} \cdots \mathbf{0} \\ \vdots & \vdots \ddots \vdots \\ -C_{\xi\gamma+m} & \mathbf{0} \cdots \mathbf{0} \end{bmatrix}, \; C_{\xi} = \begin{bmatrix} \mathbf{0} \\ \boldsymbol{c}_{ze}^T \end{bmatrix} \tag{35}$$

Note that this obtained error system with $u(t)$ as an input has relative degree of γ.

3. Adaptive PID control system design

Here we show an adaptive PID control system design scheme for the error system (32) based on system's ASPR properties.

3.1 Almost Strictly Positive Realness (ASPR-ness)

Let's consider the following nth order SISO system:

$$\begin{aligned} \dot{\boldsymbol{x}}(t) &= A\boldsymbol{x}(t) + \boldsymbol{b}u(t) \\ y(t) &= \boldsymbol{c}^T\boldsymbol{x}(t) \end{aligned} \tag{36}$$

where, $\boldsymbol{x}(t) \in R^n$ is a state vector and $u(t), y(t) \in R$ are the input and the output, respectively. The ASPR-ness (almost strictly positive real-ness) of the system (36) is defined as follows:

Definition 1. *(Bar-Kana & Kaufman, 1985; Kaufman et al., 1997) The system (36) is called ASPR if there exists a static output feedback such that the resulting closed-loop system is strictly positive real (SPR). That is, system (36) is ASPR if there exists a control input with a feedback gain θ_p^*,*

$$u(t) = -\theta_p^* y(t) + v(t), \quad \theta_p^* > 0 \tag{37}$$

such that the resulting closed-loop system from $v(t)$ to $y(t)$:

$$\begin{aligned} \dot{\boldsymbol{x}}(t) &= A_{cl}\boldsymbol{x}(t) + \boldsymbol{b}v(t) \\ y(t) &= \boldsymbol{c}^T\boldsymbol{x}(t) \end{aligned} \tag{38}$$

$$A_{cl} = A - \theta_p^* \boldsymbol{b}\boldsymbol{c}^T \tag{39}$$

is strictly positive real (SPR).

The sufficient conditions for a system to be ASPR are given as follows (Kaufman et al., 1997):

(1) The relative degree of the system is 0 or 1.

(2) The system is minimum-phase.

(3) The high frequency gain of the system is positive.

Remark 2: The system (38) with the transfer function $G_c(s) = \boldsymbol{c}^T(sI - A_{cl})^{-1}\boldsymbol{b}$ is positive real if, for $\mathrm{Re}(s) \geq 0$, $\mathrm{Re}G_c(s) \geq 0$, and it is SPR if , for some $\varepsilon > 0$, $G_c(s-\varepsilon)$ is PR. Furthermore, if the system (38) is SPR, then there exist symmetric positive definite matrices P and Q such that the following Kalman-Yakubovich-Popov Lemma is satisfied.

$$\begin{aligned} A_{cl}^T P + P A_{cl} &= -Q \\ P\boldsymbol{b} &= \boldsymbol{c} \end{aligned} \tag{40}$$

3.2 Augmented ASPR controlled system design

Since the error system (32) has a relative degree of γ and also might be nonminimum-phase (If the system(1) is nonminimum-phase, then the error system (32) is nonminimum-phase), it is not ASPR. Here, how to make an augmented ASPR controlled system is first considered in order to design an adaptive PID controller based on the ASPR properties of the controlled system.

Let's consider the following parallel feedforward compensator (PFC) of order n_f:

$$\begin{aligned} \dot{\boldsymbol{x}}_f(t) &= A_f \boldsymbol{x}_f(t) + \boldsymbol{b}_f u(t) \\ y_f(t) &= \boldsymbol{c}_f^T \boldsymbol{x}_f(t) \end{aligned} \tag{41}$$

The augmented error system with the PFC (41) can be represented by

$$\begin{aligned} \dot{\boldsymbol{x}}_a(t) &= A_a \boldsymbol{x}_a(t) + \boldsymbol{b}_a u(t) \\ e_a(t) &= \boldsymbol{c}_a^T \boldsymbol{x}_a(t) = e(t) + y_f(t) \end{aligned} \tag{42}$$

where

$$\begin{gathered} \boldsymbol{x}_a(t) = \left[e(t)^T,\ \boldsymbol{\xi}(t)^T,\ \boldsymbol{\eta}_{ze}(t)^T,\ \boldsymbol{x}_f(t)^T \right]^T \\ A_a = \begin{bmatrix} A_e & C_e & \mathbf{0} & \mathbf{0} \\ B_\xi & A_{IM} & C_\xi & \mathbf{0} \\ C & \mathbf{0} & Q_{ze} & \mathbf{0} \\ \mathbf{0} & \mathbf{0} & \mathbf{0} & A_f \end{bmatrix} \quad \boldsymbol{b}_a = \begin{bmatrix} \boldsymbol{b}_e \\ \mathbf{0} \\ \mathbf{0} \\ \boldsymbol{b}_f \end{bmatrix} \quad \boldsymbol{c}_a = \begin{bmatrix} 1 \\ \mathbf{0} \\ \mathbf{0} \\ \boldsymbol{c}_f \end{bmatrix} \end{gathered} \tag{43}$$

Thus, if PFC is designed so as to render the resulting augmented system ASPR, we can consider the ASPR augmented system as a new controlled system.

3.3 PFC design

The PFC must render the resulting augmented system ASPR. Up to now several kinds of PFC design scheme which make the augmented system ASPR have been proposed (Iwai & Mizumoto, 1994; Kaufman et al., 1997; Mizumoto & Iwai, 1996). Here, we show a simple robust PFC design scheme based on an estimated or approximated model of the controlled system.

Let $G_p^*(s)$ be a given or roughly estimated approximated model of the controlled system (We suppose that any approximated model is available). A PFC, which renders the resulting augmented system ASPR, can be designed as follows:

$$G_{PFC}(s) = \frac{1}{k}\left\{ G_{ASPR}(s) - G_p^*(s) \right\}\ ,\ k \geq 1 \tag{44}$$

where $G_{ASPR}(s)$ is a desired (or designed) ASPR model. In a general case, k is designed as $k = 1$, however, in order to expand the versatility of the PFC design, here we introduce a weight $k \geq 1$. The resulting augmented system can be represented by

$$\begin{aligned} G_a(s) &= G_p(s) + G_{PFC}(s) \\ &= G_p^*(s) + G_{PFC}(s) + G_p(s) - G_p^*(s) \\ &= \tilde{G}_{ASPR}(s)\left\{ 1 + \Delta(s) \right\} \end{aligned} \tag{45}$$

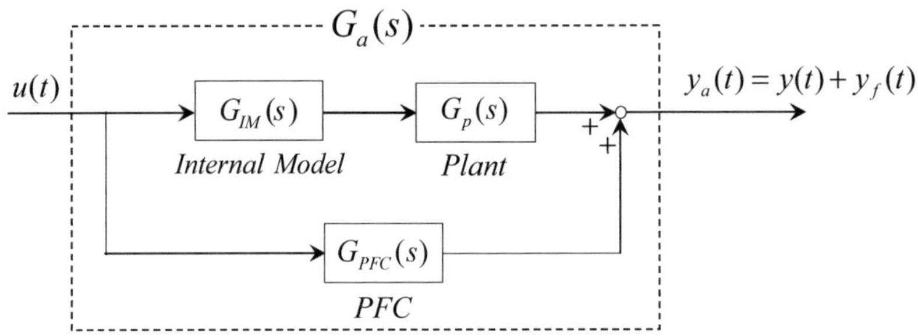

Fig. 2. Block diagram of the augmented system

where

$$\begin{aligned}\tilde{G}_{ASPR}(s) &= G_p^*(s) + G_{PFC}(s)\\ \Delta(s) &= \tilde{G}_{ASPR}(s)^{-1}\Delta G_p(s)\\ \Delta G_p(s) &= G_p(s) - G_p^*(s)\end{aligned} \tag{46}$$

$\Delta(s)$ represents an uncertain part of the augmented system.
The following lemma concerns the ASPR-ness of the resulting augmented system (45) (Mizumoto & Iwai, 1996).

Lemma 1. *The augmented system (45) is ASPR if*

(1) $G_{ASPR}(s)$ *is ASPR.*

(2) $\Delta(s) \in RH_\infty$.

(3) $\|\Delta(s)\|_\infty < 1$.
Where $\|\Delta(s)\|_\infty$ *denote the* H_∞ *norm of* $\Delta(s)$ *which is defined as* $\|\Delta(s)\|_\infty = \sup_{s \in C_{+e}} |\Delta(s)|$.

Remark 3: Theoretically, one can select any ASPR model as $G_{ASPR}(s)$. However, performance of the control system may be influenced by the given ASPR model. For example, if the time constant of the given $G_{ASPR}(s)$ is small, one can attain fast tracking of the augmented system with small input. However, since the resulting PFC might have a large gain, the tracking of the practical output $y(t)$ has delay. One the centrally, if the time constant of $G_{ASPR}(s)$ is large, one can attain quick tracking for the practical output $y(t)$. However, large control input will be required (Minami et al., 2010).

The overall block diagram of the augmented system for the system with an internal model filter $G_{IM}(s)$ can be shown as in Fig. 2. Thus, introducing an internal model filter, the PFC must be designed for a system $G_{IM}(s)G_p(s)$. Unfortunately, in the case where $G_{IM}(s)$ is not stable the PFC design conditions given in Theorem 1 are not satisfied even if the controlled system $G_p(s)$ is originally stable. For such cases, the PFC can be designed according to the following procedure.

Step 1: Introduce a PFC as shown in Figure 3.
Step 2: Consider designing a PFC $G_{PFC}(z)$ so as to render the augmented system $G_c(s) = G_p(s) + G_{PFC}(s)$ for the controlled system $G_p(z)$ ASPR.

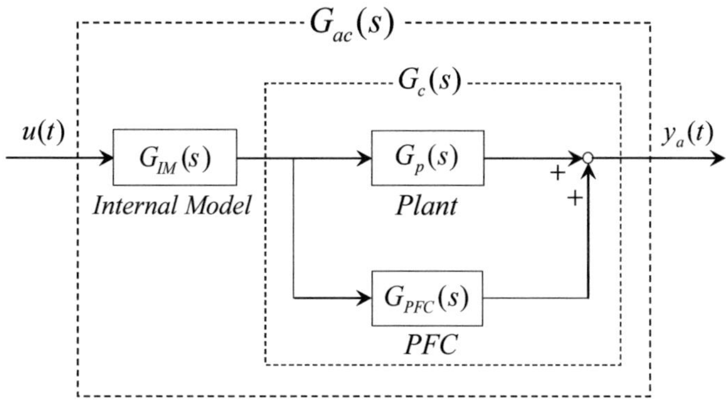

Fig. 3. Block diagram of a modified augmented system

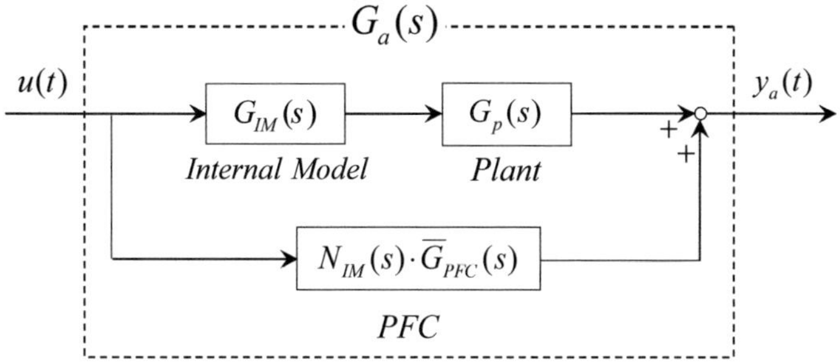

Fig. 4. Equivalent augmented system

Step 3: Design the desired ASPR model so that the obtained PFC $G_{PFC}(s)$ has $D_{IM}(s)$ as a part of the numerator. That is, the designed $G_{PFC}(z)$ must have a form of

$$G_{PFC}(s) = D_{IM}(s) \cdot \bar{G}_{PFC}(s) \, , \; \bar{G}_{PFC}(s) = \frac{\bar{N}_{PFC}(s)}{\bar{D}_{PFC}(s)} \tag{47}$$

where $D_{IM}(s)$ and $\bar{D}_{PFC}(s)$ are coprime polynomials.

In this case, the obtained augmented system $G_{ac}(z) = G_c(s)G_{IM}(s)$ is ASPR since both $G_c(s)$ is ASPR and $G_{IM}(s)$ is ASPR with relative degree of 0. Further, since the overall system given in Fig. 3 is equivalent to the system shown in Fig. 4, one can obtain an equivalent PFC that can render $G_p(s)G_{IM}(s)$ ASPR.

3.4 Adaptive PID controller design

For an ASPR controlled system with a PFC, let's consider an ideal PID control input given as follows:

$$u^*(t) = -\tilde{\theta}_p^* e_a(t) - \tilde{\theta}_i^* w(t) - \tilde{\theta}_d^* \dot{e}_a(t) \tag{48}$$

with

$$\tilde{\theta}_p^* > 0 \, , \; \tilde{\theta}_i^* > 0 \, , \; \tilde{\theta}_d^* > 0 \tag{49}$$

and

$$\dot{w}(t) = e_a(t) - \sigma_i w(t) \;\; , \;\; \sigma_i > 0 \tag{50}$$

$w(t)$ is an pseudo-integral signal of $e_a(t)$ and $\tilde{\theta}_p^*$ is the ideal feedback gain which makes the resulting closed-loop of (42) SPR. That is, for the control system with $u^*(t)$ as the control input, considering a closed-loop system:

$$\begin{aligned} \dot{\boldsymbol{x}}_a(t) &= A_c \boldsymbol{x}_a(t) + \boldsymbol{b}_a v(t) \\ e_a(t) &= \boldsymbol{c}_a^T \boldsymbol{x}_a(t) \end{aligned} \tag{51}$$

where

$$\begin{aligned} A_c &= A_a - \tilde{\theta}_p^* \boldsymbol{b}_a \boldsymbol{c}_a^T \\ v(t) &= -\tilde{\theta}_i^* w(t) - \tilde{\theta}_d^* \dot{e}_a(t) \end{aligned} \tag{52}$$

the closed-loop system $(A_c, \boldsymbol{b}_a, \boldsymbol{c}_a)$ is SPR.
This means that the resulting control system with the input (48) will be stabilized by setting sufficiently large $\tilde{\theta}_p^*$ and any $\tilde{\theta}_i^* > 0$ and $\tilde{\theta}_d^* > 0$, which can be easily confirmed using the ASPR properties of the controlled system.
Unfortunately, however, since the controlled system is unknown, one can not design ideal PID gains. Therefore, we consider designing the PID controller adaptively by adaptively adjusting the PID parameters as follows:

$$\begin{aligned} u(t) &= -\tilde{\theta}_p(t) e_a(t) - \tilde{\theta}_i(t) w(t) - \tilde{\theta}_d(t) \dot{e}_a(t) \\ &= -\tilde{\boldsymbol{\theta}}(t)^T \tilde{\boldsymbol{z}}(t) \end{aligned} \tag{53}$$

where

$$\begin{aligned} \tilde{\boldsymbol{\theta}}(t)^T &= \left[\tilde{\theta}_p(t)\ \tilde{\theta}_i(t)\ \tilde{\theta}_d(t)\right] \\ \tilde{\boldsymbol{z}}(t) &= \left[e_a(t)\ \ w(t)\ \ \dot{e}_a(t)\right]^T \end{aligned} \tag{54}$$

and $\tilde{\boldsymbol{\theta}}(t)$ is adaptively adjusting by the following parameter adjusting law.

$$\begin{aligned} \dot{\tilde{\theta}}_p(t) &= \gamma_p e_a^2(t), \quad \gamma_p > 0 \\ \dot{\tilde{\theta}}_i(t) &= \gamma_i w(t) e_a(t), \quad \gamma_i > 0 \\ \dot{\tilde{\theta}}_d(t) &= \gamma_d \dot{e}_a(t) e_a(t), \quad \gamma_d > 0 \end{aligned} \tag{55}$$

The resulting closed-loop system can be represented as

$$\begin{aligned} \dot{\boldsymbol{x}}_a(t) &= A_c \boldsymbol{x}_a(t) + \boldsymbol{b}_a \{\Delta u(t) + v(t)\} \\ e_a(t) &= \boldsymbol{c}_a^T \boldsymbol{x}_a(t) \end{aligned} \tag{56}$$

where

$$\Delta u(t) = u(t) - u^*(t) \tag{57}$$

$$= -\Delta\tilde{\boldsymbol{\theta}}(t)^T \tilde{\boldsymbol{z}}(t) \tag{58}$$

with

$$\Delta\tilde{\boldsymbol{\theta}}(t) = \begin{bmatrix} \tilde{\theta}_p(t) - \tilde{\theta}_p^* \\ \tilde{\theta}_i(t) - \tilde{\theta}_i^* \\ \tilde{\theta}_d(t) - \tilde{\theta}_d^* \end{bmatrix} \tag{59}$$

3.5 Stability analysis

Considering the ideal proportional gain $\tilde{\theta}_p^*$, the closed-loop system $(A_c, \boldsymbol{b}_a, \boldsymbol{c}_a)$ is SPR. Then there exist symmetric positive definite matrices $P = P^T > 0, Q = Q^T > 0$, such that the following Kalman-Yakubovich-Popov Lemma is satisfied

$$\begin{aligned} A_c^T P + PA_c &= -Q \\ P\boldsymbol{b}_a &= \boldsymbol{c}_a \end{aligned} \tag{60}$$

Now, consider the following positive definite function $V(t)$:

$$V(t) = V_1(t) + V_2(t) + V_3(t) \tag{61}$$

$$V_1(t) = \boldsymbol{x}_a(t)^T P \boldsymbol{x}_a(t) \tag{62}$$

$$V_2(t) = \tilde{\theta}_i^* w(t)^2 + \tilde{\theta}_d^* e_a(t)^2 \tag{63}$$

$$V_3(t) = \Delta\tilde{\boldsymbol{\theta}}(t)^T \Gamma^{-1} \Delta\tilde{\boldsymbol{\theta}}(t) \tag{64}$$

The time derivative of $V_1(t)$ can be expressed by

$$\begin{aligned} \dot{V}_1(t) &= \dot{\boldsymbol{x}}_a(t)^T P\, \boldsymbol{x}_a(t) + \boldsymbol{x}_a(t)^T P\, \dot{\boldsymbol{x}}_a(t) \\ &= \boldsymbol{x}_a(t)^T \left(A_c^T P + PA_c\right) \boldsymbol{x}_a(t) + 2\boldsymbol{b}_a^T P \boldsymbol{x}_a(t)\{\Delta u(t) + v(t)\} \\ &= -\boldsymbol{x}_a(t)^T Q \boldsymbol{x}_a(t) + 2e_a(t)\{\Delta u(t) + v(t)\} \end{aligned} \tag{65}$$

Further, the derivative of $V_2(t)$ is obtained as

$$\begin{aligned} \dot{V}_2(t) &= 2\tilde{\theta}_i^* w(t)\dot{w}(t) + 2\tilde{\theta}_d^* e_a(t)\dot{e}_a(t) \\ &= 2\tilde{\theta}_i^* w(t)\{e_a(t) - \sigma_i w(t)\} + 2\tilde{\theta}_d^* e_a(t)\dot{e}_a(t) \\ &= 2\tilde{\theta}_i^* w(t)e_a(t) + 2\tilde{\theta}_d^* \dot{e}_a(t)e_a(t) - 2\sigma_i\tilde{\theta}_i^* w(t)^2 \\ &= -2e_a(t)v(t) - 2\sigma_i\tilde{\theta}_i^* w(t)^2 \end{aligned} \tag{66}$$

and the time derivative of $V_3(t)$ can be obtained by

$$\begin{aligned} \dot{V}_3(t) &= \Delta\dot{\tilde{\boldsymbol{\theta}}}(t)^T \Gamma^{-1} \Delta\tilde{\boldsymbol{\theta}}(t) + \Delta\tilde{\boldsymbol{\theta}}(t)^T \Gamma^{-1} \Delta\dot{\tilde{\boldsymbol{\theta}}}(t) \\ &= \frac{2}{\gamma_p}\Delta\tilde{\theta}_p(t)\Delta\dot{\tilde{\theta}}_p(t) + \frac{2}{\gamma_i}\Delta\tilde{\theta}_i(t)\Delta\dot{\tilde{\theta}}_i(t) + \frac{2}{\gamma_d}\Delta\tilde{\theta}_d(t)\Delta\dot{\tilde{\theta}}_d(t) \\ &= 2\Delta\tilde{\theta}_p(t)e_a(t)^2 + 2\Delta\tilde{\theta}_i(t)w(t)e_a(t) + 2\Delta\tilde{\theta}_d(t)\dot{e}_a(t)e_a(t) \\ &= -2\Delta u(t)e_a(t) \end{aligned} \tag{67}$$

Finally, we have

$$\dot{V}(t) = -\boldsymbol{x}_a(t)^T Q \boldsymbol{x}_a(t) \leq 0 \tag{68}$$

and thus we can conclude that $\|\boldsymbol{x}_a(t)\|$ is bounded and L_2 and all the signals in the control system are also bounded. Furthermore, form (42) and boundedness of all the signals in the control system, we have $\|\dot{\boldsymbol{x}}_a(t)\| \in L_\infty$. Thus, using Barbalat's Lemma (Sastry & Bodson, 1989), we obtain

$$\lim_{t\to\infty} \boldsymbol{x}_a(t) = 0 \tag{69}$$

and then we can conclude that

$$\lim_{t\to\infty} e(t) = 0 \tag{70}$$

Remark 4: It should be noted that if there exist undesired disturbance and/or noise, one can not ensure the stability of the control system with the parameter adjusting law (55). In such case, one can design parameter adjusting laws as follows using σ-modification method:

$$\begin{aligned}
\dot{\tilde{\theta}}_p(t) &= \gamma_p e_a^2(t) - \sigma_P \tilde{\theta}_p(t), \quad \gamma_p > 0,\ \sigma_P > 0 \\
\dot{\tilde{\theta}}_i(t) &= \gamma_i w(t) e_a(t) - \sigma_I \tilde{\theta}_i(t), \quad \gamma_i > 0,\ \sigma_I > 0 \\
\dot{\tilde{\theta}}_d(t) &= \gamma_d \dot{e}_a(t) e_a(t) - \sigma_D \tilde{\theta}_d(t), \quad \gamma_d > 0,\ \sigma_D > 0
\end{aligned} \tag{71}$$

In this case, we only confirm the boundedness of all the signals in the control system.

Remark 5: If the exosystem (2) has unstable characteristic polynomial, then since $w_d(t)$ and/or $r(t)$ are not bounded, one cannot guarantee the boundedness of the signals in the control system, although it is attained that $\lim_{t\to\infty} e(t) = 0$.

4. Application to control of unsaturated highly accelerated stress test system

4.1 Unsaturated highly accelerated stress test system

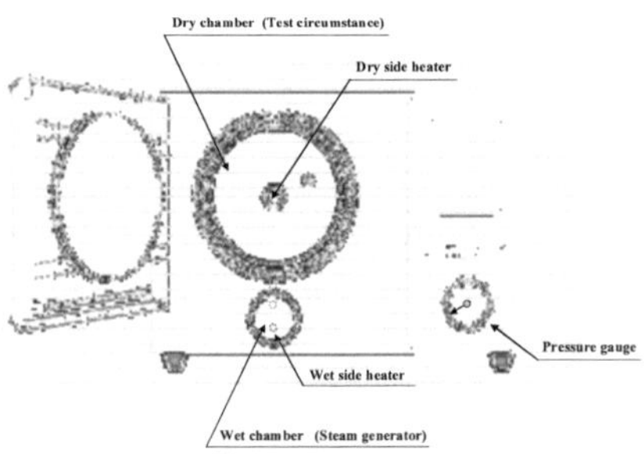

Fig. 5. Schematic view of the unsaturated HAST system

We consider to apply the ASPR based adaptive PID method to the control of an unsaturated HAST (Highly Accelerated Stress Test) system. Fig. 5 shows a schematic view of the unsaturated HAST system. In this system the temperature in the dry chamber has to raise quickly at a set point within 105.0 to 144.4 degree and must be kept at set point with 100 % or 85 % or 75% RH (relative humidity). To this end, we control the temperature in the dry chamber and wet chamber by heaters setting in the chambers.

In the general unsaturated HAST system, the system is controlled by a conventional PID scheme with static PID gains. However, since the HAST system has highly nonlinearities and the system might be changed at higher temperature area upper than 100 degree and furthermore, the dry chamber and the wet chamber cause interference of temperatures each other, it was difficult to control this system by static PID. Fig. 6 shows the experimental result with a packaged PID under the control conditions of 120 degree in the dry chamber at 85 % RH (The result shows the performance of the HAST which is available in the market). The temperature in the dry chamber was oscillating and thus the relative humidity was also oscillated, and it takes long time to reach the set point stably. The requirement from the user is to attain a faster rising time and to maintain the steady state quickly.

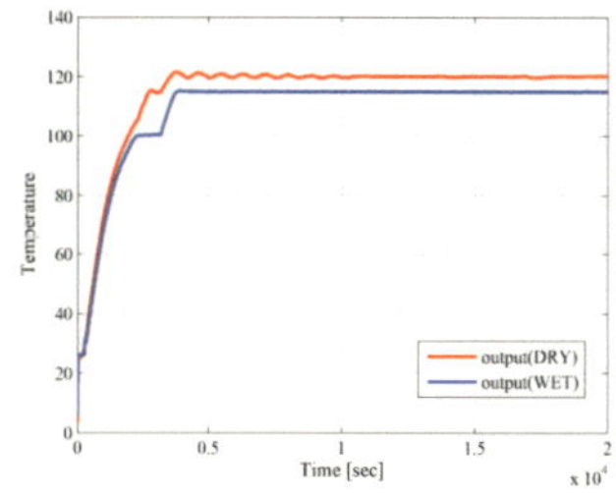

Fig. 6. Temperature in the dry chamber with a packaged PID: set point at 120 degree

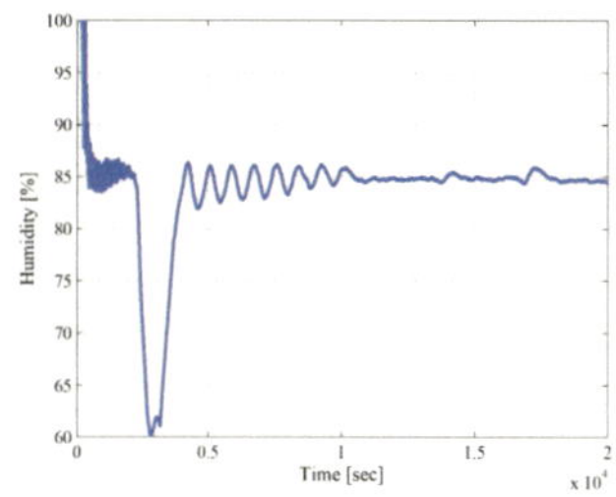

Fig. 7. Relative humidity with a packaged PID: 85 % RH

4.2 System's approximated model

Using a step response under 100 degree, we first identify system models of dry chamber and wet chamber respectively (see Figs. 8).

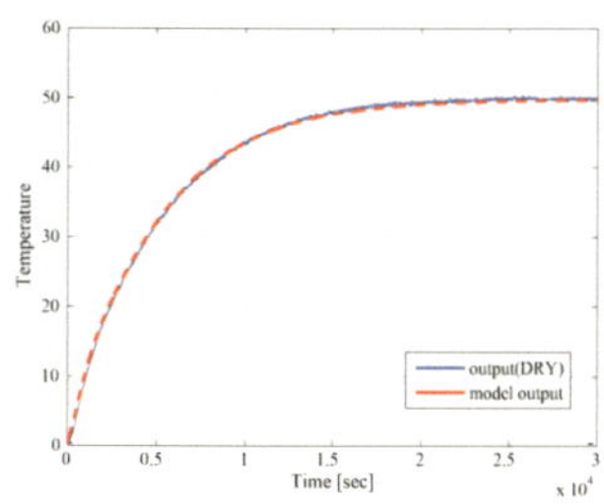

(a) Temperature in the dry chamber

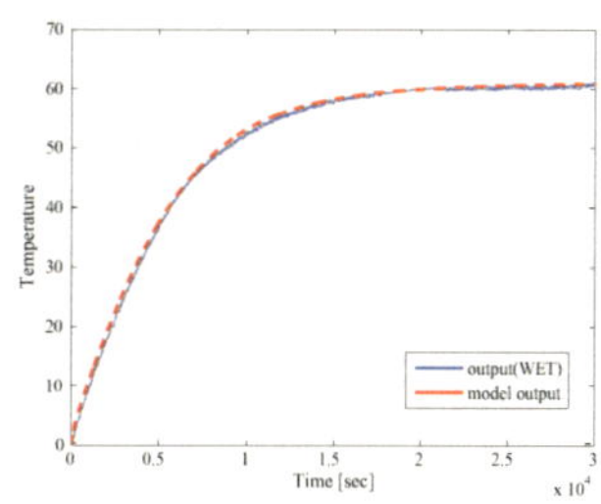

(b) Temperature in the wet chamber

Fig. 8. Step response

The identified models were obtained as follows by using Prony's Method (Iwai et al., 2005):
For dry chamber:

$$G_{P-DRY}(s) = \frac{a_1 s^4 + b_1 s^3 + c_1 s^2 + d_1 s + e_1}{s^5 + f_1 s^4 + g_1 s^3 + h_1 s^2 + i_1 s + j_1} \tag{72}$$

$$\begin{aligned} a_1 &= 0.02146\,,\ b_1 = 0.000185\,,\ c_1 = 1.344 \times 10^{-6}\,,\ d_1 = 1.656 \times 10^{-9} \\ e_1 &= 1.068 \times 10^{-12}\,,\ f_1 = 0.02373\,,\ g_1 = 0.0001138 \\ h_1 &= 1.778 \times 10^{-7}\,,\ i_1 = 1.357 \times 10^{-10}\,,\ j_1 = 2.146 \times 10^{-14} \end{aligned} \tag{73}$$

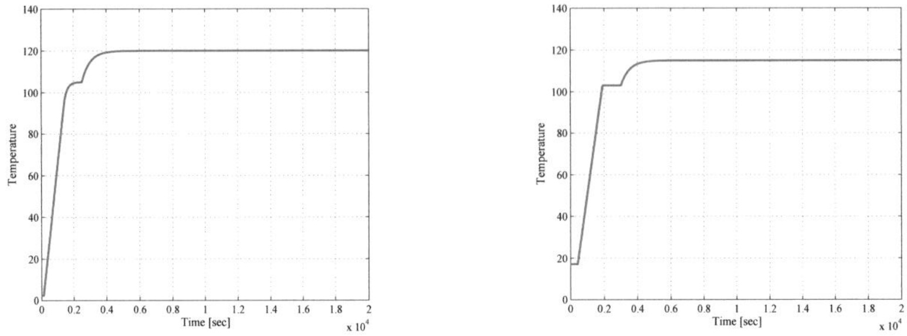

(a) Temperature in dry chamber (b) Temperature in wet chamber

Fig. 9. Reference signals

For wet chamber:

$$G_{P-WET}(s) = \frac{a_2 s^3 + b_2 s^2 + c_2 s + d_2}{s^4 + e_2 s^3 + f_2 s^2 + g_2 s + h_2} \tag{74}$$

$$\begin{aligned} a_2 &= 0.02122\,,\ b_2 = 7.078\times 10^{-5}\,,\ c_2 = 3.906\times 10^{-8} \\ d_2 &= 9.488\times 10^{-12}\,,\ e_2 = 0.006775\,,\ f_2 = 4.493\times 10^{-6} \\ g_2 &= 1.424\times 10^{-9}\,,\ h_2 = 1.555\times 10^{-13} \end{aligned} \tag{75}$$

It is noted that the HAST system is a two-input/two-output system so that we would have the following system representation.

$$\begin{bmatrix} y_{DRY}(t) \\ y_{WET}(t) \end{bmatrix} = \begin{bmatrix} G_{11}(s)\ G_{12}(s) \\ G_{21}(s)\ G_{22}(s) \end{bmatrix} \begin{bmatrix} u_{DRY}(t) \\ u_{WET}(t) \end{bmatrix} \tag{76}$$

For this system, we consider designing a decentralized adaptive PID controller to each control input $u_{DRY}(t)$ and $u_{WET}(t)$. Therefore, in order to design PFCs for each subsystem, we only identified subsystems $G_{11}(s) = G_{P-DRY}(s)$ and $G_{22}(s) = G_{P-WET}(s)$.

4.3 Control system design

The control objective is to have outputs $y_{DRY}(t)$ and $y_{WET}(t)$, which are temperatures in the dry chamber and the wet chamber respectively, track a desired reference signal to attain a desired temperature in dry chamber and desired relative humidity. For example, if one would like to attain a test condition with the temperature in dry chamber of 120 degree with 85 % RH, the reference signals shown in Fig. 9 will be set.

In order to attain control objective, we first design internal model filters as follows:

$$G_{IM-DRY}(s) = \frac{100s+1}{s}\,,\ G_{IM-WET}(s) = \frac{170s+1}{s} \tag{77}$$

Further, for each controlled subsystem with the internal models, we set desired ASPR models as follows in order to design PFCs for each subsystems.

$$G_{ASPR-DRY}(s) = \frac{49.8}{250s+1}\,,\ G_{ASPR-WET}(s) = \frac{61.0}{100s+1} \tag{78}$$

Then the PFCs were designed according to the model-based PFC design scheme given in (44) using obtained approximated model $G_{P-DRY}(s)$ and $G_{P-WET}(s)$ as follows:

$$G_{PFC-DRY}(s) = \frac{1}{k_{DRY}}\left\{G_{ASPR-DRY}(s) - G_{P-DRY}(s)\right\}\ ,\ k_{DRY} = 100 \quad (79)$$

$$G_{PFC-WET}(s) = \frac{1}{k_{WET}}\left\{G_{ASPR-WET}(s) - G_{P-WET}(s)\right\}\ ,\ k_{WET} = 170 \quad (80)$$

For the obtained ASPR augmented subsystems with PFCs, the adaptive PID controllers are designed as in (53) with parameter adjusting laws given in (71). The designed parameters in (71) are given as follows:

$$\Gamma_{DRY} = \Gamma_{WET} = \mathrm{diag}[\gamma_d, \gamma_i, \gamma_d] = \mathrm{diag}[1 \times 10^{-2}, 1 \times 10^{-5}, 1 \times 10^{-8}] \quad (81)$$

$$\sigma_D = \sigma_I = \sigma_D = = 1.0 \times 10^{-10} \quad (82)$$

$$\sigma_i = 0 \quad (83)$$

4.4 Experimental results

We performed the following 4 types experiments.

(1) Quickly raise the temperature up to 120 degree and keep the relative humidity at 85 % RH.

(2) Quickly raise the temperature up to 130 degree and keep the relative humidity at 85 % RH.

(3) Quickly raise the temperature up to 121 degree and keep the relative humidity at 100 % RH.

(4) Quickly raise the temperature up to 120 degree and change the temperature to 130 and again 120 with keeping the relative humidity at 85 % RH.

Figs. 10 to 13 show the results for Experiment (1). Fig. 10 shows the temperature in the dry and wet chambers and the relative humidity. It can be seen that temperatures quickly reached to the desired values and the relative humidity was kept at set value. Fig. 11 shows the results with the given reference signal. Both temperatures in dry and wet chamber track the reference signal well. Fig. 12 are control inputs and Fig. 13 shows adaptively adjusted PID parameters. Figs. 14 to 17 show the resilts for Experiment (2), Figs. 18 to 21 show the resilts for Experiment (3) and Figs. 22 to 25 show the resilts for Experiment (4). All cases attain satisfactory performance.

5. Conclusion

In this Chapter, an ASPR based adaptive PID control system design strategy for linear continuous-time systems was presented. The adaptive PID scheme based on the ASPR property of the system can guarantee the asymptotic stability of the resulting PID control system and since the method presented in this chapter utilizes the characteristics of the ASPR-ness of the controlled system, the stability of the resulting adaptive control system can be guaranteed with certainty. Furthermore, by adjusting PID parameters adaptively, the method maintains a better control performance even if there are some changes of the system properties. In order to illustrate the effectiveness of the presented adaptive PID design scheme for real world processes, the method was applied to control of an unsaturated highly accelerated stress test system.

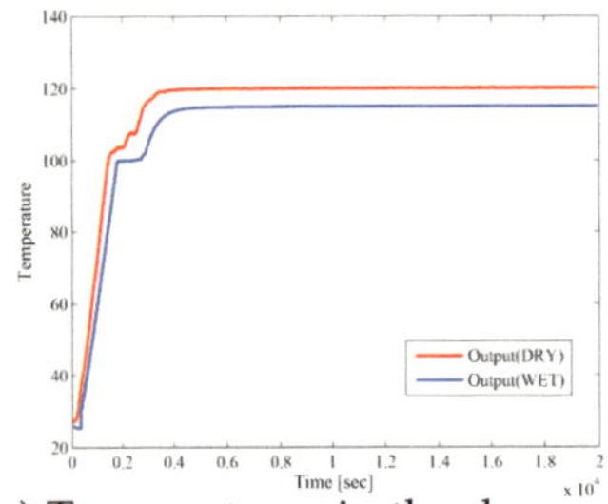

(a) Temperatures in the dry and wet chambers

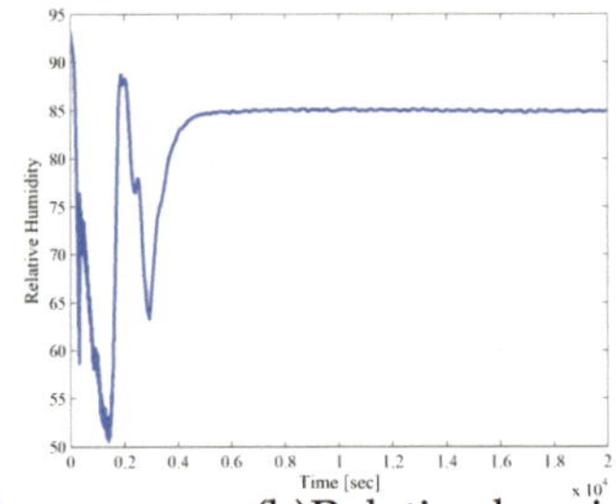

(b)Relative humidity

Fig. 10. Experimental results of outputs: 120 degree and 85 % RH

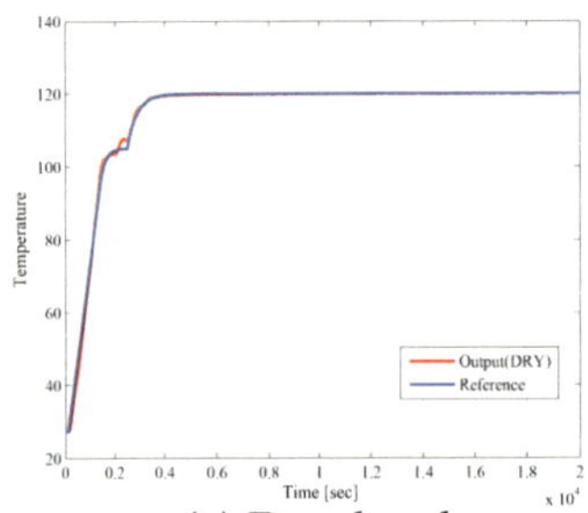

(a) Dry chamber

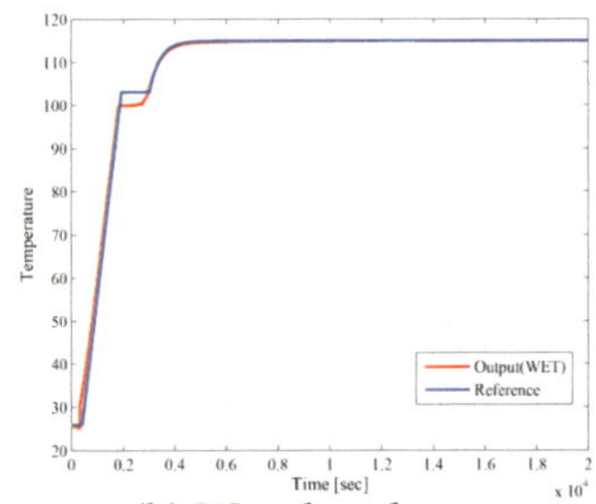

(b) Wet chamber

Fig. 11. Comparison between Output and Reference signal: 120 degree and 85 % RH

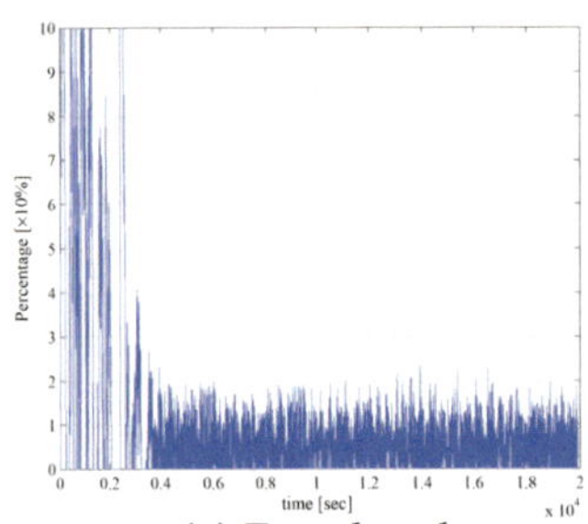

(a) Dry chamber

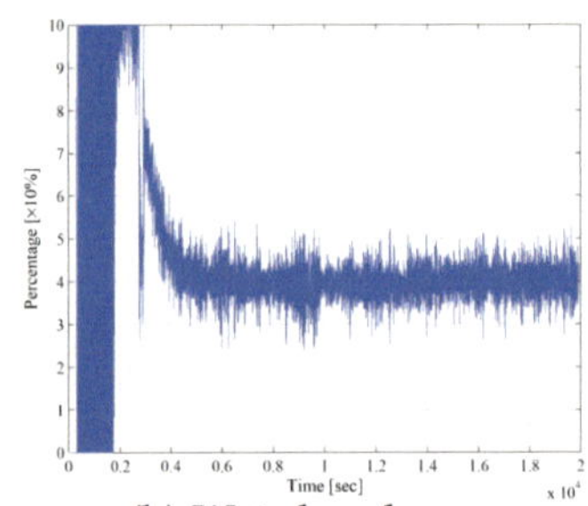

(b) Wet chamber

Fig. 12. Control Input: 120 degree and 85 % RH

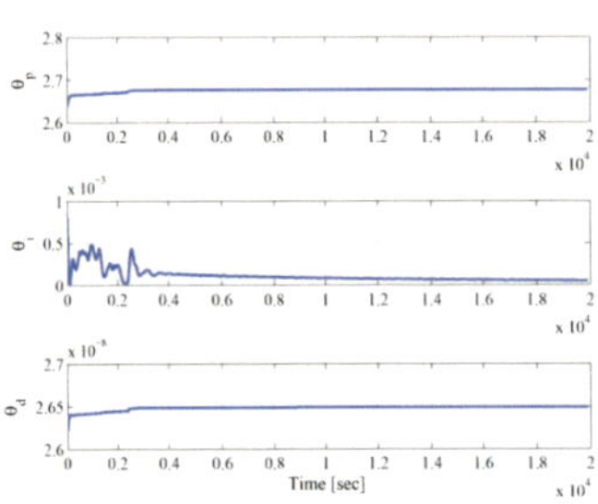

(a) Dry chamber

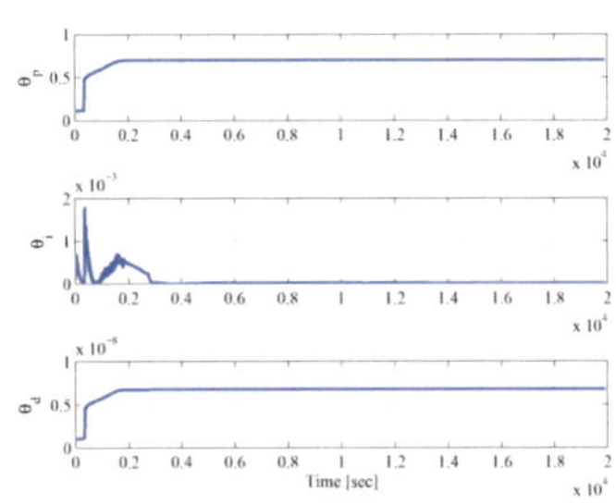

(b) Wet chamber

Fig. 13. Adaptively adjusted PID gains: 120 degree and 85 % RH

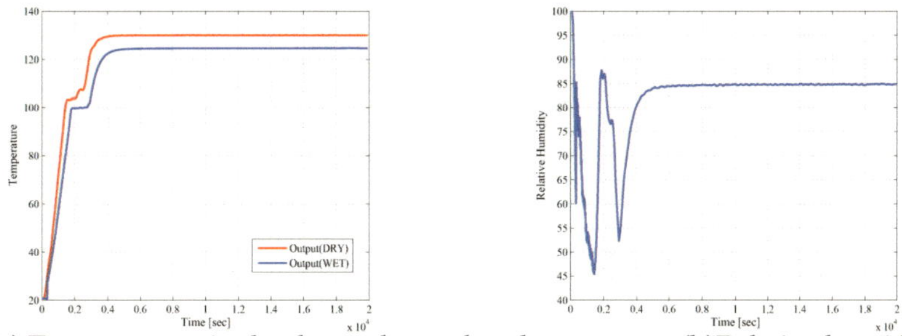

(a) Temperatures in the dry and wet chambers (b)Relative humidity

Fig. 14. Experimental results of outputs: 130 degree and 85 % RH

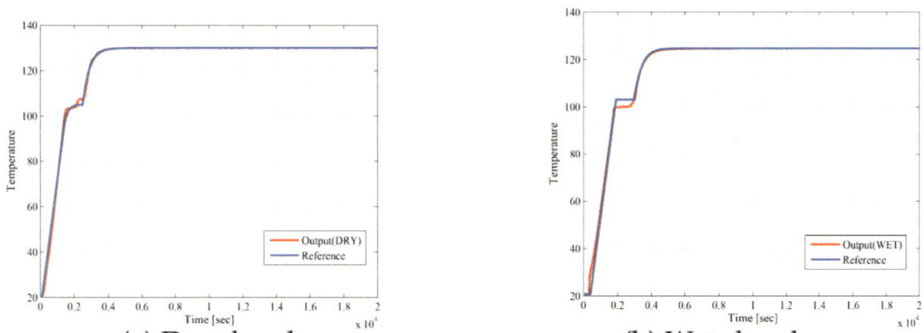

(a) Dry chamber (b) Wet chamber

Fig. 15. Comparison between Output and Reference signal: 130 degree and 85 % RH

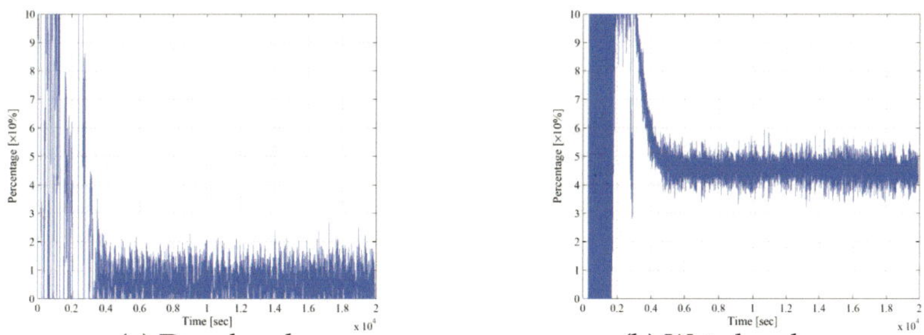

(a) Dry chamber (b) Wet chamber

Fig. 16. Control Input: 130 degree and 85 % RH

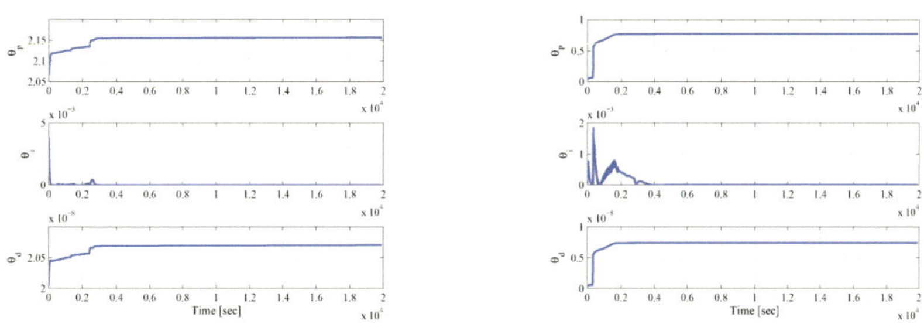

(a) Dry chamber (b) Wet chamber

Fig. 17. Adaptively adjusted PID gains: 130 degree and 85 % RH

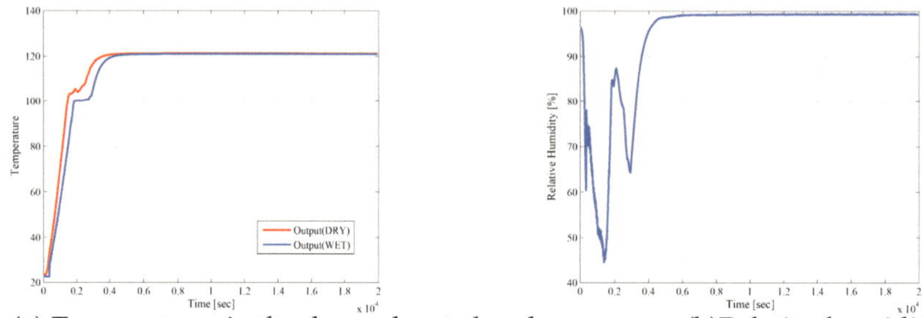

(a) Temperatures in the dry and wet chambers (b)Relative humidity

Fig. 18. Experimental results of outputs: 121 degree and 100 % RH

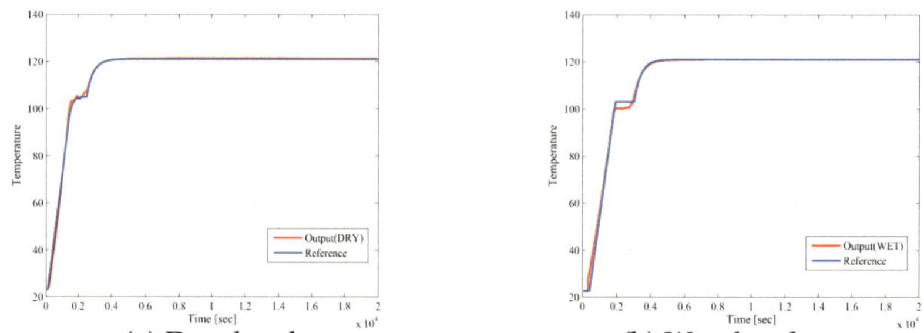

(a) Dry chamber (b) Wet chamber

Fig. 19. Comparison between Output and Reference signal: 121 degree and 100 % RH

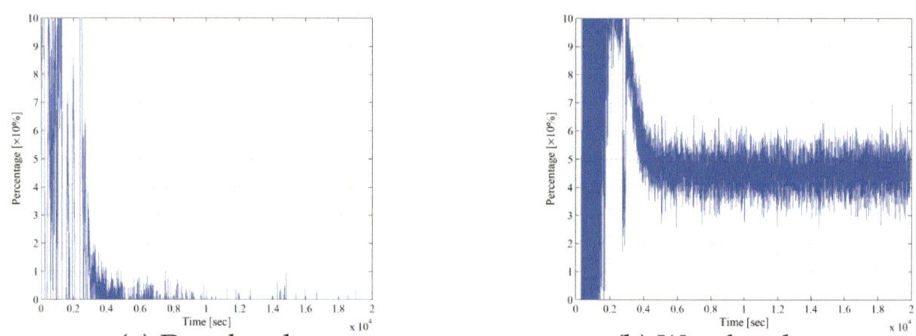

(a) Dry chamber (b) Wet chamber

Fig. 20. Control Input: 121 degree and 100 % RH

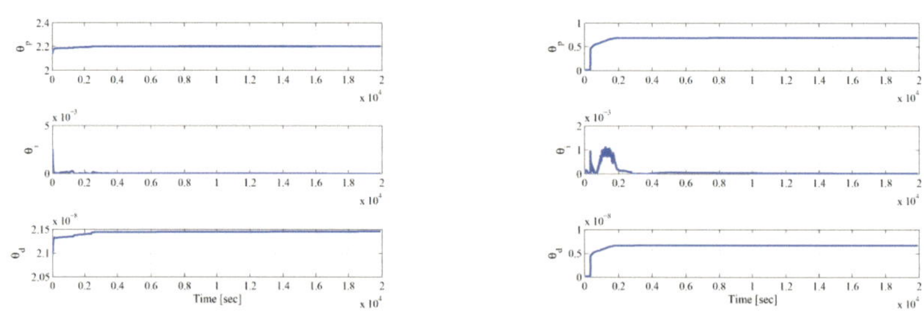

(a) Dry chamber (b) Wet chamber

Fig. 21. Adaptively adjusted PID gains: 121 degree and 100 % RH

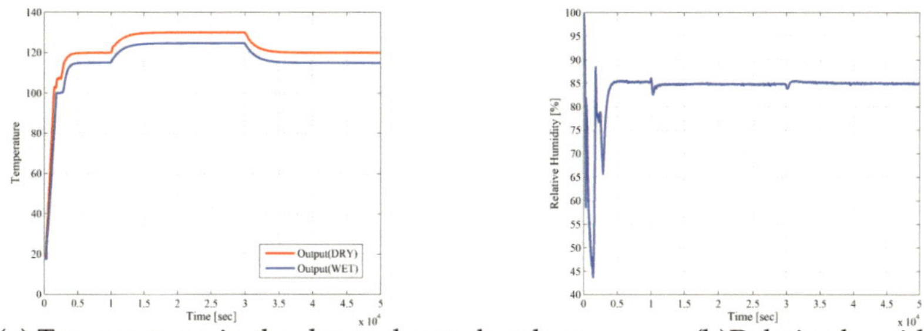

(a) Temperatures in the dry and wet chambers (b)Relative humidity

Fig. 22. Experimental results of outputs: 120 $\rightarrow$ 130 $\rightarrow$ 120 degree with 85 % RH

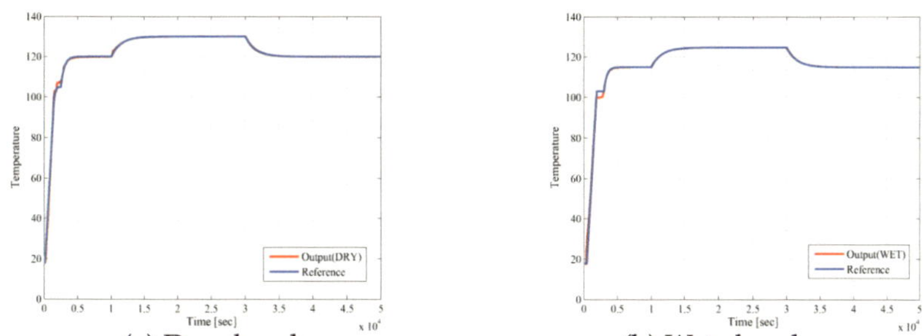

(a) Dry chamber (b) Wet chamber

Fig. 23. Comparison between Output and Reference signal: 120 $\rightarrow$ 130 $\rightarrow$ 120 degree with 85 % RH

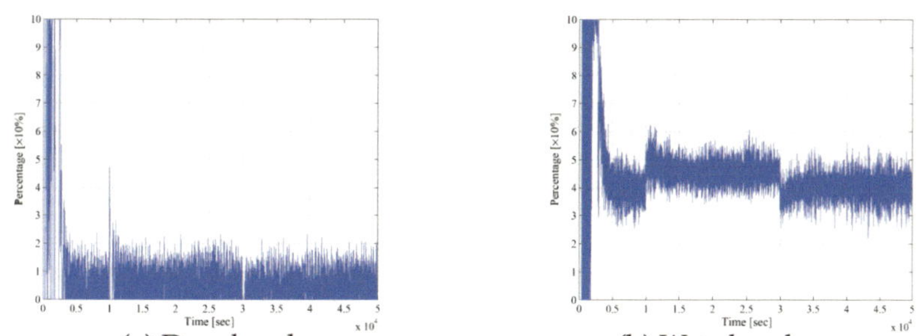

(a) Dry chamber (b) Wet chamber

Fig. 24. Control Input: 120 $\rightarrow$ 130 $\rightarrow$ 120 degree with 85 % RH

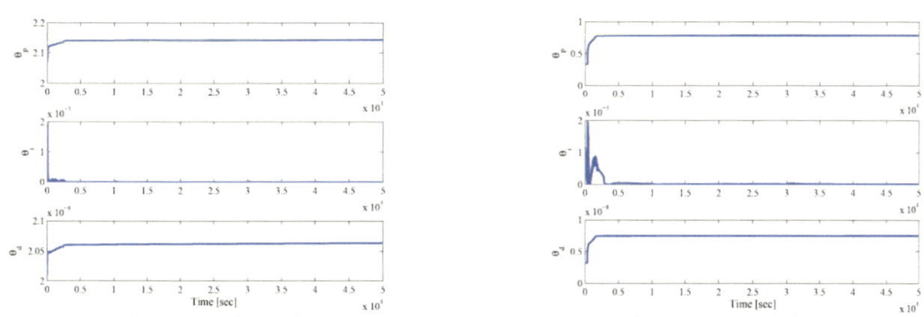

(a) Dry chamber (b) Wet chamber

Fig. 25. Adaptively adjusted PID gains: 120 $\rightarrow$ 130 $\rightarrow$ 120 degree with 85 % RH

6. References

Astrom, K. & Hagglund, T. (1995). Pid control, theory, design and tuning, Instrument Society of America, USA, second Ed.

Bar-Kana & Kaufman, H. (1985). Global stability and performance of a simplified adaptive algorithm, *International Journal of Control* 42(6): 1491–1505.

Chang, W.-D., Hwang, R.-C. & Hsieh, J.-G. (2003). A multivariable on-line adaptive pid controller using auto-tuning neurons, *Engineering Application of Artificial Intelligence* 16: 57–63.

Isidori, A. (1995). Nonlinear control systems, Springer-Verlag, third Ed.

Iwai, Z. & Mizumoto, I. (1994). Realization of simple adaptive control by using parallel feedforward compensator, *Int. J. of Control* 59(6): 1543–1565.

Iwai, Z., Mizumoto, I., Liu, L., Shah, S. & Jiang, H. (2006). Adaptive stable pid controller with parallel feedforward compensator, *Proc. of 9th Int. Conf. on Control, Automation, Robotics and Vision, Singapore* pp. 1253–1258.

Iwai, Z., Mizumoto, I., Nagata, M., Kumon, M. & Kubo, Y. (2005). Accuracy of identification and control performance in 3 parameter process model approximations (identification by prony's method and examination through model-driven pid control system design), *Trans. of the Japan Society of Mechanical Engineers (Ser. C)* 71(702): 589–596.

Kaufman, H., Barkana, I. & Sobel, K. (1997). *Direct Adaptive Control Algorithms*, 2nd edn, Springer.

Kono, T., Yamamoto, T., Hinamoto, T. & Shah, S. (2007). Design of a data-driven performance-adaptive pid controller, *Proc. of 9th IFAC Workshop on Adaptive and Learning in Control and Signal Processing, St. Petersburg, Russia* CD-ROM.

Minami, A., Mizumoto, I. & Iwai, Z. (2010). Model-based pfc design based on time-varying aspr model for anti-windup adaptive pid control, *SICE Annual Conference 2010* CD-ROM: 18–21.

Mizumoto, I. & Iwai, Z. (1996). Simplified adaptive model output following control for plants with unmodelled dynamics, *Int. J. of Control* 64(1): 61–80.

Ren, T.-J., Chen, T.-C. & Chen, C.-J. (2008). Motion control for a two-wheeled vehicle using a self-tuning pid controller, *Control Engineering Practice* 16(3): 365–375.

Sastry, S. & Bodson, M. (1989). *Adaptive Control –Stability, Convergence, and Robustness–*, Prentice Hall.

Tamura, K. & Ohmori, H. (2007). Auto-tuning method of expanded pid control for mimo systems, *Proc. of 9th IFAC Workshop on Adaptive and Learning in Control and Signal Processing, St. Petersburg, Russia* CD-ROM.

Yamamoto, T. & Shah, S. (2004). Design and experimental evaluation of multivariable self-tuning pid controller, *IEE Proc. of Control Theory and Applications* 151(5): 645–652.

Yu, D., Chang, T. & Yu, D. (2007). A stable self-learning pid sontrol for multivariable time varying systems, *Control Engineering Practice* 15(12): 1577–1587.

3

Analysis via Passivity Theory of a Class of Nonlinear PID Global Regulators for Robot Manipulators

Jose Luis Meza[1], Víctor Santibáñez[1], Rogelio Soto[2], Jose Perez[3] and Joel Perez[3]
[1]*Instituto Tecnológico de la Laguna*
[2]*Tecnológico de Monterrey, Campus Monterrey*
[3]*Universidad Autónoma de Nuevo León*
Mexico

1. Introduction

We present a simple global asymptotic stability analysis, by using passivity theory for a class of nonlinear PID regulators for robot manipulators. Nonlinear control structures based on the classical PID controller, which assure global asymptotic stability of the closed-loop system, have emerged. Some works that deal with global nonlinear PID regulators based on Lyapunov theory have been reported by (Arimoto, 1995a), (Kelly, 1998) and (Santibáñez & Kelly, 1998). Recently, a particular case of the class of nonlinear PID global regulators originally proposed in (Santibáñez & Kelly, 1998) was presented by (Sun et al., 2009). Few saturated PID controllers (that is, bounded PID controllers taking into account the actuator torque constraints) have been reported: for the case of semiglobal asymptotic stability, a saturated linear PID controller was presented in (Alvarez et al., 2003) and (Alvarez et al., 2008); for the case of global asymptotic stability, saturated nonlinear PID controllers were introduced in (Gorez, 1999), (Meza et al., 2005), (Santibáñez et al., 2008). The work introduced by (Gorez, 1999) was the first bounded PID-like controller in assuring global regulation; the latter works, introduced in (Meza et al., 2005) and (Santibáñez et al., 2008), also guarantee global regulation, but with the advantage of a controller structure which is simpler than that presented in (Gorez, 1999). A local adaptive bounded regulator was presented by (Laib, 2000). Recently a new saturated nonlinear PID regulator for robots has been proposed in (Santibáñez et al., 2010), the controller structure considers the saturation phenomena of the control computer, the velocity servo-drivers and the torque constraints of the actuators. The work in (Orrante et al., 2010) presents a variant of the work presented by (Santibáñez et al., 2010), where now the controller is composed by a saturated velocity proportional (P) inner loop, provided by the servo-driver, and a saturated position proportional-integral (PI) outer loop, supplied by the control computer.

In this chapter we use a passivity based approach to explain the results of global regulation of a class of nonlinear PID controllers proposed by (Santibáñez & Kelly, 1998), that include the particular cases reported by (Arimoto, 1995a) and (Kelly, 1998). At the end of the 80's, it was established in (Kelly & Ortega, 1988) and (Landau & Horowitz, 1988) that the nonlinear

dynamics of rigid robots describe a passivity mapping from torque input to velocity output. This property is know as the passive structure of rigid robots (Ortega & Spong, 1989). The controller design methodology for robot manipulators introduced by (Takegaki & Arimoto, 1981), also called energy shaping plus damping injection technique, allows to naturally split the controller tasks into potential energy shaping for stabilization at the desired equilibrium, and damping injection, to make this equilibrium attractive (Ortega et al., 1995b). As a feature of this kind of controllers, it can be shown that the passivity property, from a new input torque to output velocity, is preserved for robots in closed-loop with the energy shaping term and damping injection term of the controller. Furthermore, considering that corresponding feedback to the integral action define a passive mapping, then it is possible to use a passivity theorem of interconnected systems to explain the stability of a class of nonlinear PID global regulators for robots. The theorem used allows to conclude global asymptotic stability of the origin of an unforced feedback system, which is composed by the feedback interconnection of state strictly passive dynamic systems with a passive and zero state observable system.
The objective of this chapter is to present in a simple framework the global asymptotic stability analysis, by using passivity theory for a class of nonlinear PID regulators for robot manipulators. The remainder of this chapter is organized as follows: Section 2 summarizes the dynamics for rigid robots and also recalls some of their important properties. The rationale behind the energy shaping plus damping injection technique for rigid robots are given in Section 3. The class of nonlinear PID regulators is given in Section 4. In Section 5 we recall the definition of the passivity concepts for dynamical systems and we present the passivity theorem useful for asymptotic stability analysis of interconnected systems. In Section 6 we present a passivity analysis and application of passivity theorem to conclude global asymptotic stability. An Evaluation in simulation to verify the theoretical results is presented in Section 7. Finally, our conclusions are shown in Section 8. Throughout this chapter, the norm of a vector x is defined as $\|x\| = \sqrt{x^T x}$ and that of a matrix A is defined as the corresponding induced norm $\|A\| = \sqrt{\lambda_{\mathrm{M}}\{A^T A\}}$. L_2^n and L_{2e}^n denote the space of n–dimensional square integrable functions and its extension, respectively.

2. Robot dynamics

For control design purposes, it is necessary to have a mathematical model that reveals the dynamical behavior of a system. Robots manipulators are articulated mechanical systems composed of links connected by joints. Links and joints are usually made as rigid as possible so as to achieve high precision in robot positioning. The joints are mainly of two types: revolute and prismatic. Its dynamic model is characterized by nonlinear coupled second-order differential equations, which describe the temporal interactions of the joint motions in response to the inertial, centrifugal and Coriolis, gravitational and actuating torques or forces. The most commonly used equations to model the dynamics of a robot are the Euler-Lagrange and Newton-Euler formulations. Here we use the Euler - Lagrange formulation. In this section we consider robot manipulators formed by an open kinematic chain. We assume that all the links are joined together by revolute joints. In the absence of friction and other disturbances, the Lagrangian $\mathcal{L}(q,\dot{q})$ of a mechanical system is defined by

$$\mathcal{L}(q,\dot{q}) = \mathcal{K}(q,\dot{q}) - \mathcal{U}(q),$$

where $\mathcal{K}(q,\dot{q})$ y $\mathcal{U}(q)$ represent the kinetic and potential energy of the system respectively. The equation of motion of Euler - Lagrange for a manipulator of n degrees of freedom are given by:

$$\frac{d}{dt}\left(\frac{\partial \mathcal{L}(q,\dot{q})}{\partial \dot{q}}\right) - \frac{\partial \mathcal{L}(q,\dot{q})}{\partial q} = \tau \tag{1}$$

where $\tau \in \mathbb{R}^n$ is the vector of external generalized forces acting on each joint of robot, $q \in \mathbb{R}^n$ represents the vector of generalized coordinates of the system, $\dot{q} = \frac{d}{dt}q$ is the vector of generalized velocities. In the case of a robot manipulator of n degrees of freedom, kinetic energy is a quadratic function of the velocity vector $\dot{q}$ of the form:

$$\mathcal{K}(q,\dot{q}) = \frac{1}{2}\dot{q}^T M(q)\dot{q}$$

where $M(q) \in \mathbb{R}^{n\times n}$ is the manipulator inertia matrix, which is symmetric and positive definite. The equations of motion of Euler - Lagrange provide the following dynamic model (Spong et al., 2006):

$$M(q)\ddot{q} + \dot{M}(q)\dot{q} - \frac{1}{2}\frac{\partial}{\partial q}(\dot{q}^T M(q)\dot{q}) + \frac{\partial \mathcal{U}(q)}{\partial q} = \tau$$

or in compact form, the dynamics of a serial n-link rigid robot can be written as:

$$M(q)\ddot{q} + C(q,\dot{q})\dot{q} + g(q) = \tau \tag{2}$$

where $C(q,\dot{q})\dot{q}$ and $g(q)$ are given by:

$$C(q,\dot{q})\dot{q} = \dot{M}(q)\dot{q} - \frac{1}{2}\frac{\partial}{\partial q}(\dot{q}^T M(q)\dot{q})$$

$$g(q) = \frac{\partial \mathcal{U}(q)}{\partial q}. \tag{3}$$

$C(q,\dot{q})\dot{q} \in \mathbb{R}^n$ is the vector of centrifugal and Coriolis forces, and $g(q) \in \mathbb{R}^n$ is the vector of gravitational forces or torques obtained as the gradient of the potential energy of robot $\mathcal{U}(q)$. In terms of the state vector $[q^T \;\; \dot{q}^T]^T$, the robot dynamics given by equation (2) can be written as:

$$\frac{d}{dt}\begin{bmatrix} q \\ \dot{q} \end{bmatrix} = \begin{bmatrix} \dot{q} \\ M(q)^{-1}(q)[\tau(t) - C(q,\dot{q})\dot{q} - g(q)]. \end{bmatrix} \tag{4}$$

2.1 Planar robot of 2 degrees of freedom

The equation of motion of Euler-Lagrange (1) for a robot of n degrees of freedom can be equivalently written as:

$$\frac{d}{dt}\left(\frac{\partial \mathcal{L}(q,\dot{q})}{\partial \dot{q}_i}\right) - \frac{\partial \mathcal{L}(q,\dot{q})}{\partial \dot{q}_i} = \tau_i \qquad i = 1,\cdots,n.$$

In the particular case of a planar robot of $n = 2$ degrees of freedom rotational joints, consider the diagram shown in Fig. 1 (Reyes & Kelly, 2001). It is known that the compact form (2) of the dynamics of a robot for two degrees of freedom with rigid links can be written as:

$$\begin{bmatrix} M_{11}(\boldsymbol{q}) & M_{11}(\boldsymbol{q}) \\ M_{11}(\boldsymbol{q}) & M_{11}(\boldsymbol{q}) \end{bmatrix} \begin{bmatrix} \ddot{q}_1 \\ \ddot{q}_2 \end{bmatrix} + \begin{bmatrix} C_{11}(\boldsymbol{q},\dot{\boldsymbol{q}}) & C_{12}(\boldsymbol{q},\dot{\boldsymbol{q}}) \\ C_{21}(\boldsymbol{q},\dot{\boldsymbol{q}}) & C_{22}(\boldsymbol{q},\dot{\boldsymbol{q}}) \end{bmatrix} \begin{bmatrix} \dot{q}_1 \\ \dot{q}_2 \end{bmatrix} + \begin{bmatrix} g_1(\boldsymbol{q}) \\ g_2(\boldsymbol{q}) \end{bmatrix} = \begin{bmatrix} \tau_1 \\ \tau_2 \end{bmatrix} \tag{5}$$

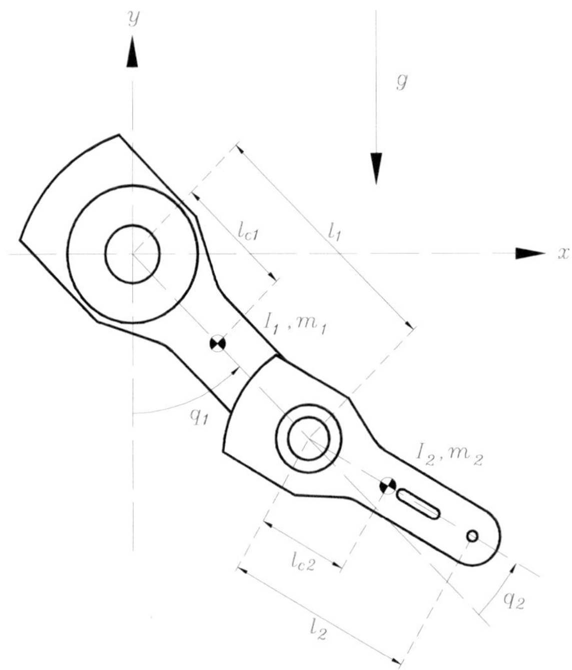

Fig. 1. Diagram of the prototype planar robot with 2 degrees of freedom.

The meaning of the parameters of the prototype planar robot is shown in Table 1.

Description	*notation*
Angular displacement of link 1	q_1
Angular displacement of link 2	q_2
Length link 1	l_1
Length link 2	l_2
Link (1) center of mass	l_{c1}
Link (2) center of mass	l_{c2}
Mass link 1	m_1
Mass link 2	m_2
Inertia link 1	I_1
Inertia link 2	I_2
Gravity acceleration	g

Table 1. Parameters of prototype planar robot with 2 degrees of freedom

The elements of the inertia matrix $M(\boldsymbol{q})$ in terms of the parameters of the robot, are given by

$$
\begin{aligned}
M_{11}(\boldsymbol{q}) &= m_1 l_{c1}^2 + m_2 \left(l_1^2 + l_{c2}^2 + 2 l_1 l_{c2} \cos(q_2) \right) + I_1 + I_2 \\
M_{12}(\boldsymbol{q}) &= m_2 \left(l_{c2}^2 + l_1 l_{c2} \cos(q_2) \right) + I_2 \\
M_{21}(\boldsymbol{q}) &= m_2 \left(l_{c2}^2 + l_1 l_{c2} \cos(q_2) \right) + I_2 \\
M_{22}(\boldsymbol{q}) &= m_2 l_{c2}^2 + I_2.
\end{aligned}
$$

The matrix elements $C_{ij}(\boldsymbol{q}, \dot{\boldsymbol{q}})$ $(i, j = 1, 2)$ of centrifugal and Coriolis forces $C(\boldsymbol{q}, \dot{\boldsymbol{q}})$ are

$$
\begin{aligned}
C_{11}(\boldsymbol{q}, \dot{\boldsymbol{q}}) &= -m_2 l_1 l_{c2} \sin(q_2) \dot{q}_2 \\
C_{12}(\boldsymbol{q}, \dot{\boldsymbol{q}}) &= -m_2 l_1 l_{c2} \sin(q_2) \left(\dot{q}_1 + \dot{q}_2 \right) \\
C_{21}(\boldsymbol{q}, \dot{\boldsymbol{q}}) &= m_2 l_1 l_{c2} \sin(q_2) \dot{q}_1 \\
C_{22}(\boldsymbol{q}, \dot{\boldsymbol{q}}) &= 0.
\end{aligned}
$$

Finally the elements of the vector of gravitational torques $\boldsymbol{g}(\boldsymbol{q})$ are given by:

$$
\begin{aligned}
g_1(\boldsymbol{q}) &= (m_1 l_{c1} + m_2 l_1) g \sin(q_1) + m_2 l_{c2} g \sin(q_1 + q_2) \\
g_2(\boldsymbol{q}) &= m_2 l_{c2} g \sin(q_1 + q_2).
\end{aligned}
$$

2.2 Properties of the robot dynamics

Although the equation of motion (2) is complex, it has several fundamental properties which can be exploited to facilitate the analysis of stability. Three important properties of the robot dynamics are the following:

Property 1. (Koditschek, 1984) The matrix $C(\boldsymbol{q}, \dot{\boldsymbol{q}})$ and the time derivative $\dot{M}(\boldsymbol{q})$ of the inertia matrix satisfy:

$$
\dot{\boldsymbol{q}}^T \left[\frac{1}{2} \dot{M}(\boldsymbol{q}) - C(\boldsymbol{q}, \dot{\boldsymbol{q}}) \right] \dot{\boldsymbol{q}} = \mathbf{0} \quad \forall\, \boldsymbol{q}, \dot{\boldsymbol{q}} \in \mathbb{R}^n.
$$

Property 2. (Tomei, 1991) There exists a positive constant k_g such that

$$
k_g \geq \left\| \frac{\partial \boldsymbol{g}(\boldsymbol{q})}{\partial \boldsymbol{q}} \right\| \qquad \forall\, \boldsymbol{q} \in \mathbb{R},
$$

and

$$
\| \boldsymbol{g}(\boldsymbol{x}) - \boldsymbol{g}(\boldsymbol{y}) \| \leq k_g \| \boldsymbol{x} - \boldsymbol{y} \| \qquad \forall\, \boldsymbol{x}, \boldsymbol{y} \in \mathbb{R}^n.
$$

Property 3. (Passive structure of rigid robots) In relation to the dynamic model (2). The operator

$$
\begin{aligned}
H_R &: L_{2e}^n \to L_{2e}^n \\
&: \boldsymbol{\tau} \mapsto \dot{\boldsymbol{q}}
\end{aligned}
$$

is passive (Kelly & Ortega, 1988; Landau & Horowitz, 1988; Ortega & Spong, 1989), i.e.:

$$
\underbrace{\int_0^T \dot{\boldsymbol{q}}(t)^T \boldsymbol{\tau}(t)\, dt}_{applied-energy} = \underbrace{V_1(T) - V_1(0)}_{stored-energy}, \tag{6}
$$

which can be expressed as

$$\int_0^T \dot{q}(t)^T \tau(t)\, dt \geq -V_a(0), \tag{7}$$

where $V_a(t)$ is the total energy of robot plus a suitable constant which is introduced so that $V_a(t)$ is a nonnegative function (Ortega et al., 1995b):

$$V_a(t) = \frac{1}{2}\dot{q}(t)^T M(q(t))\dot{q}(t) + \mathcal{U}(q(t)) - k_u \tag{8}$$

being $\frac{1}{2}\dot{q}^T(t)M(q(t))\dot{q}(t)$ energy kinetic, $\mathcal{U}(q(t))$ the potential energy of robot due to gravity, y

$$k_u = \min_{q} \mathcal{U}(q(t)).$$

3. Energy shaping methodology

Energy is one of the fundamental concepts in control of mechanical systems with multi-degrees-of-freedom. The action of a controller can be understood in energy terms as a dynamical system called "actuator" that supplies energies to the controlled system, upon interconnection, to modify desirably the behavior of the closed-loop (interconnected) system. This idea has its origin in (Takegaki & Arimoto, 1981) and is later called the "energy-shaping" approach, which is now known as a basic controller design technique common in control of mechanical systems. Its systematic interpretation is called "passivity-based control" (Arimoto, 2009).

The main idea of this methodology is to reshape the robot system's natural energy and inject damping via velocity feedback, for asymptotic stabilization purposes, such that a regulation objective is reached . This is achieved by choosing a controller structure such that, first, the total potential energy function of the closed–loop system due to gravity and the controller is radially unbounded function in the position error with a unique and global minimum at zero position error, and second, it injects damping via velocity feedback. The resulting closed loop system is an autonomous one which has the nice property that zero position error and zero velocity form always the unique equilibrium point. By using the total energy, i.e., the kinetic plus total potential energy, as a Lyapunov function, it follows that this equilibrium is stable. In order to prove that the equilibrium is in fact globally asymptotically stable, the final key step is to exploit the autonomous nature of the closed loop system to invoke the Krasovskii–LaSalle's theorem. This approach has been continued by their colleagues and several researchers (Arimoto, 1995a; Nijmeijer & Van der Schaft, 1990; Wen & Bayard, 1988), who have offered extensions and improvements (Ailon & Ortega, 1993; Berghuis & Nijmeijer, 1993b; Kelly, 1993; Ortega et al., 1998) and (Ortega & Garcia-Canseco, 2004; Ortega et al., 2008; Sepulchre et al., 1997; Vander, 1999).

As a feature of this kind of controllers, it can be shown that the passivity property, from a new input torque to output velocity, is preserved either for rigid robots or elastic joint robots in closed loop with the energy shaping term of the controller. The damping injection term, via velocity feedback or by way of a suitable filtering of position, defines an input, output or state strictly passive mapping from velocity input to damping output, and asymptotically stabilizes the desired equilibrium of the closed loop system.

We broach the regulation problem whose goal is to find $\tau(t)$ such that

$$\lim_{t\to\infty} q(t) = q_d$$

where $q_d \in \mathbb{R}^n$ is a vector of constant desired joint displacements. The features of the system can be enhanced by reshaping its total potential energy. This can be done by constructing a controller to meet a desired energy function for the closed-loop system, and inject damping, via velocity feedback, for asymptotic stabilization purposes (Nijmeijer & Van der Schaft, 1990). To this end, in this section we consider controllers whose control law can be written by

$$\tau = \frac{\partial \mathcal{U}_a(q_d, \tilde{q})}{\partial \tilde{q}} - \frac{\partial \mathcal{F}(\dot{q})}{\partial \dot{q}} \tag{9}$$

where $\mathcal{F}(\dot{q})$ is some kind of dissipation function from which the damping force can be derived, an example is the so called Rayleigh dissipative function $\mathcal{F}(\dot{q}) = \frac{1}{2}\dot{q}^T K_v \dot{q}$, where K_v is the matrix of coefficient of viscous friction, $\tilde{q} = q_d - q \in \mathbb{R}^n$ denotes the joint position error and $\mathcal{U}_a(q_d, \tilde{q})$ is some kind of artificial potential energy provided by the controller whose properties will be established later. The first right hand side term of (9) corresponds to the energy shaping part and the other one to the damping injection part.
We assume the dissipation function $\mathcal{F}(\dot{q})$ satisfies the following conditions:

$$\frac{\partial \mathcal{F}(\dot{q})}{\partial \dot{q}} = \mathbf{0} \Leftrightarrow \dot{q} = \mathbf{0} \tag{10}$$

$$\dot{q}^T \frac{\partial \mathcal{F}(\dot{q})}{\partial \dot{q}} > 0 \ \forall\, \dot{q} \neq \mathbf{0}. \tag{11}$$

The closed-loop system equation obtained by substituting the control law (9) into the robot dynamics (2) leads to

$$\frac{d}{dt}\begin{bmatrix}\tilde{q}\\ \dot{q}\end{bmatrix} = \tag{12}$$

$$\begin{bmatrix} -\dot{q} \\ M^{-1}[\frac{\partial}{\partial \tilde{q}}\{\mathcal{U}(q_d - \tilde{q}) + \mathcal{U}_a(q_d, \tilde{q})\} - \frac{\partial \mathcal{F}(\dot{q})}{\partial \dot{q}} - C(q, \dot{q})\dot{q}] \end{bmatrix} \tag{13}$$

where (3) has been used. If the total potential energy $\mathcal{U}_T(q_d, \tilde{q})$ of the closed-loop system, defined as the sum of the potential energy $\mathcal{U}(q)$ due to gravity plus the artificial potential energy $\mathcal{U}_a(q_d, \tilde{q})$ introduced by the controller

$$\mathcal{U}_T(q_d, \tilde{q}) = \mathcal{U}(q_d - \tilde{q}) + \mathcal{U}_a(q_d, \tilde{q}), \tag{14}$$

is radially unbounded in $\tilde{q}$, and $\tilde{q} = \mathbf{0} \in \mathbb{R}^n$ is an unique minimum, which is global for all q_d, then the origin $\left[\tilde{q}^T \ \dot{q}^T\right]^T = \mathbf{0} \in \mathbb{R}^{2n}$ of the closed–loop system (13) is global and asymptotically stable (Takegaki & Arimoto, 1981).

4. A class of nonlinear PID global regulators

4.1 Classical PID regulators

Conventional proportional-integral-derivative PID regulators have been extensively used in industry due to their design simplicity, inexpensive cost, and effectiveness. Most of the present industrial robots are controlled through PID regulators (Arimoto, 1995a). The classical version of the PID regulator can be described by the equation:

$$\tau = K_p\tilde{q} - K_v\dot{q} + K_i \int_0^t \tilde{q}(\sigma)\, d\sigma \tag{15}$$

where K_p, K_v and K_i are positive definite diagonal $n \times n$ matrices, and $\tilde{q} = q_d - q$ denotes the position error vector. Even though the PID controller for robot manipulators has been very used in industrial robots (Arimoto, 1995a), there still exist open problems, that make interesting its study. A open problem is the lack of a proof of global asymptotic stability (Arimoto, 1994). The stability proofs shown until now are only valid in a local sense (Arimoto, 1994; Arimoto et al., 1990; Arimoto & Miyazaki, 1983; Arimoto, 1996; Dorsey, 1991; Kelly, 1995; Kelly et al., 2005; Rocco, 1996; Wen, 1990) or, in the best of the cases, in a semiglobal sense (Alvarez et al., 2000; Meza et al., 2007). In (Ortega et al., 1995a), a so–called PI^{2}D controller is introduced, which is based on a PID structure but uses a filter of the position in order to estimate the velocity of the joints, and adds a term which is the integral of such an estimate of the velocity (this added term motivates the name PI^{2}D); for this controller, semiglobal asymptotic stability was proved. To solve the global positioning problem, some globally asymptotically stable PID–like regulators have also been proposed (Arimoto, 1995a; Gorez, 1999; Kelly, 1998; Santibáñez & Kelly, 1998), such controllers, however, are nonlinear versions of the classical linear PID. We propose a new global asymptotic stability analysis, by using passivity theory for a class of nonlinear PID regulators for robot manipulators. For the purpose of this chapter, it is convenient to recall the following definition presented in (Kelly, 1998).

Definition 1. $\mathcal{F}(m, \varepsilon, \boldsymbol{x})$ with $1 \geq m > 0$, $\varepsilon > 0$ and $\boldsymbol{x} \in \mathbb{R}^n$ denotes the set of all continuous differentiable increasing functions $\mathbf{sat}(\boldsymbol{x}) = [\,\mathrm{sat}(x_1) \;\; \mathrm{sat}(x_2) \;\; \cdots \;\; \mathrm{sat}(x_n)]^T$ such that

- $|x| \geq |\,\mathrm{sat}(x)| \geq m\,|x| \quad \forall\, x \in \mathbb{R} : |x| < \varepsilon$
- $\varepsilon \geq |\,\mathrm{sat}(x)| \geq m\varepsilon \quad \forall\, x \in \mathbb{R} : |x| \geq \varepsilon$
- $1 \geq \frac{d}{dx}\mathrm{sat}(x) \geq 0 \quad \forall\, x \in \mathbb{R}$

where $|\cdot|$ stands for the absolute value.

◇

For instance, the nonlinear vector function $\mathbf{sat}(\tilde{\boldsymbol{q}}) = [\,\mathrm{sat}(\tilde{q}_1) \;\; \mathrm{sat}(\tilde{q}_2) \;\; \cdots \;\; \mathrm{sat}(\tilde{q}_n)]^T$, considered in Arimoto (Arimoto, 1995a) whose entries are given by

$$\mathrm{sat}(x) = \mathrm{Sin}(x) = \begin{cases} \sin(x) & \text{if } |x| < \pi/2 \\ 1 & \text{if } x \geq \pi/2 \\ -1 & \text{if } x \leq -\pi/2 \end{cases} \tag{16}$$

belongs to set $\mathcal{F}(\sin(1), 1, \boldsymbol{x})$.

◇

4.2 A class of nonlinear PID controllers

The class of nonlinear PID global regulators under study was proposed in (Santibáñez & Kelly, 1998). The structure is based on the gradient of a $\mathcal{C}^1$ artificial potential function $\mathcal{U}_a(\tilde{\boldsymbol{q}})$ satisfying some typical features required by the energy shaping methodology (Takegaki & Arimoto, 1981). The PID control law can be written by (see Fig. 2).

$$\boldsymbol{\tau} = \frac{\partial \mathcal{U}_a(\tilde{\boldsymbol{q}})}{\partial \tilde{\boldsymbol{q}}} - K_v \dot{\boldsymbol{q}} + K_i \int_0^t [\alpha\,\mathrm{sat}(\tilde{\boldsymbol{q}}(\sigma)) + \dot{\tilde{\boldsymbol{q}}}(\sigma)]\, d\sigma \tag{17}$$

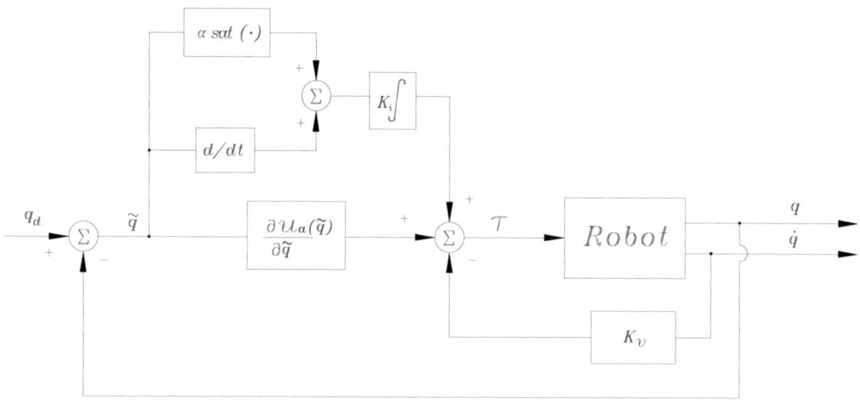

Fig. 2. Block diagram of nonlinear PID control

where

- $\mathcal{U}_a(\tilde{q})$ is a kind of $\mathcal{C}^1$ artificial potential energy induced by a part of the controller.
- K_v and K_i are diagonal positive definite $n \times n$ matrices
- $\dot{\tilde{q}}$ is the velocity error vector
- $\text{sat}(\tilde{q}) \in \mathcal{F}(m, \varepsilon, \tilde{q})$,
- α is a small constant, satisfying (Santibáñez & Kelly, 1998)

By defining z as:

$$z(t) = \int_0^t [\alpha\, \text{sat}(\tilde{q}(\sigma)) + \dot{\tilde{q}}(\sigma)]\, d\sigma - K_p^{-1} g(q_d), \tag{18}$$

we can describe the closed-loop system by

$$\frac{d}{dt}\begin{bmatrix} \tilde{q} \\ \dot{q} \\ z \end{bmatrix} = \tag{19}$$

$$\begin{bmatrix} -\dot{q} \\ M(q)^{-1}\left[\nabla_{\tilde{q}} \mathcal{U}_T(q_d, \tilde{q}) - K_v \dot{q} - C(q, \dot{q})\dot{q} + K_i z\right] \\ \alpha\, \text{sat}(\tilde{q}) - \dot{q} \end{bmatrix} \tag{20}$$

which is an autonomous nonlinear differential equation whose origin $\left[\tilde{q}^T \; \dot{q}^T \; z^T\right]^T = \mathbf{0} \in \mathbb{R}^{3n}$ is the unique equilibrium.

4.3 Some examples

Some examples of this kind of nonlinear PID regulators

$$\tau = \frac{\partial \mathcal{U}_a(\tilde{q})}{\partial \tilde{q}} - K_v \dot{q} + K_i \int_0^t [\alpha\, \text{sat}(\tilde{q}(\sigma)) + \dot{\tilde{q}}(\sigma)]\, d\sigma \tag{21}$$

are:

- (Kelly, 1998) $\tau = K'_p\tilde{q} - K_v\dot{q} + K'_i\int_0^t \text{sat}(\tilde{q}(\sigma))\, d\sigma$
  where $K'_p = K_p + K_{pa}$, K_{pa} is a diagonal positive definite $n \times n$ matrix with $\lambda_{\text{m}}\{K_{pa}\} > k_g$, $K_p = K_i$, $K'_i = \alpha K_i$. This controller has associated an artificial potential energy $\mathcal{U}_a(\tilde{q})$ given by

$$\mathcal{U}_a(\tilde{q}) = \frac{1}{2}\tilde{q}^T K_{pa}\tilde{q}.$$

- (Arimoto et al., 1994a) $\tau = K_{pa}\text{Sin}[\tilde{q}] - K_v\dot{q} + K_i\int_0^t[\alpha\, \text{sat}(\tilde{q}(\sigma)) + \dot{\tilde{q}}(\sigma)]\, d\sigma$
  where K_{pa} is a diagonal positive definite $n \times n$ matrix whose entries are k_{pai} and $\mathbf{sat}(\tilde{q}) = \mathbf{Sin}[\tilde{q}] = [\text{Sin}(\tilde{q}_1)\ \ \text{Sin}(\tilde{q}_2)\ldots\ \ \text{Sin}(\tilde{q}_n)]^T$ with Sin(.) defined in (16).
  This controller has associated a $\mathcal{C}^2$ artificial potential energy $\mathcal{U}_a(\tilde{q})$ given by

$$\mathcal{U}_a(\tilde{q}) = \sum_{=1}^{n} k_{pai}[1 - \text{Cos}(\tilde{q})],$$

  where

$$\text{Cos}(x) = \begin{cases} \cos(x) & \text{if } |x| < \pi/2 \\ -x + \pi/2 & \text{if } x \geq \pi/2 \\ x + \pi/2 & \text{if } x \leq -\pi/2 \end{cases}$$

- $\tau = K_{pa}\tanh[\tilde{q}] - K_v\dot{q} + K_i\int_0^t[\alpha\, \text{sat}(\tilde{q}(\sigma)) + \dot{\tilde{q}}(\sigma)]\, d\sigma$
  where K_{pa} is a diagonal positive definite $n \times n$ matrix whose entries are k_{pai} and $\mathbf{sat}(\tilde{q}) = \mathbf{tanh}[\tilde{q}] = [\tanh(\tilde{q}_1)\ \ \tanh(\tilde{q}_2)\ldots\ \ \tanh(\tilde{q}_n)]^T$. This controller has associated a $\mathcal{C}^\infty$ artificial potential energy $\mathcal{U}_a(\tilde{q})$ given by

$$\mathcal{U}_a(\tilde{q}) = \sum_{i=1}^{n} k_{pai} \ln[\cosh(\tilde{q}_i)]$$

- $\tau = K_{pa}\text{Sat}[\tilde{q}] - K_v\dot{q} + K_i\int_0^t[\alpha\, \text{sat}(\tilde{q}(\sigma)) + \dot{\tilde{q}}(\sigma)]\, d\sigma$
  where K_{pa} is a diagonal positive definite $n \times n$ matrix whose entries are k_{pai} and $\mathbf{sat}(\tilde{q}) = \mathbf{Sat}[\tilde{q}] = [\text{Sat}(\tilde{q}_1)\ \ \text{Sat}(\tilde{q}_2)\ldots\ \ \text{Sat}(\tilde{q}_n)]^T$. This controller has associated a $\mathcal{C}^1$ artificial potential energy $\mathcal{U}_a(\tilde{q})$ given by

$$\mathcal{U}_a(q_d, \tilde{q}) = \sum_{i=1}^{n}\left[\int_0^{\tilde{q}_i} k_{pai}\text{Sat}(\sigma_i; \lambda_i)\, d\sigma_i\right]$$

where $\text{Sat}(x;\lambda)$ stands for the well known hard saturation function

$$\text{Sat}(x;\lambda) = \begin{cases} x & \text{if } |\tilde{q}_i| < \lambda \\ \lambda & \text{if } \tilde{q}_i \geq \lambda \\ -\lambda & \text{if } \tilde{q}_i \leq -\lambda \end{cases}.$$

Following the ideas given in (Santibáñez & Kelly, 1995) and (Loria et al., 1997) it is possible to demonstrate, for all above mentioned regulators, that $\mathcal{U}_a(\tilde{q})$ leads to a radially unbounded virtual total potential function $\mathcal{U}_T(q_d, \tilde{q})$.

5. Passivity concepts

In this chapter, we consider dynamical systems represented by

$$\dot{x} = f(x, u) \tag{22}$$

$$y = h(x, u) \tag{23}$$

where $u \in \mathbb{R}^n$, $y \in \mathbb{R}^n$, $x \in \mathbb{R}^m$, $f(0,0) = 0$ and $h(0,0) = 0$. Moreover f, h are supposed sufficiently smooth such that the system is well–defined, i.e., $\forall\ u \in L^n_{2e}$ and $x(0) \in \mathbb{R}^m$ we have that the solution $x(\cdot)$ is unique and $y \in L^n_{2e}$.

Definition 2. (Khalil, 2002) The system (22)–(23) is said to be passive if there exists a continuously differentiable positive semidefinite function $V(x)$ (called the storage function) such that

$$u^T y \geq \dot{V}(x) + \epsilon\|u\|^2 + \delta\|y\|^2 + \rho\psi(x) \tag{24}$$

where ϵ, δ, and ρ are nonnegative constants, and $\psi(x) : \mathbb{R}^m \to \mathbb{R}$ is a positive definite function of x. The term $\rho\psi(x)$ is called the state dissipation rate. Furthermore, the system is said to be

- lossless if (24) is satisfied with equality and $\epsilon = \delta = \rho = 0$; that is, $u^T y = \dot{V}(x)$
- input strictly passive if $\epsilon > 0$ and $\delta = \rho = 0$,
- output strictly passive if $\delta > 0$ and $\epsilon = \rho = 0$,
- state strictly passive if $\rho > 0$ and $\epsilon = \delta = 0$,

If more than one of the constants ϵ, δ, ρ are positive we combine names.

□

Now we recall the definition of an observability property of the system (22)–(23).

Definition 3. (Khalil, 2002) The system (22)–(23) is said to be zero state observable if

$$u(t) \equiv 0 \text{ and } y(t) \equiv 0 \Rightarrow x(t) \equiv 0.$$

Equivalently, no solutions of $\dot{x} = f(x, 0)$ can stay identically in $S = \{x \in \mathbb{R}^m : h(x, 0) = 0\}$, other than the trivial solution $x(t) \equiv 0$.

Right a way, we present a theorem that allows to conclude global asymptotic stability for the origin of an unforced feedback system, which is composed by the feedback interconnection of a state strictly passive system with a passive system, which is an adaptation of a passivity theorem useful for asymptotic stability analysis of interconnected system presented in (Khalil, 2002).

Theorem 1. Consider the feedback system of Fig. 3 where H_1 and H_2 are dynamical systems of the form

$$\dot{x}_i = f_i(x_i, e_i)$$

$$y_i = h_i(x_i, e_i)$$

for $i = 1, 2$, where $f_i : \mathbb{R}^{m_i} \times \mathbb{R}^n \to \mathbb{R}^{m_i}$ and $h_i : \mathbb{R}^{m_i} \times \mathbb{R}^n \to \mathbb{R}^n$ are supposed sufficiently smooth such that the system is well–defined. $f_1(0, e_1) = 0 \Rightarrow e_1 = 0$, $f_2(0,0) = 0$, y $h_i(0,0) = 0$.

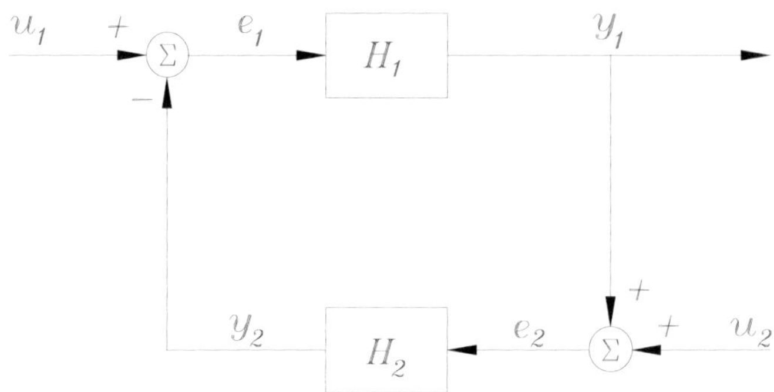

Fig. 3. Feedback connection

The system has the same number of inputs and outputs. Suppose the feedback system has a well–defined state–space model

$$\dot{x} = f(x,u)$$
$$y = h(x,u)$$

where

$$x = \begin{bmatrix} x_1 \\ x_2 \end{bmatrix}, u = \begin{bmatrix} u_1 \\ u_2 \end{bmatrix}, y = \begin{bmatrix} y_1 \\ y_2 \end{bmatrix}$$

f and h are sufficiently smooth, $f(\mathbf{0},\mathbf{0}) = \mathbf{0}$, and $h(\mathbf{0},\mathbf{0}) = \mathbf{0}$. Let H_1 be a state strictly passive system with a positive definite storage function $V_1(x_1)$ and state dissipation rate $\rho_1\psi_1(x_1)$ and H_2 be a passive and zero state observable system with a positive definite storage function $V_2(x_2)$; that is,

$$e_1^T y_1 \geq \dot{V}_1(x_1) + \rho_1\psi_1(x_1)$$
$$e_2^T y_2 \geq \dot{V}_2(x_2)$$

Then the origin $x = \mathbf{0}$ of

$$\dot{x} = f(x,\mathbf{0}) \tag{25}$$

is asymptotically stable. If $V_1(x_1)$ and $V_2(x_2)$ are radially unbounded then the origin of (25) will be globally asymptotically stable.

Proof. Take $u_1 = u_2 = \mathbf{0}$. In this case $e_1 = -y_2$ and $e_2 = y_1$. Using $V(x) = V_1(x_1) + V_2(x_2)$ as a Lyapunov function candidate for the closed–loop system, we have

$$\begin{aligned} \dot{V}(x) &= \dot{V}_1(x_1) + \dot{V}_2(x_2) \\ &\leq e_1^T y_1 - \rho_1\psi_1(x_1) + e_2^T y_2 \\ &= -\rho_1\psi_1(x_1) \leq 0, \end{aligned}$$

which shows that the origin of the closed-loop system is stable. To prove asymptotic stability we use the LaSalle's invariance principle and the zero state observability of the system H_2. It remains to demonstrate that $x = \mathbf{0}$ is the largest invariant set in $\Omega = \{x \in \mathbb{R}^{m_1+m_2} : \dot{V}(x) =$

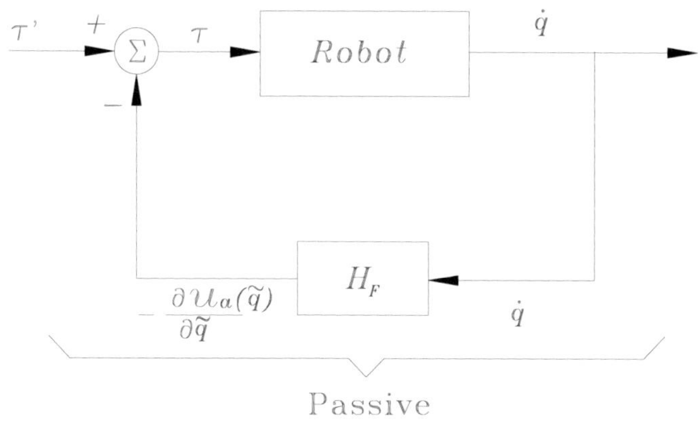

Fig. 4. Passivity structure of rigid robots in closed-loop

$0\}$. To this end, in the search of the largest invariant set, we have that $\dot{V}(x) = 0 \Rightarrow 0 \leq -\rho_1\psi_1(x_1) \leq 0 \Rightarrow -\rho_1\psi_1(x_1) = 0$. Besides

$$\rho_1 > 0 \Rightarrow \psi_1(x_1) \equiv 0 \Rightarrow x_1 \equiv 0$$

Now, as $x_1 \equiv 0 \Rightarrow \dot{x}_1 = f_1 \equiv 0$ and in agreement with the assumption about f_1 in the sense that $f_1(\mathbf{0}, e_1) = \mathbf{0} \Rightarrow e_1 = \mathbf{0}$, we have $e_1 \equiv \mathbf{0} \Rightarrow y_2 \equiv \mathbf{0}$. Also $x_1 \equiv \mathbf{0}$, $e_1 \equiv \mathbf{0} \Rightarrow y_1 \equiv \mathbf{0}$ (owing to assumption $h_1(\mathbf{0},\mathbf{0}) = \mathbf{0}$). Finally, $y_1 \equiv \mathbf{0} \Rightarrow e_2 \equiv \mathbf{0}$, and

$$e_2 \equiv \mathbf{0} \text{ and } y_2 \equiv \mathbf{0} \Rightarrow x_2 \equiv \mathbf{0}$$

in agreement with the zero state observability of H_2. This shows that the largest invariant set in Ω is the origin, hence, by using the Krasovskii–LaSalle's theorem, we conclude asymptotic stability of the origin of the unforced closed-loop system (25). If $V(x)$ is radially unbounded then the origin will be globally asymptotically stable.

$\nabla\nabla\nabla$

6. Analysis via passivity theory

In this section we present our main result: the application of the passivity theorem given in Section 5, to prove global asymptotic stability of a class of nonlinear PID global regulators for rigid robots. First, we present two passivity properties of rigid robots in closed-loop with energy shaping based controllers.

Property 4. Passivity structure of rigid robots in closed-loop with energy shaping based controllers (see Fig. 4). The system (2) in closed-loop with

$$\tau = \frac{\partial \mathcal{U}_a(q_d, \tilde{q})}{\partial \tilde{q}} + \tau' \tag{26}$$

is passive, from input torque τ' to output velocity $\dot{q}$, with storage function

$$\begin{aligned} V(\dot{q}, \tilde{q}) = {} & \frac{1}{2}\dot{q}^T M(q)\dot{q} + \mathcal{U}_T(q_d, \tilde{q}) \\ & -\mathcal{U}_T(q_d, \mathbf{0}), \end{aligned} \tag{27}$$

This is,

$$\int_0^T \dot{q}(t)^T \tau' \, dt \geq -V(\dot{q}(0), \tilde{q}(0)), \tag{28}$$

where $\mathcal{U}_a(q_d, \tilde{q})$ is the artificial potential energy introduced by the controller with properties requested by the energy shaping methodology and $\mathcal{U}_T(q_d, \tilde{q})$ is the total potential energy of the closed-loop system, which has an unique minimum that is global.
Furthermore the closed-loop system is zero state observable.

Proof. The system (2) in closed-loop with control law (26) is given by

$$\frac{d}{dt}\begin{bmatrix}\tilde{q}\\ \dot{q}\end{bmatrix} = \tag{29}$$

$$\begin{bmatrix} -\dot{q} \\ M^{-1}(q)\left[\frac{\partial \mathcal{U}_T(q_d - \tilde{q})}{\partial \tilde{q}} - C(q, \dot{q})\dot{q}\right] + \left[M^{-1}(q)\tau'\right] \end{bmatrix} \tag{30}$$

where (3) and (14) have been used. In virtue of Property 1, the time derivate of the storage function (27) along the trajectories of the closed-loop system (30) yields

$$\dot{V}(\dot{q}(t), \tilde{q}(t)) = \dot{q}^T \tau'$$

where integrating from 0 to T, in a direct form we obtain (28), thus, passivity from τ' to $\dot{q}$ has been proved.

◇

The zero state observability property of the system (30) can be proven, by taking the output as $y = \dot{q}$ and the input as $u = \tau'$, because

$$\dot{q} \equiv \mathbf{0}, \; \tau' \equiv \mathbf{0} \Rightarrow \tilde{q} \equiv \mathbf{0}.$$

The robot passive structure is preserved in closed-loop with the energy shaping based controllers, because this kind of controllers also have a passive structure. Passivity is invariant for passive systems which are interconnected in closed-loop, and the resulting system is also passive.

◇

Property 5. State strictly passivity of rigid robots in closed-loop with the energy shaping plus damping injection based regulators (see Fig. 5). The system (2) in closed-loop with

$$\tau = \frac{\partial \mathcal{U}_a(q_d, \tilde{q})}{\partial \tilde{q}} - K_v \dot{q} + \tau'' \tag{31}$$

is state strictly passive, from input torque τ'' to output $(\dot{q} - \alpha\,\mathrm{sat}(\tilde{q}))$, with storage function

$$\begin{aligned} V(\dot{q}, \tilde{q}) = {} & \frac{1}{2}\dot{q}^T M(q)\dot{q} + \mathcal{U}_T(q_d, \tilde{q}) \\ & -\mathcal{U}_T(q_d, \mathbf{0}) - \alpha\,\mathrm{sat}(\tilde{q})^T M(q)\dot{q}, \end{aligned} \tag{32}$$

where $\frac{1}{2}\dot{q}^T M(q)\dot{q}$ is the kinetic energy, $\mathcal{U}_T(q_d, \tilde{q})$ is the total potential energy of the closed-loop system, and $\alpha\,\mathrm{sat}(\tilde{q}) M(q)\dot{q}$ is a cross term which depends on position error and velocity,

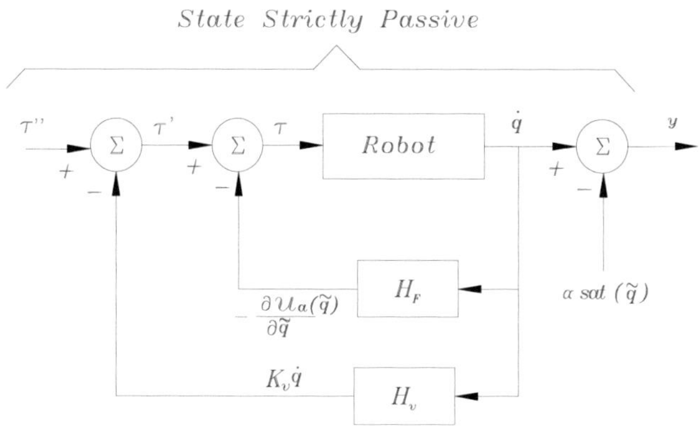

Fig. 5. State strictly passivity of rigid robots in closed-loop with energy shaping plus damping injection based regulator

and α is a small constant (Santibáñez & Kelly, 1998). In this case $K_v\dot{q}$ is the damping injection term. The State dissipation rate is given by :

$$\begin{aligned}\varphi(\dot{q},\tilde{q}) = & \dot{q}^T K_v \dot{q} + \alpha\ \dot{\text{sat}}(\tilde{q}) M(q)\dot{q} \\ & -\alpha\,\text{sat}(\tilde{q})\, C(\tilde{q},\dot{q})^T \dot{q} \\ & -\alpha\,\text{sat}(\tilde{q}) K_p \tilde{q} + \alpha\,\text{sat}(\tilde{q}) K_v\, \dot{q}.\end{aligned} \tag{33}$$

Consequently the inner product of the input τ'' and the output $y = (\dot{q} - \alpha\,\text{sat}(\tilde{q}))$ is given by:

$$(\dot{q} - \alpha\,\text{sat}(\tilde{q}))^T \tau'' \geq \dot{V}(\dot{q},\tilde{q}) + \varphi(\dot{q},\tilde{q}), \tag{34}$$

Proof. The closed-loop system(2) with control law (31) is

$$\frac{d}{dt}\begin{bmatrix}\tilde{q}\\ \dot{q}\end{bmatrix} = \tag{35}$$

$$\begin{bmatrix} -\dot{q} \\ M^{-1}(q)\left[\frac{\partial \mathcal{U}_T(q_d - \tilde{q})}{\partial \tilde{q}} - K_v\dot{q} - C(q,\dot{q})\dot{q}\right] + \left[M^{-1}(q)\tau''\right]\end{bmatrix} \tag{36}$$

where (3) and (14) have been used. In virtue of property 1, the time derivate of the storage function (32) along the trajectories of the closed-loop system (36) yields to

$$\dot{V}(\dot{q}(t),\tilde{q}(t)) = (\dot{q} - \alpha\,\text{sat}(\tilde{q}))^T \tau'' - \varphi(\dot{q}(t),\tilde{q}(t)),$$

from which we get (34), so state strictly passivity from input τ'' to output $(\dot{q} - \alpha\,\text{sat}(\tilde{q}))$ is proven.

◊

The robot dynamics enclosed loop with the energy shaping plus damping injection based controllers defines a state strictly passive mapping, from torque input τ'' to output $y = (\dot{q} - \alpha\,\text{sat}(\tilde{q}))$

$$y^T \tau'' \geq \dot{V}_1(\dot{q}(t),\tilde{q}(t)) + \varphi(\dot{q}(t),\tilde{q}(t)), \tag{37}$$

where $\varphi(\dot{q},\tilde{q})$ is called the state dissipation rate given by (33) with a storage function

$$V_1(\tilde{q},\dot{q}) = \frac{1}{2}\dot{q}^T M(q)\dot{q} + \mathcal{U}_T(q_d,\tilde{q}) - \mathcal{U}_T(q_d,\mathbf{0})$$
$$-\alpha\,\mathrm{sat}(\tilde{q})^T M(q)\dot{q}, \tag{38}$$

which is positive definite function and radially unbounded (Santibáñez & Kelly, 1995).
The integral action defines a zero state observable passive mapping with a radially unbounded and positive definite storage function

$$V_2(z) = \frac{1}{2}z^T K_i z.$$

By considering the robot dynamics in closed loop with the energy shaping plus damping injection based control action, in the forward path and the integral action in the feedback path (see Fig. 6), then, the feedback system satisfies in a direct way the theorem 1 conditions and we conclude global asymptotic stability of the closed loop system.

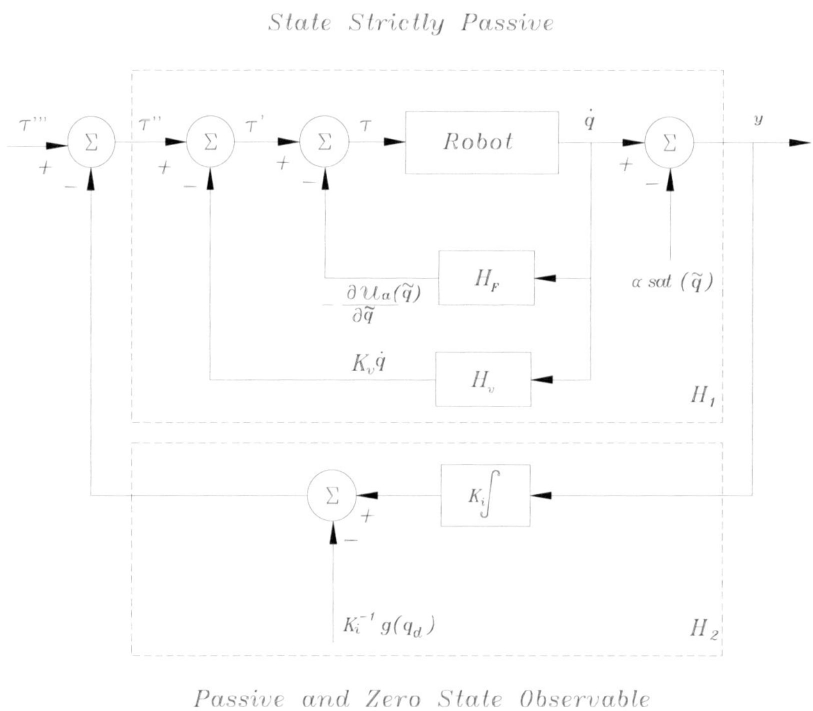

Fig. 6. Robot dynamics with Nonlinear PID controller

So we have proved the following:

Proposition 1.
Consider the class of nonlinear PID regulators (17) in closed-loop with robot dynamics (2). The closed-loop system can be represented by an interconnected system, which satisfies the following conditions

- A1. The system in the forward path defines a state strictly passive mapping with a radially unbounded positive definite storage function.
- A2. The system in the feedback path defines a zero state observable passive mapping with a radially unbounded positive definite storage function.

Besides, the equilibrium $\left[\tilde{q}^T \; \dot{q}^T \; z^T\right]^T = \mathbf{0} \in \mathbb{R}^{3n}$ of the closed-loop system (20) is globally asymptotically stable.

7. Simulation results

Computer simulations have been carried out to illustrate the performance of a class of nonlinear PID global regulators for robot manipulators. A example of this kind of nonlinear PID regulators is given by (Kelly, 1998)

$$\tau = K_p\tilde{q} - K_v\dot{q} + K_i\int_0^t \left[\alpha\,\mathrm{sat}(\tilde{q}(\sigma)) + \dot{\tilde{q}}(\sigma)\right] d\sigma \tag{39}$$

where artificial potential energy is given by $\mathcal{U}_a(\tilde{q}) = \frac{1}{2}\tilde{q}^T K_p \tilde{q}$, hence $\frac{\partial \mathcal{U}_a(\tilde{q})}{\partial \tilde{q}} = K_p\tilde{q}$.
The manipulator used for simulation is a two revolute joined robot (planar elbow manipulator), as show in Fig. 1. The meaning of the symbols is listed in Table 2 whose numerical values have been taken from (Reyes & Kelly, 2001).

Parameters	*Notation*	*Value*	*Unit*
Length link 1	l_1	0.45	m
Length link 2	l_2	0.45	m
Link (1) center of mass	l_{c1}	0.091	m
Link (2) center of mass	l_{c2}	0.048	m
Mass link 1	m_1	23.902	kg
Mass link 2	m_2	3.88	kg
Inertia link 1	I_1	1.266	Kg m^2/rad
Inertia link 2	I_2	0.093	Kg m^2/rad
Gravity acceleration	g	9.81	m/s^2

Table 2. Physical parameters of the prototype planar robot with 2 degrees of freedom

The entries of the dynamics of this two degrees–of–freedom direct–drive robotic arm are given by (Meza et al., 2007):

$$M(q) = \begin{bmatrix} 2.351 + 0.168\cos(q_2) & 0.102 + 0.084\cos(q_2) \\ 0.102 + 0.084\cos(q_2) & 0.102 \end{bmatrix}$$

$$C(q,\dot{q}) = \begin{bmatrix} -0.084\sin(q_2)\dot{q}_2 & -0.084\sin(q_2)(\dot{q}_1 + \dot{q}_2) \\ 0.084\sin(q_2)\dot{q}_1 & 0 \end{bmatrix}$$

$$g(q) = 9.81\begin{bmatrix} 3.921\sin(q_1) + 0.186\sin(q_1 + q_2) \\ 0.186\sin(q_1 + q_2) \end{bmatrix}$$

The PID tuning method is based on the stability analysis presented in (Santibáñez & Kelly, 1998). The tuning procedure for the PID controller gains can be written as:

$$\lambda_{\mathrm{M}}\{K_i\} \geq \lambda_{\mathrm{m}}\{K_i\} > 0$$
$$\lambda_{\mathrm{M}}\{K_v\} \geq \lambda_{\mathrm{m}}\{K_v\} > 0$$
$$\lambda_{\mathrm{M}}\{K_p\} \geq \lambda_{\mathrm{m}}\{K_p\} > k_g$$

where K_p denotes a diagonal positive definite $n \times n$ gain matrix resulting of the artificial potential energy $\mathcal{U}_a(\tilde{q}) = \frac{1}{2}\tilde{q}^T K_p \tilde{q}$ of the controller. The PID tuning requires to compute k_g. Using property 2 and the above expressions of the gravitational torque vector, we obtain that $k_g = 80.578 \quad [\mathrm{kg\ m^2/sec^2}]$.
The gain was tuned as $K_p = \mathrm{diag}\{130, 81\}$ [Nm/rad], $K_i = \mathrm{diag}\{30, 5\}$ [Nm/rad sec] and $K_v = \mathrm{diag}\{31, 18\}$ [Nm sec/rad] and $\alpha = 1$. The maximum torques supplied by the actuators are $\tau_1^{\max} = 150$ [Nm] and $\tau_2^{\max} = 15$ [Nm]. With the end of supporting the effectiveness of the proposed controller we have used a squared signal whose amplitude is decreased in magnitude every two seconds. More specifically, the robot task is coded in the following desired joint positions

$$\mathbf{q}_{\mathbf{d}_1}(t) = \begin{cases} 45 & \text{degrees if } 0 \leq t < 2 \text{ sec} \\ 30 & \text{degrees if } 2 \leq t < 4 \text{ sec} \\ 20 & \text{degrees if } 4 \leq t < 6 \text{ sec} \\ 0 & \text{degrees if } 6 \leq t < 8 \text{ sec} \end{cases}$$

$$\mathbf{q}_{\mathbf{d}_2}(t) = \begin{cases} 15 & \text{degrees if } 0 \leq t < 2 \text{ sec} \\ 10 & \text{degrees if } 2 \leq t < 4 \text{ sec} \\ 5 & \text{degrees if } 4 \leq t < 6 \text{ sec} \\ 0 & \text{degrees if } 6 \leq t < 8 \text{ sec} \end{cases}$$

Above position references are piecewise constant and really demand large torques to reach the amplitude of the respective requested step. In order to evaluate the effectiveness of the proposed controller. The proposed Nonlinear PID control scheme has been tuned to get their best performance in the presence of a step input whose amplitude is 45 deg for link 1 and 15 deg for link 2. The simulations results are depicted in Figs. (7)-(10), they show the desired and actual joint positions and the applied torques for the nonlinear PID control. From Figs. (7)-(8), one can observe that the transient for the nonlinear PID in each change of the step magnitude, of the links are really good and the accuracy of positioning is satisfactory.
Applied torque τ_1 and τ_2 are sketched in Figs. (9)-(10) these figures show the evolution of the applied torques to the robot joints during the execution of the simulations. Notice that initial torque peaks fit to the nominal torque limits.

8. Conclusions

In this chapter we have given sufficient conditions for global asymptotic stability of a class of nonlinear PID type controllers for rigid robot manipulators. By using a passivity approach, we have presented the asymptotic stability analysis based on the energy shaping methodology. The analysis has been done by using an adaptation of a passivity theorem presented in the literature. This passivity theorem, deals with systems composed by the feedback interconnection of a state strictly passive system with a passive system. Simulation results confirm that the class of nonlinear PID type controllers for rigid robot manipulators

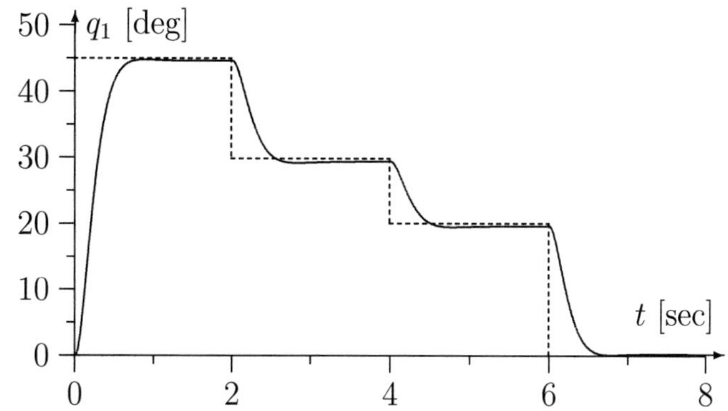

Fig. 7. Desired and actual positions 1 for the Nonlinear PID control

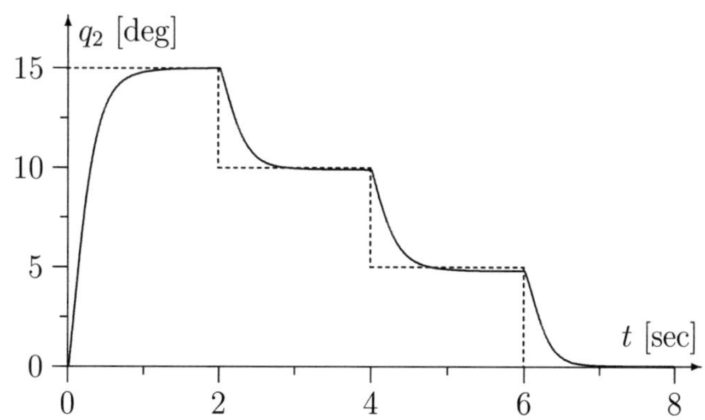

Fig. 8. Desired and actual positions 2 for the Nonlinear PID control

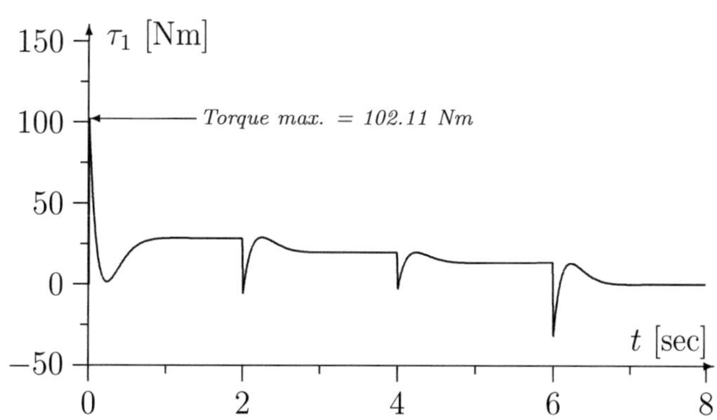

Fig. 9. Applied torque τ_1 Nonlinear PID

have a good precision. The performance of the nonlinear PID type controllers has been

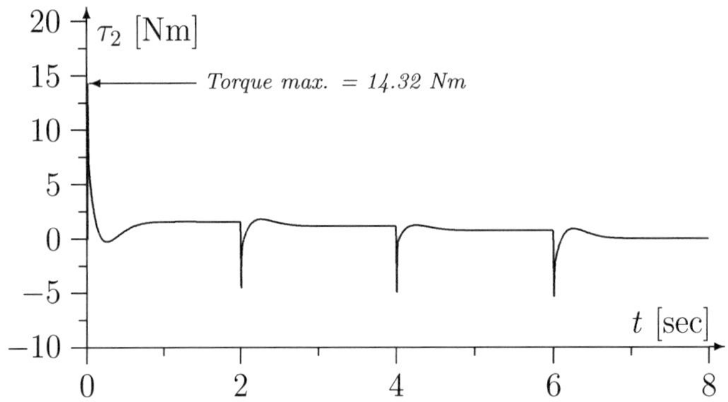

Fig. 10. Applied torque τ_2 Nonlinear PID

verified on a two degree of freedom direct drive robot arm.

9. Acknowledgment

The authors would like to thank CONACYT(México) Grant No. 134534, Promep, Cátedra e-robots Tecnológico de Monterrey, DGEST for their support.

10. References

Ailon, A. & Ortega, R. (1993) "An observer-based set-point controller for robot manipulators with flexible joints", *Systems and Control Letters*, North Holland, 21, pp. 329-335.

Alvarez, J.; Cervantes, I. & Kelly, R. (2000). PID regulation of robot manipulators: stability and performance *Systems and Control Letters, 41*, 73-83.

Alvarez, J.; Kelly, R. & Cervantes, I. (2003). Semiglobal stability of saturated linear PID control for robot manipulators. Automatica, Vol. 39, pp. 989-995.

Alvarez, J.; Santibáñez, V. & Campa, R. (2008). Stability of robot manipulators under saturated PID compensation. IEEE Transactions on Control Systems Technology, Vol. 16, No. 6, pp. 1333-1341.

Arimoto, S.; Naniwa, T.; Parra–Vega, V. & Whitcomb, L. (1994), A quasi–natural potential and its role in design of hyper–stable PID servo–loop for robotic systems, *In Proc. of the CAI Pacific Symposium'94 on Control and Industrial Automation Application*, Hong Kong, pp. 110–117.

Arimoto, S. (1994). State of the art and future research directions of robot control, *4Th IFAC Symposium on Robot Control*, Italy.

Arimoto, S.; Naniwa, T. & Suzuki H.(1990), Asymptotic stability and robustness of PID local feedback for position control of robot manipulators, *proceeding of ICARCV 90*, Singapore 1990, 19-27.

Arimoto, S. & Miyazaki, F. (1983). Stability and robustness of PID feedback control for robot manipulators of sensory capability, *Robotics Researchs, First International Symposium* MIT press, Cambridge, MA.

Arimoto, S. (1995a) "Fundamental problems of robot control: Part I, Innovations in the realm of robot servo–loops", *Robotica*, Vol. 13, pp. 19–27.

Arimoto, S. (1996) *Control theory of non-linear mechanical systems: A passivity-based and circuit-theoretic approach*, Oxford University Press.

Arimoto, S. (2009) Control of mechanical systems. Scholarpedia, 4(4):6520.

Berghuis, H. & Nijmeijer, H. (1993b) "Global regulation of robots using only position measurement", *Systems and Control Letters*, Vol. 21, pp. 289–293.

Qu, Z. & Dorsey, J. (1991) Robust PID control of robots, *Int. J. Robot. Automat.,6*, 1991, 228-235.

Gorez, R. (1999) Globally stable PID-like control of mechanical systems, *Systems and Control Letters, 38*,pp. 61-72.

Khalil, H. K. (2002). *Nonlinear Systems*, 3rd edition. Prentice-Hall, Englewood Cliffs, NJ.

Kelly, R. (1993) "A simple set–point robot controller by using only position measurements", Proc. of the IFAC'93 World Congress, Sydney, Australia, Vol. 6, pp. 173–176.

Kelly, R. (1995) A Tuning procedure for stable PID control of robot manipulators, *Robotica, 13*(2), pp. 141-148.

Kelly, R.; Santibáñez, V. & Loria, A. (2005). *Control of Robot Manipulators in Joint Space*, Springer–Verlag.

Kelly, R. (1998). "Global positioning of robot manipulators via PD control plus a class of nonlinear integral actions ", *IEEE Transactions on Automatic Control*,Vol. 43, No. 7,pp 934–938.

Kelly, R. & Ortega, R. (1988). "Adaptive control of robot manipulators: an input–output approach" IEEE International Conference on Robotics and Automation, Philadelphia, PA.

Koditschek, D. (1984). "Natural motion for robot arms" Proceedings of the 1984 IEEE Conference on Decision and Control, Las Vegas CA., Dec., pp. 733-735.

Laib, A. (2000). Adaptive output regulation of robot manipulators under actuator constraints. IEEE Transactions on Robotics and Automation, Vol. 16, pp. 29-35.

Landau, I. D. & Horowitz, R. (1988). "Synthesis of adaptive controllers for robot manipulators using a Passive feedback system approach", IEEE International Conference on Robotics and Automation, Philadelphia, PA.

Loria, A.; Kelly, R.; Ortega & Santibáñez, V. (1997). On global output feedback regulation of Euler–Lagrange systems with bounded inputs, *IEEE Transactions on Automatic Control*", Vol. 42, No. 8, pp. 1138–1143.

Meza, J.L.; Santibáñez V. & Hernandez, V. (2005). Saturated nonlinear PID global regulator for robot manipulators: Passivity based analysis. Proceedings of the 16th IFAC World Congress, Prague, Czech Republic.

Meza, J.L.; Santibáñez V. & Campa, R.(2007). An estimate of the domain of attraction for the PID regulator of manipulators, *Int. J. Robotics Automat. 22*, 187-195.

Nijmeijer, H. & Van der Schaft, A. (1990) *"Nonlinear Dynamical Control Systems"*, Springer–Verlag, NY.

Ortega, R. & Spong, M. (1989) "Adaptive motion control of rigid robots: a tutorial", *Automatica*, Vol. 25, No. 6, pp. 877–888.

Ortega, R.; Loria, A. & Kelly, R. (1995a) A semiglobally stable output feedback PI^{2}D regulator for robot manipulators, *IEEE Transactions on Automatic Control, 40* (8), pp. 1432–1436.

Ortega, R.; Loría, A.; Kelly, R. & Praly, L. (1995b) "On passivity–based output feedback global stabilization of Euler–Lagrange systems" *International Journal of Robust and Nonlinear Control*, Vol. 5, pp. 313–323.

Orrante, J.; Santibáñez V. & Campa, R. (2010). On Saturated PID Controllers for Industrial Robots: the PA10 Robot Arm as Case of Study, Advanced Strategies for Robot Manipulators, S. Ehsan Shafiei (Ed.), ISBN: 978-953-307-099-5, InTech, Available from: http://www.intechopen.com/articles/show/title/on-saturated-pid-controllers-for-industrial-robots-the-pa10-robot-arm-as-case-of-study.

Ortega, R.; A. Loria, P. J. Nicklasson, P. J. & Sira-Ramirez, H. (1998). Passivity- based control of Euler-Lagrange systems, in Communications and Control Engineering. Berlin, Germany: Springer.

Ortega, R. & Garcia-Canseco, E. (2004) Interconnection and damping assignment passivity-based control: A survey, Eur. J Control, vol. 10, pp. 432–450.

Ortega, R.; Vander, A.; Castaños, F. & Astolfi, A. (2008). Control by Interconnection and Standard Passivity-Based Control of Port-Hamiltonian Systems, IEEE Transaction on Automatic Control, Vol.53,No. 11, pp. 2527- 2542.

Reyes, F & Kelly, R. (2001). Experimental evaluation of model–based controllers on a direct–drive robot arm, *Mechatronics,11*, 267–282.

Rocco, P. (1996) Stability of PID control for industrial robot arms, *IEEE Transactions on Robotics and Automation, 12*(4), pp. 606-614.

Santibáñez, V. & Kelly, R. (1998) "A class of Nonlinear PID Global regulators for robot manipulators", Proc. of 1998 IEEE International Conference on Robotics and Automation, Leuven, Belgium.

Santibáñez, V. & Kelly, R. (1997) "Strict Lyapunov functions for global regulation of robot manipulators", AUTOMATICA, Vol. 33, No. 4, pp. 675-682.

Santibáñez, V.; Kelly, R.; Zavala-Rio, A. & Parada, P. (2008). A new saturated nonlinear PID global regulator for robot manipulators, Proceedings of the 17th IFAC World Congress,Seoul, Korea.

Santibáñez, V.; Camarillo, K.; Moreno-Valenzuela, J. & Campa, R. (2010). A practical PID regulator with bounded torques for robot manipulators. International Journal of Control Automation and Systems, Vol. 8, No. 3, pp. 544-555.

Sepulchre, R.; Jankovic, M. & Kokotovic, P. (1997). Constructive Nonlinear Control. New York: Springer.

Spong, M.; Hutchinson, M. & Vidyasagar, M.(2006). *Robot Modeling and Control,* John Wiley and Sons.

Sun, D.; Hu, S.; Shao, X. & Liu, C., (2009) Global stability of a saturated nonlinear PID controller for robot manipulators. IEEE Transactions on Control Systems Technology, Vol. 17, No. 4, pp. 892-899.

Takegaki, M. & Arimoto, S. (1981) "A new feedback method for dynamic control of manipulators", *ASME Journal of Dynamic Systems, Measurement, and Control,* Vol. 103, pp. 119–125.

Tomei, P. (1991). Adaptive PD controller for robot manipulators, *IEEE Transactions on Robotics and Automation, 7*(4), 565-570.

Van der Schaft, A. (1999). Gain and Passivity Techniques in Nonlinear Control. Berlin, Germany: Springer.

Wen, J. T. & Bayard, D. S. (1988) "New class of control laws for robotic manipulators. Part 1. Non–adaptive case", *International Journal of Control,* Vol 47, No. 5, pp. 1361–1385.

Wen, J. T. & Murphy, S. (1990). *PID control for robot manipulators,* CIRSSE Document 54, Rensselaer Polytechnic Institute.

4

A PI^{2}D Feedback Control Type for Second Order Systems

América Morales Díaz and Alejandro Rodríguez-Angeles
Center for Research and Advanced Studies, CINVESTAV-IPN
Robotics and Advanced Manufacturing Group, Unidad Saltillo
Mechatronics Group, Electrical Engineering Deparment, Unidad Zacatenco
México

1. Introduction

Second order autonomous systems are key systems in the study of non linear systems because their solution trajectories can be represented by curves in the plane (Khalil, 2002), which helps in the development of control strategies through the understanding of their dynamical behaviour. Such autonomous systems are often obtained when considering feedback control strategies, because the closed loop system might be rewritten in terms of the state system and perturbation terms, which are function of the state as well. Thus, analyses of stability properties of second order autonomous systems and their convergence are areas of interest on the control community.

Moreover, several applications consider nonlinear second order systems; there are various examples of this:

1. In mechanical systems the pendulum, the inverted pendulum, the translational oscillator with rotational actuator (TORA) and the mass-spring systems;
2. In electrical systems there are examples such as the tunnel diode circuit, some electronic oscillators as the negative-resistance twin-tunnel-diode circuit; and finally
3. Other type of these systems are mechanical-electrical-electronic combinations, for example a two degree of freedom (DOF) robot arm or a mobile planar robot and among every degree of freedom on a robotic structure can be represented by a second order nonlinear system.

Therefore, due to the wide applications in second order nonlinear systems, several control laws have been proposed, which comprises from simple ones, like linear controllers, to the more complex, like sliding mode, backstepping approach, output-input feedback linearization, among others (Khalil, 2002).

Despite the development of several control strategies for nonlinear second order systems, it is not surprising that for several years and even nowadays the classical PID controllers have been widely used in technical and industrial applications and even on research fields. This is due to the good understanding that engineers have of them. Moreover, the PID controllers have several important functions: provide feedback, has the ability to eliminate steady state offset through integral action, and it can anticipate the future through derivative action.

PID controllers are sufficient for many control problems, particularly when system dynamics are favourable and the performance requirements are moderate. These types of

controllers are important elements of distributed control system. Many useful features of PID control are considered trade secrets, (Astrom and Hagglund, 1995). To build complicated automation systems in widely production systems as energy, transportation and manufacturing, PID control is often combined with logic, sequential machines, selectors and simple function blocks. And even advanced techniques as model predictive control is encountered to be organized in hierarchically, where PID control is used in the lower level. Therefore, it can be inferred that PID control is a key ingredient in control engineering.
For the above reasons several authors have developed PID control strategies for nonlinear systems, this is the case of (Ortega, Loria and Kelly, 1995) that designed an asymptotically stable proportional plus integral regulator with position feedback for robots with uncertain payload that results in a PI²D regulator. In the work of (Kelly, 1998), the author proposed a simple PD feedback control plus integral action of a nonlinear function of position errors of robot manipulators, that resulted effective on the control of this class of second nonlinear systems and it is known as PD control with gravity compensation. Also PID modifications for control of robot manipulators are proposed at the work of (Loria, Lefeber and Nijmeijer, 2000), where global asymptotic stability is proven. In process control a kind of PI² compensator was developed in the work of (Belanger and Luyben, 1997) as a low frequency compensator, due to the additional double integral compensation rejects the effects of ramp-like disturbances; and in the work of (Monroy-Loperena, Cervantes, Morales and Alvarez-Ramirez, 1999), a parametrization of the PI² controller in terms of a nominal closed-loop and disturbance estimation constants is obtained, despite both works are on the process control field, their analysis comprises second order plants.
In the present work a class of nonlinear second order system is consider, where the control input can be consider as result of state feedback, that in the case of second order systems is equivalent to a PD controller, meanwhile double integral action is provided when the two state errors are consider, both regulation and tracking cases are considered.
Stability analysis is developed and tuning gain conditions for asymptotic convergence are provided. A comparison study against PID type controller is presented for two examples: a simple pendulum and a 2 DOF robot arm. Simulation results confirm the stability and convergence properties that are predicted by the stability analysis, which is based on Lyapunov theory. Finally, the chapter closes with some conclusions.

2. Problem formulation

Two cases are considered in this work, first regulation to a constant reference is boarded, second tracking a time varying reference is studied; in both cases stability and tuning gain conditions are provided.

2.1 Regulation

Consider the following type of second order system:

$$\begin{aligned} \dot{x}_1 &= x_2 \\ \dot{x}_2 &= -f(x) + g(x)u \end{aligned} \tag{1}$$

Where $x \in \mathbb{R}^n$ is the state, $u \in \mathbb{R}^n$ is the control input, such that fully actuated systems are considered, $g(x) \in \mathbb{R}^{nxn}$ is a non linear function that maps the input to the system dynamics, and it is assumed that such function is known and invertible along all solutions of the

system, $f(x)$ is a nonlinear function that is continuously differentiable, and locally Lipschitz. It is assumed that the state is measurable and that $f(x)$ is known.
The control objective is to regulate the state $x = [x_1 \quad x_2]^T$ to a constant value $x_{ref} = \left[x_{1ref} \quad 0\right]^T$.
The proposed dynamic control considers full cancellation of the system dynamics, and it is given by

$$u = g(x)^{-1}\left(f(x) + u_n\right) \tag{2}$$

where u_n represents a nominal feedback control that would be designed to ensure the regulation of (1) to x_{ref}.
The nominal control is designed as a feedback state control plus a type of double integral control and is provided in the following equation

$$u_n = -K_P\left(x_1 - x_{1ref}\right) - K_D x_2 - K_I \int\left(\left(x_1 - x_{1ref}\right) + x_2\right)dt \tag{3}$$

Control (3) provides an extra integral action with the integration of the state x_2. The constant gains are K_P, K_D and K_I and must be positive. The integral action provides an augmented state, therefore system (1) in closed loop with control (2) and (3) is re-written as

$$\begin{aligned} \dot{x}_1 &= x_2 \\ \dot{x}_2 &= -K_P\left(x_1 - x_{1ref}\right) - K_D x_2 - K_I x_3 \\ \dot{x}_3 &= \left(x_1 - x_{1ref}\right) + x_2 \end{aligned} \tag{4}$$

The closed-loop system (4) has a unique equilibrium point in $x_{ref} = \left[x_{1ref} \quad 0 \quad 0\right]^T$.
In the following a stability analysis for the regulation case is determined.

2.1.1 Stability analysis for the regulation case

Consider the following position error vector $e = [e_1 \quad e_2 \quad e_3]^T$, with $e_1 = x_1 - x_{1ref}$, $e_2 = x_2$, $e_3 = \int(e_1 + e_2)dt$, such that the closed loop error dynamics (4), which corresponds to an autonomous system, might be rewritten as

$$\begin{aligned} \dot{e}_1 &= e_2 \\ \dot{e}_2 &= -K_P e_1 - K_D e_2 - K_I e_3 \\ \dot{e}_3 &= e_1 + e_2 \end{aligned} \tag{5}$$

Provided that the gains K_P, K_D and K_I are different from zero and positive, it is immediate to obtain that the equilibrium of system (5) corresponds to $e^* = [0 \quad 0 \quad 0]^T$. On the following stability conditions and tuning guidelines for the control gains K_P, K_D and K_I will be presented.

Theorem 1

Consider the autonomous dynamic second order system given by (5), which represents the closed loop error dynamics obtained from system (1) with the control law (2), and the nominal PI²D controller given by (3). The autonomous dynamic system (5) converge asymptotically to its equilibrium point $e^* = [0 \quad 0 \quad 0]^T$, if the positive control gains K_P, K_D and K_I satisfy the following conditions

$$\begin{aligned} &K_I > 8 \\ &K_D > 3K_I + 2\sqrt{2K_I^2 - K_I - 1} \\ &K_P > K_I + K_D \end{aligned} \tag{6}$$

Proof:

Consider the position error vector $e = [e_1 \quad e_2 \quad e_3]^T$ and the Lyapunov function

$$V_e = \tfrac{1}{2} e^T M(K_P, K_D, K_I) e \tag{7}$$

where $M(K_P, K_D, K_I) \in \mathbb{R}^{n \times n}$ is a symmetric positive definite matrix, with all entries $m_{i,j}$ real and positive for all i,j; in order to simplify the Lyapunov function computation the following conditions are introduced

$$\begin{aligned} &m_{1,3} = m_{3,1} = 0 \\ &m_{1,2} = m_{2,3} \end{aligned}$$

The time derivative of the Lyapunov function (7) is function of the closed loop error dynamics (5), and it is given by

$$\begin{aligned} \dot{V}(e) &= e^T M \dot{e} \\ &= e_1 e_2 \left(m_{1,1} + m_{1,2} - m_{2,2} K_P - m_{1,2} K_D \right) + e_1 e_3 \left(m_{3,3} - m_{1,2} K_P - m_{1,2} K_I \right) + \\ &+ e_2 e_3 \left(m_{3,3} - m_{1,2} K_D - m_{2,2} K_I \right) - e_1^2 m_{1,2} K_P + e_2^2 \left(2m_{1,2} - m_{2,2} K_D \right) - e_3^2 m_{1,2} K_I \end{aligned}$$

Thus, a straightforward simplification of the time derivative of the Lyapunov function is to cancel the crossed error terms e_1e_2, e_1e_3, e_2e_3, which results in $\dot{V}(e)$ given by quadratic error terms. First, in order to cancel the crossed term e_1e_3 conditions on $m_{3,3}$ can be obtained, and then to cancel the crossed term e_2e_3 the matrix entry $m_{2,2}$ is defined appropriately, finally by defining $m_{1,1}$ the crossed error term e_1e_2 is eliminated. So far the conditions on matrix $M(K_P, K_D, K_I)$ are summarized as follows

$$\begin{aligned} &m_{1,3} = m_{3,1} = 0 \\ &m_{1,2} = m_{2,3} \\ &m_{1,1} = \frac{m_{1,2}}{K_I} \left(K_P^2 + K_P(K_I - K_D) + K_I(K_D - 1) \right) \\ &m_{2,2} = \frac{m_{1,2}}{K_I} (K_P + K_I - K_D) \\ &m_{3,3} = m_{12} (K_P + K_I) \end{aligned} \tag{8}$$

On the other hand, to guarantee that $m_{1,1}$, $m_{2,2}$ and $m_{3,3}$ of the matrix $M(K_P, K_D, K_I)$ are positive, it is necessary to satisfy the following conditions. For the matrix entry $m_{3,3}$ to be positive, it is enough to have that $K_P > 0$ and $K_I > 0$ which are satisfied by conditions (6) of Theorem 1. For $m_{2,2} > 0$ it follows that $K_P > K_D - K_I$, that is satisfy by the conditions $K_D > 0$ and $K_P > K_D + K_I$ as stated at (8). Finally, for $m_{1,1} > 0$ it follows that $K_P^2 + K_P(K_I - K_D) + K_I(K_D - 1) > 0$, which implies conditions on K_P and K_D, to find out such conditions, the solutions of equation $K_P^2 + K_P(K_I - K_D) + K_I(K_D - 1) = 0$ are computed as follows

$$K_P = \frac{(K_D - K_I) \pm \sqrt{(K_I - K_D)^2 - 4K_I(K_D - 1)}}{2}$$

For K_P to be positive, it is required that $K_D > K_I$, that is satisfied by conditions (6). Then for K_P to be purely real, it is required that the argument of the squared root being positive, i.e. $(K_I - K_D)^2 - 4K_I(K_D - 1) > 0$, which when it is equal to zero implies the solutions $K_D = 3K_I \pm \sqrt{2K_I^2 - K_I}$. Thus, by taking the positive part of the solution and considering that $K_D > 3K_I + \sqrt{2K_I^2 - K_I}$ it is guaranteed that K_P is real, and condition $K_I > 1/2$ implies that K_D is real. Notice that all these conditions are satisfied by those stated at Theorem 1, equations (6).

At this point, it is guaranteed that the solutions K_P are real and positive, those to ensure that $m_{1,1} > 0$ it is considered that K_P must satisfied

$$K_P > \frac{(K_D - K_I) + \sqrt{(K_I - K_D)^2 - 4K_I(K_D - 1)}}{2}$$

Such a condition is clearly over satisfied by the condition $K_P > K_D + K_I$ given at Theorem 1, equations (6).

Therefore, if conditions given by (6) at Theorem 1 are satisfied, it implies that $m_{1,1}$, $m_{2,2}$ and $m_{3,3}$ of the matrix $M(K_P, K_D, K_I)$ are positive.

Furthermore, the definition of $m_{1,1}$, $m_{2,2}$ and $m_{3,3}$ stated by (8), yield a time derivative Lyapunov function given by quadratic error terms. Nonetheless it is necessary to check positive definitiveness of the Lyapunov function (7), which after conditions (8) are considered is given by

$$V(e) = m_{1,2}\left\{ \frac{e_1^2}{K_I}\left[K_P^2 + K_P(K_I - K_D) + K_I(K_D - 1)\right] + 2e_1e_2 + e_2^2\frac{K_P + K_I - K_D}{K_I} + 2e_2e_3 + e_3^2(K_P + K_I) \right\}$$

Notice that the positive condition on the coefficient of the term e_1^2, i.e. $K_P^2 + K_P(K_I - K_D) + K_I(K_D - 1) > 0$ has already been considered for positive definitiveness of $m_{1,1}$. Therefore, the conditions on K_P, K_D and K_I given by (6) imply that all coefficients of the quadratic terms of $V(e)$ are positive. Now to guarantee $V(e) > 0$ note that the cross error terms $2e_1e_2$ and $2e_2e_3$ can be rewritten as part of a quadratic form, for that it is required

$$K_P^2 + K_P(K_I - K_D) + K_I(K_D - 1) > 1$$
$$\frac{K_P + K_I - K_D}{K_I} > 2$$
$$K_P + K_I > 1$$

The last two conditions imply that $K_P > 1 - K_I$ and $K_P > K_I + K_D$ which are satisfied by conditions (6). And the first condition implies to solve the equation $K_P^2 + K_P(K_I - K_D) + K_I(K_D - 1) = 1$ for K_P. Similar to the way in which conditions for positive value of the entries of matrix $M(K_P, K_D, K_I)$, it follows that

$$K_P > \frac{(K_D - K_I) + \sqrt{(K_I - K_D)^2 - 4[K_I(K_D - 1) - 1]}}{2}$$

That is conservatively satisfied by the condition $K_P > K_D + K_I$ given at Theorem 1, equation (6). On the other hand for K_P to be real, it is necessary that $K_D > 3K_I + 2\sqrt{2K_I^2 - K_I - 1}$, and for K_D to be real it is required that $K_I > 1$; all these conditions are clearly satisfied by those stated at Theorem 1, equations (6).
Therefore, if the conditions given by (6) are satisfied, the Lyapunov function results on a sum of quadratic terms

$$V(e) = m_{1,2}\left\{(e_1 + e_2)^2 + (e_2 + e_3)^2 + k_1 e_1^2 + k_2 e_2^2 + k_3 e_3^2\right\} \tag{9}$$

for positive parameters k_1, k_2, k_3; thus concluding that $V(e) > 0$ for $e \neq 0$, and $V(e) = 0$ for $e = 0$.
Since the definition of the matrix entries (8) allows cancellation of all cross error terms on the time derivative of the Lyapunov function (7), then along the position error solutions, it follows that

$$\dot{V}(e) = e^T M \dot{e} = -m_{1,2}\left[K_P e_1^2 + K_I e_3^2 + \frac{K_P K_D + K_I(K_D - 2) - K_D^2}{K_I} e_2^2\right] \tag{10}$$

To ensure that $\dot{V}(e) < 0$, it is required that $K_P K_D + K_I(K_D - 2) - K_D^2 > 0$, which implies that

$$K_P > \frac{2K_I - K_D K_I + K_D^2}{K_D}$$

which is satisfied by the condition $K_P > K_D + K_I$ given at Theorem 1, equations (6). Nonetheless to guaranteed that K_P is real, it follows that $2K_I - K_D K_I + K_D^2 > 0$, that implies when considering equal to zero, that the solutions are

$$K_D = \frac{K_I \pm \sqrt{K_I(K_I - 8)}}{2}$$

Thus for K_D to be real it is required that $K_I > 8$ and finally the condition on K_D results on

$$K_D > \frac{K_I + \sqrt{K_I(K_I - 8)}}{2}$$

Such that, the above conditions are satisfied by considering those of Theorem 1, equation (6). Therefore, by satisfying conditions (6) it can be guaranteed that all coefficients of the derivative of the Lyapunov function $\dot{V}(e)$ are positive, such that $\dot{V}(e) < 0$ for $e \neq 0$, and $\dot{V}(e) = 0$ for $e = 0$.
Thus, it can be concluded that the closed loop system dynamic (5) is stable and the error vector e converges globally asymptotically to its equilibrium $e^* = [0 \quad 0 \quad 0]^T$.

■

Remark 1

The conditions stated at Theorem 1, equations (6) are rather conservative in order to guarantee stability and asymptotic convergence of the closed loop errors. The conditions (6) are only sufficient but not necessary to guarantee the stability of the system.

Remark 2

Because full cancellation of the system dynamics function $f(x)$ in (1) is assumed by the control law (2), in order to obtain the closed loop error dynamics (5), then the auxiliary polynomial $P(s) = s^3 + s^2 K_D + s(K_P + K_I) + K_I$ can be considered to obtain a Hurwitz polynomial, and to characterize some properties of the closed loop system.

2.1.2 Stability analysis for the regulation case with non vanishing perturbation

In case that no full cancellation of $f(x)$ in (1) can be guaranteed, either because of uncertainties on $f(x)$, $g(x)$, or in the system parameters, convergence of the system to the equilibrium point $e^* = [0 \quad 0 \quad 0]^T$ is not guaranteed. Nonetheless, the Lipschitz condition on $f(x)$, and assuming that $f(x)$ is bounded in terms of x, i.e $\|f(x)\| \leq \gamma \|x\|$ for positive γ, then locally uniformly ultimate boundedness might be proved for large enough control gains K_P, K_D and K_I, see (Khalil, 2002).

2.2 Tracking

In the case of tracking, the problem statement is now to ensure that the sate vector $x = [x_1 \quad x_2]^T$ follows a time varying reference $x_{ref}(t) = \left[x_{1ref}(t) \quad \dot{x}_{1ref}(t)\right]^T$; this trajectory is at least twice differentiable, smooth and bounded. For this purpose the control proposed in (2) is considered, but with the nominal controller u_n given by

$$u_n = -K_P\left(x_1 - x_{1ref}\right) - K_D\left(x_2 - \dot{x}_{1ref}\right) - K_I \int\left(\left(x_1 - x_{1ref}\right) + \left(x_2 - \dot{x}_{1ref}\right)\right)dt + \ddot{x}_{1ref} \tag{11}$$

2.2.1 Stability analysis for the tracking case

Similar to the regulation case, the following position error vector $e = [e_1 \quad e_2 \quad e_3]^T$ is defined, with $e_1 = x_1 - x_{1ref}$, $e_2 = x_2 - \dot{x}_{1ref}$, $e_3 = \int\left(\left(x_1 - x_{1ref}\right) + \left(x_2 - \dot{x}_{1ref}\right)\right)dt$, such that the closed loop error dynamics of system (1), with the controller (2) and (11) results in the same dynamic systems given by (5), such that Theorem 1 applies for the tracking case.

Remark 3

The second integral action proposed in the nominal controllers, (3) for regulation, and (11) for tracking case, can be interpreted as a composed measured output function, such that this action helps the controller by integrating the velocity errors. When all non linearity is cancelled the integral action converges to zero, yielding asymptotic stability of the complete state of the system. If not all nonlinear dynamics is cancelled, or there is perturbation on the system, which depends on the state, then it is expected that the integral action would act as estimator of such perturbation, and combined with suitable large control gains, it would render ultimate uniformly boundedness of the closed loop states.

3. Results

In this section two systems are consider, a simple pendulum with mass concentrated and a 2 DOF planar robot. First the pendulum system results are showed.

3.1 Simple pendulum system at regulation

Consider the dynamic model of a simple pendulum, with mass concentrated at the end of the pendulum and frictionless, given by

$$\begin{aligned} \dot{x}_1 &= x_2 \\ \dot{x}_2 &= -f(x) + c\,u \end{aligned} \tag{12}$$

where $f(x) = a\sin(x_1) + bx_2$ with $a = g/l > 0$, $b = k/m > 0$ and $c = 1/ml^2 > 0$, with the notation m for the mass, k for the spring effects, l the length of the pendulum, and g the gravity acceleration. The values of the model parameters are presented at Table 1, and the initial condition of the pendulum is $x(0) = [1 \quad 0]^T$.

The proposed PI²D is applied and compared against a PID control that also considers full dynamic compensation, i.e. the classical PID is programmed as follows

$$\begin{aligned} u &= g(x)^{-1}\left(f(x) + u_n\right) \\ u_n &= -K_P\left(x_1 - x_{1ref}\right) - K_D\left(x_2 - \dot{x}_{1ref}\right) - K_I \int \left(x_1 - x_{1ref}\right) dt \end{aligned}$$

The comparative results are shown in Figure 1. The control gains were tuned accordingly to conditions given by (6), see Table 1, such that it was considered that: $K_I > 8$, thus for the selected K_I value, it was obtained that $K_D > 57.49$, and after selection of K_D, it was finally obtained that $K_P > 70$. For the tuned gains listed at Table 1, it follows that the eigenvalues of the closed loop system (5) are the roots of the characteristic polynomial $P(s) = s^3 + s^2 K_D + s(K_P + K_I) + K_I$, such that $s_1 = -0.1208$, $s_2 = -1.4156$, and $s_3 = -58.4635$. Therefore, the closed loop system behaves as an overdamped system as shown in Figure 1. The behaviour of the closed-loop system for the PID and PI²D controllers is shown in Figure 1; the performance of the double integral action on the PID proposed by the nominal controller (3) shows faster and overdamped convergence to the reference $x_{ref} = \left[\pi/4 \quad 0\right]^T$ than the PID controller, in which performance it is observed overshoot. Notice however that both input controls are similar in magnitude and shape; this implies better performance of the PI²D controller without increasing the control action significantly.

$a = 10$	$K_I = 10$
$b = 0.1$	$K_D = 60$
$c = 10$	$K_P = 80$

Table 1. Pendulum parameters and control gains.

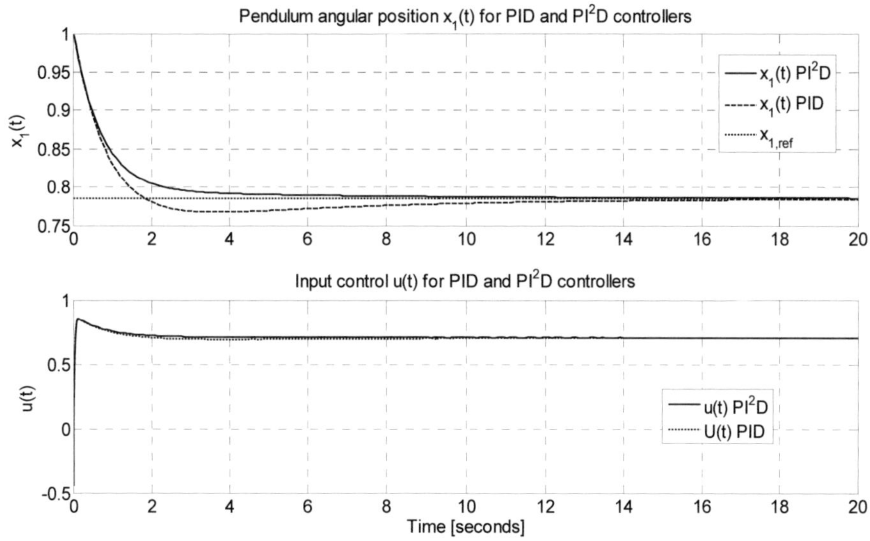

Fig. 1. Comparison study for PID vs PI²D controllers for a simple pendulum system.

For the sake of comparison another simulation is developed considering imperfect model cancellation, in this case due to pendulum parameters uncertainty considered for the definition of the controller (2). The nominal model parameters are those of Table 1, while the control parameters are $\bar{a} = 11.5$, $\bar{b} = 0.01$, $\bar{c} = 11$. The control gains and initial conditions are the same as for the case of perfect cancellation.
The obtained simulation results are shown in Figure 2, where also a change in reference signal is considered from $x_{ref} = \left[\pi/4 \;\; 0\right]^T$ [rad] in $0 \le t \le 30$ seconds to $x_{ref} = \left[\pi/3 \;\; 0\right]^T$ in $30 < t \le 60$ seconds. In the case of non complete dynamic cancellation due to uncertain parameters, it can be seen that the PI²D controller proposed by (2) and (3) also responds faster that the classical PID with dynamic cancellation, besides the control actions are similar in magnitude and shape as shown in Figure 2.

3.2 Simple pendulum system at tracking

A periodic reference given by $x_{1ref} = \sin\left(\pi t/5\right)$ [rad] is considered. The simulation results are shown in Figure 3; the control gains are the same as listed at Table 1. In Figure 3 is depicted both behaviour of the PID and PI²D with perfect dynamic compensation, the PI²D controller shows faster convergence to the desired trajectory than the PID control, nonetheless both control actions are similar in magnitude and shape, this shows that a small change on the control action might render better convergence performance, in such a case the double integral action of the PI²D controller plays a key role in improving the closed loop system performance

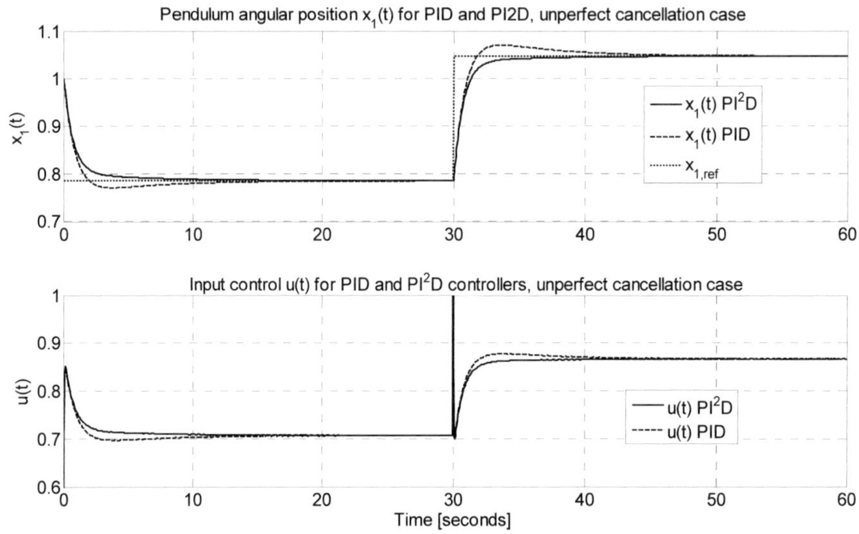

Fig. 2. Comparison study for PID vs PI²D controllers for a simple pendulum system with model parameter uncertainty.

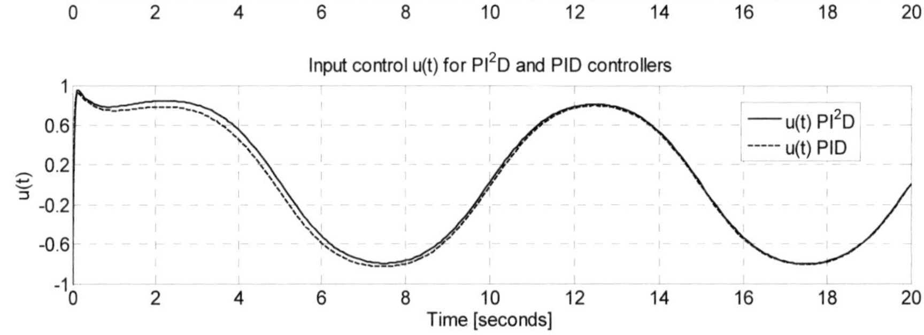

Fig. 3. Tracking response of pendulum system (1) for PID and PI²D controllers.

To close with the pendulum example, uncertainty on the parameters is considered, such that there is no cancellation of the function $f(x)=a\sin(x_1)+bx_2$, i.e. the parameters of the controller $u(t)$ given by (2) are set as $\bar{a}=0$, $\bar{b}=0$, and $\bar{c}=1$; and the controller gains are the same as listed at Table 1. In Figure 4 the comparison results are showed, despite there is no model cancellation, the PI²D controller shows better performance that the PID case, i.e faster convergence (less than 4 seconds), requiring minimum changes on the control action magnitude and shape, as shown on the below plot of Figure 4, where the control actions are similar to those of Figure 3, which implies that the control gains absorbed the model parameter uncertainties on parameter $\overline{c}=1$ as well as the non model cancellation. Notice that the control actions present a sort of chattering that is due to the effort to compensate the no model cancellation

3.3 A 2 DOF planar robot at regulation

The dynamic model of a 2 DOF serial rigid robot manipulator without friction is considered, and it is represented by

$$D(q)\ddot{q}+C(q,\dot{q})\dot{q}+g(q)=\tau \tag{13}$$

Where $q,\ \dot{q},\ \ddot{q}\in\mathbb{R}^2$ are respectively, the joint position, velocity and acceleration vectors in generalized coordinates, $D(q)\in\mathbb{R}^{2\times2}$ is the inertia matrix, $C(q,\dot{q})\in\mathbb{R}^{2\times2}$ is the Coriolis and centrifugal matrix, $g(q)\in\mathbb{R}^2$ is the gravity vector and $\tau\in\mathbb{R}^2$ is the input torque vector. The system (13) presents the following properties (Spong and Vidyasagar, 1989).

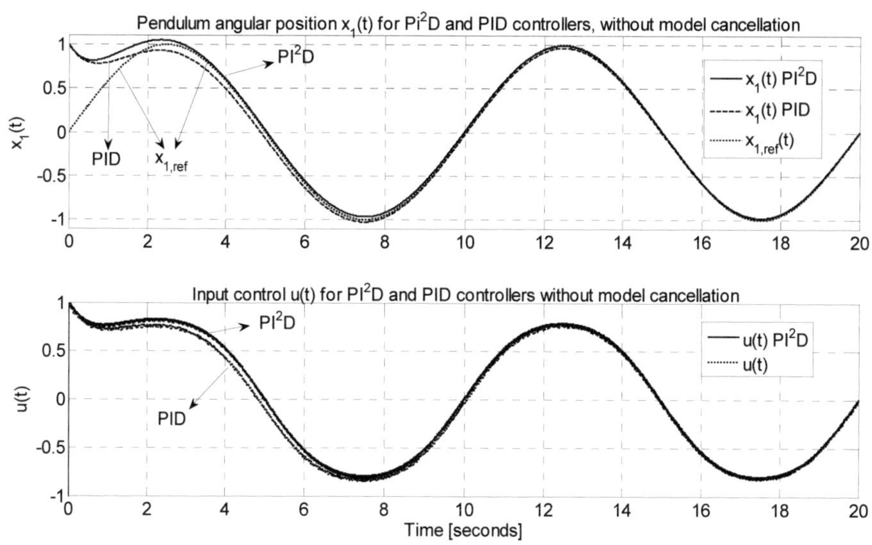

Fig. 4. Tracking response of pendulum system (1) for PID and PI²D controllers without model cancellation.

Property 1.- The inertia matrix is a positive symmetric matrix satisfying $\lambda_{\min}I\leq D(q)\leq\lambda_{\max}I$, for all $q\in\mathbb{R}^2$, and some positive constants $\lambda_{\min}\leq\lambda_{\max}$, where I is the 2-dimensional identity matrix.

Property 2.- The gravity vector $g(q)$ is bounded for all $q\in\mathbb{R}^2$. That is, there exist $n=2$ positive constants γ_i such that $\sup_{q\in\mathbb{R}^2}|g_i(q)|\leq\gamma_i$ for all $i=1,\ldots,n$.

From the generalized 2 DOF dynamic system, eq. (13), each DOF is rewritten as a nonlinear second order system as follows.

$$\begin{aligned} \dot{x}_{1,i} &= x_{2,i} \\ \dot{x}_{2,i} &= -f_i(x) + g_i(x)\, u_i \end{aligned} \tag{14}$$

With $f_i(x)$ and $g_i(x)$ obtained from rewritten system (13), solving for the acceleration vector and considering the inverse of the inertia matrix. As for the pendulum case a PI²D controller of the form given by (2) and (3) is designed and compared against a PID, similar to section 3.1, for both regulation and tracking tasks.

From Figure (5) to Figure (7), the closed loop with dynamic compensation is presented, where the angular position, the regulation error and the control input, are depicted. The PI²D controller shows better behaviour and faster response than the PID. The controller gains for both DOF of the robot are listed at Table 1. The desired reference is $x_d = \begin{bmatrix} \pi/2 & \pi/4 \end{bmatrix}^T$.

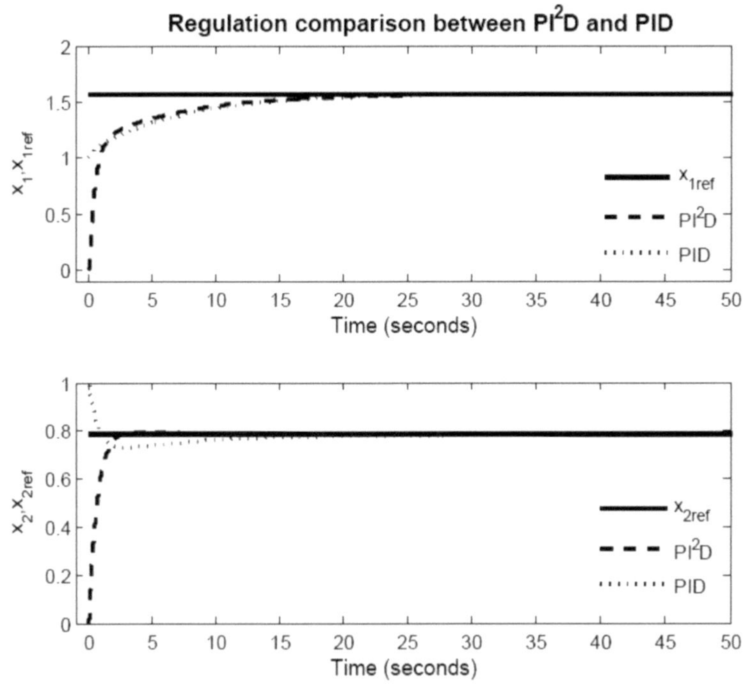

Fig. 5. Robot angular position for PI²D and PID controllers with perfect cancellation.

To test the proposed controller robustness against model and parameter uncertainty, it was considered unperfected dynamic compensation, for both links a sign change on the inertia terms corresponding to the function $g(x)$ is considered and no gravitational compensation was made, meaning that $f(x) = 0$ at the controller. The control gains remained the same as for all previous cases. Figures (8) to (10) show the simulation results. Although the inexact compensation, the proposed PI²D controller behaves faster and with a smaller control effort than the PID control.

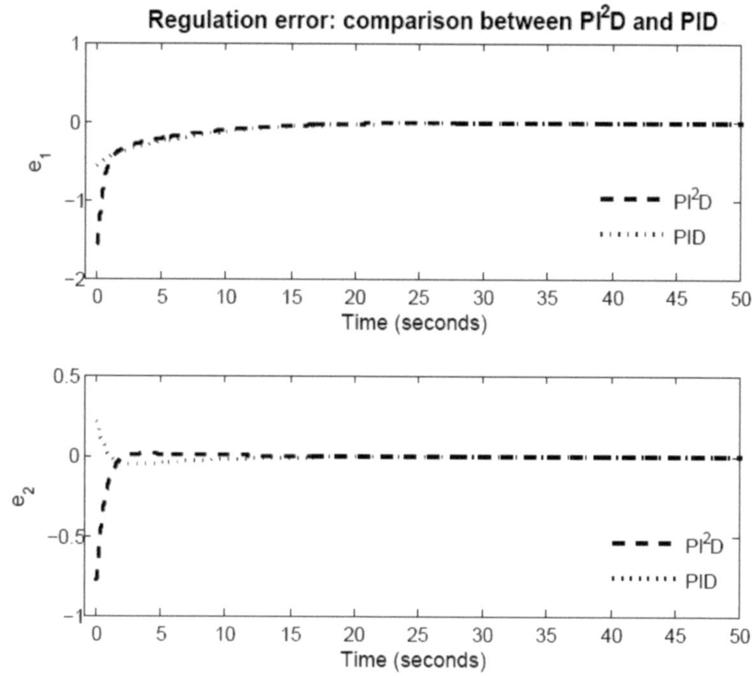

Fig. 6. Robot regulation error for PI²D and PID controllers with perfect cancellation.

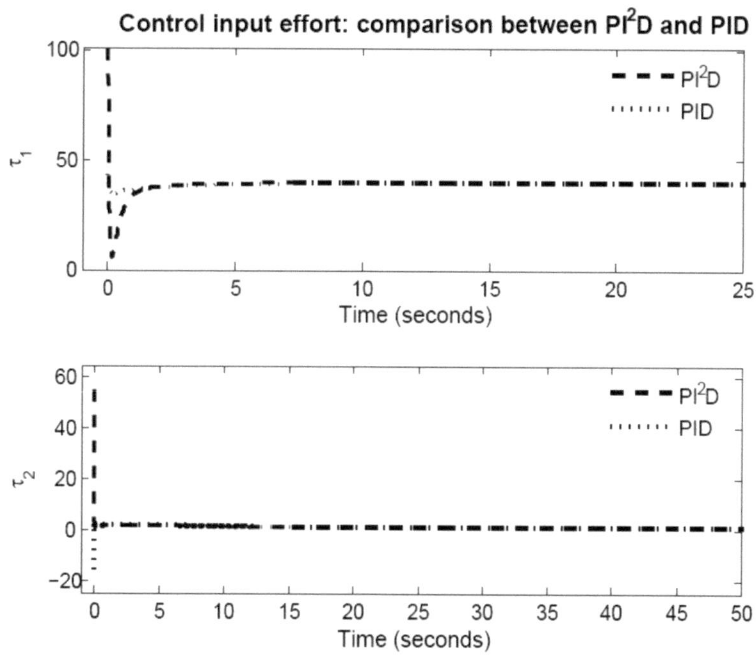

Fig. 7. Robot input torque for PI²D and PID controllers with perfect cancellation.

3.4 A 2 DOF planar robot at tracking

For the tracking case study a simple periodical signal given by $x_d(t) = \left[\sin\left(\pi t/_{40}\right) \quad \sin\left(\pi t/_{20}\right)\right]$ is tested. First perfect cancellation is considered, and then unperfected cancellation of the robot dynamics is taken into account. The control gains are the same as those listed at Table 1. Figures (11) to (13) show the system closed loop performance with perfect dynamic compensation, where the angular position, the regulation error and the control input, respectively, are depicted. The PI²D controller shows a better behaviour and faster response than the PID, both with dynamical compensation.

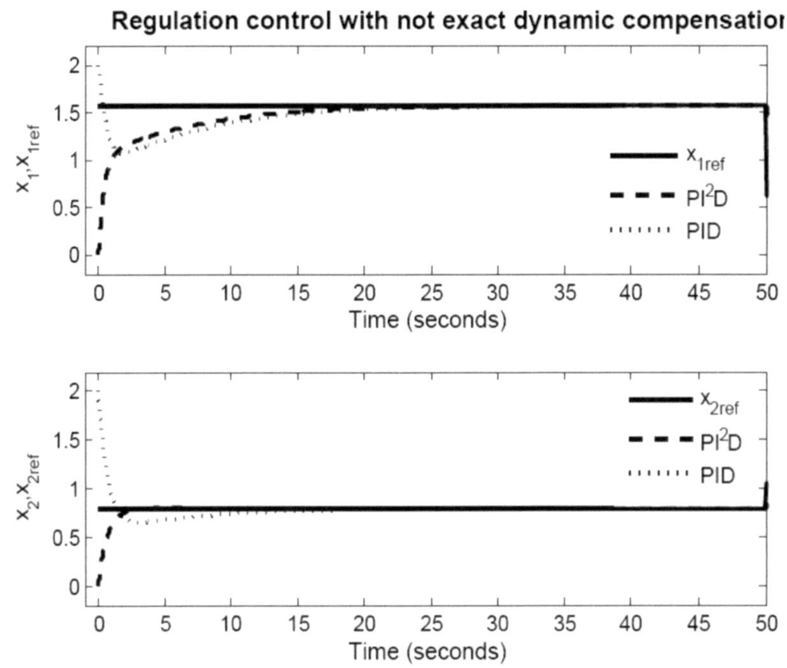

Fig. 8. Robot angular position for PI²D and PID controllers without perfect cancellation.

To test the proposed controller robustness against model and parameter uncertainty, it was considered imperfect dynamic compensation considering as in the regulation case a sign change in $g(x)$, and no compensation on $f(x)$. The control gains remained the same as for all previous cases. Figures (14) to (16) show the simulation results. Although the inexact compensation, the proposed PI²D controller behaves faster and with a smaller control effort than the PID control.

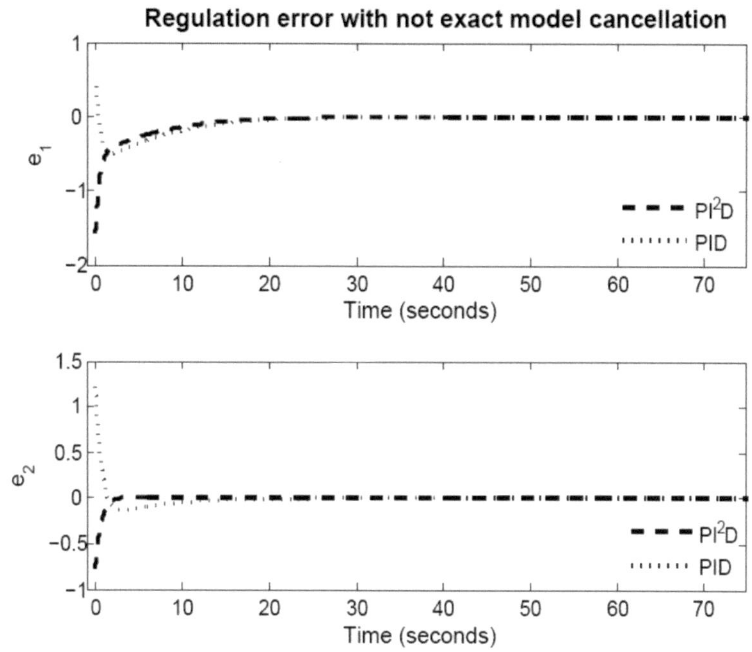

Fig. 9. Robot regulation error for PI^{2}D and PID controllers without perfect cancellation.

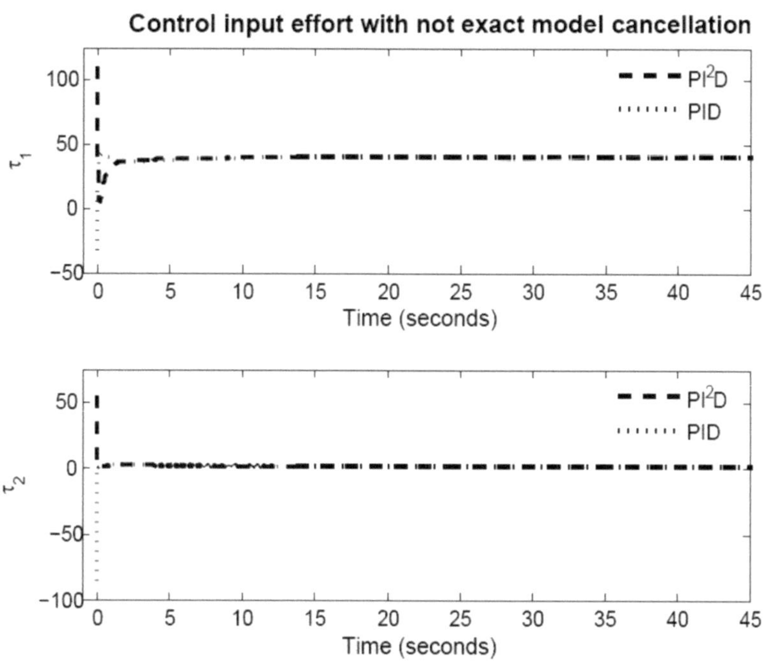

Fig. 10. Robot input torque for PI^{2}D and PID controllers without perfect cancellation.

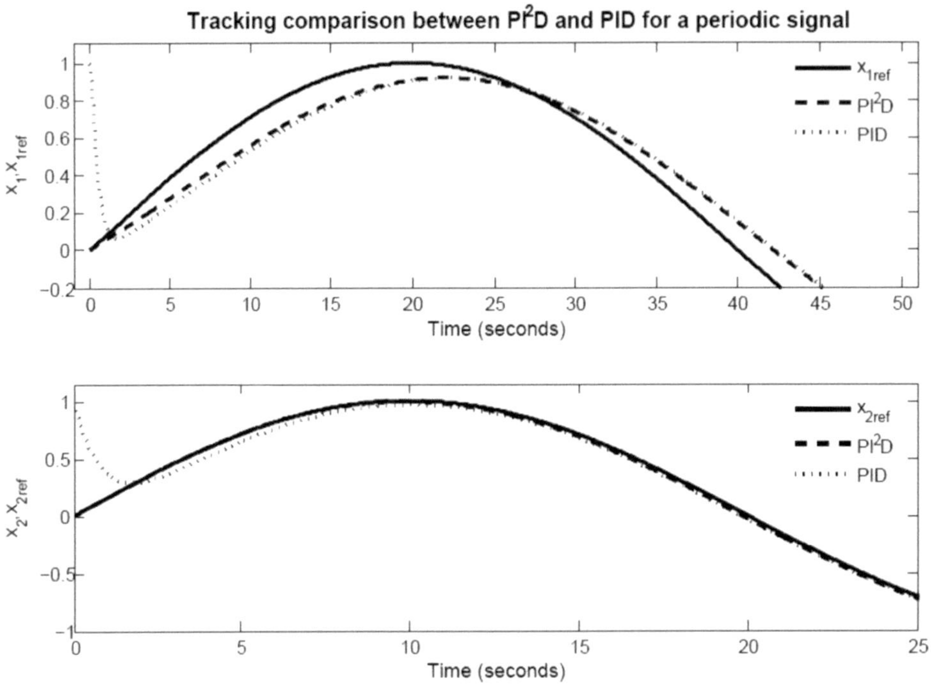

Fig. 11. Robot angular position for PI²D and PID controllers with perfect cancellation.

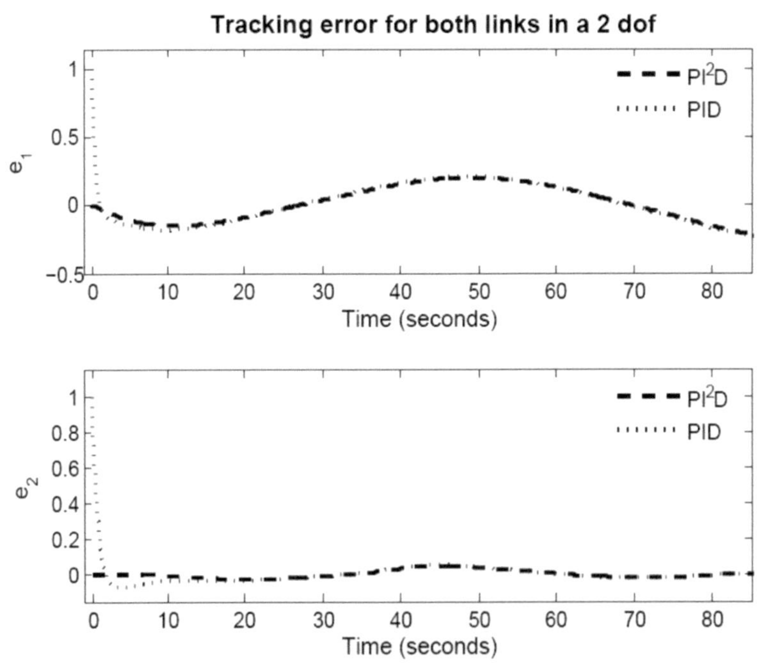

Fig. 12. Robot tracking error for PI²D and PID controllers with perfect cancellation.

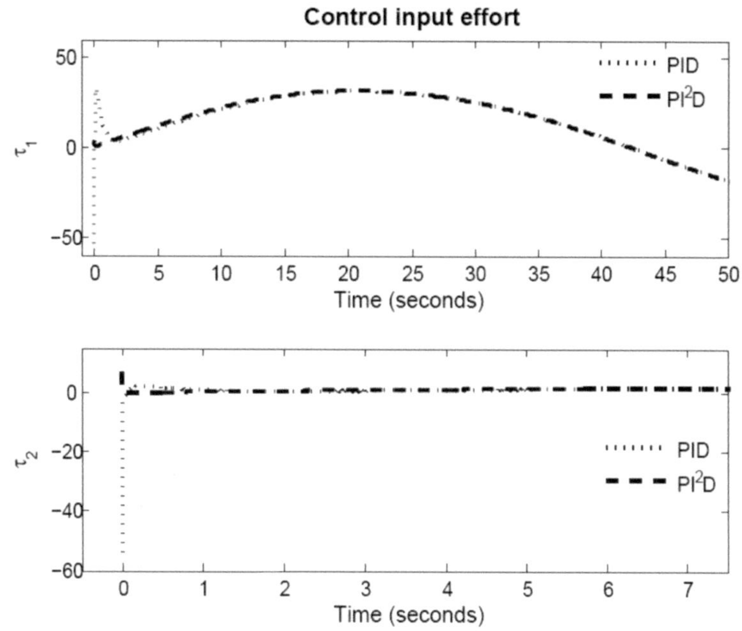

Fig. 13. Robot input torque for PI²D and PID controllers with perfect cancellation.

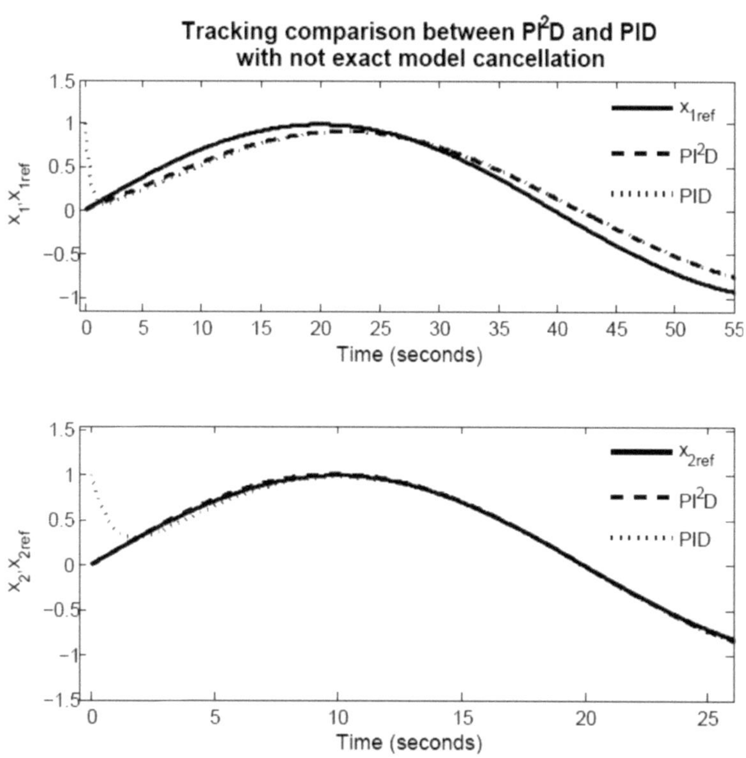

Fig. 14. Robot angular position for PI²D and PID controllers without perfect cancellation.

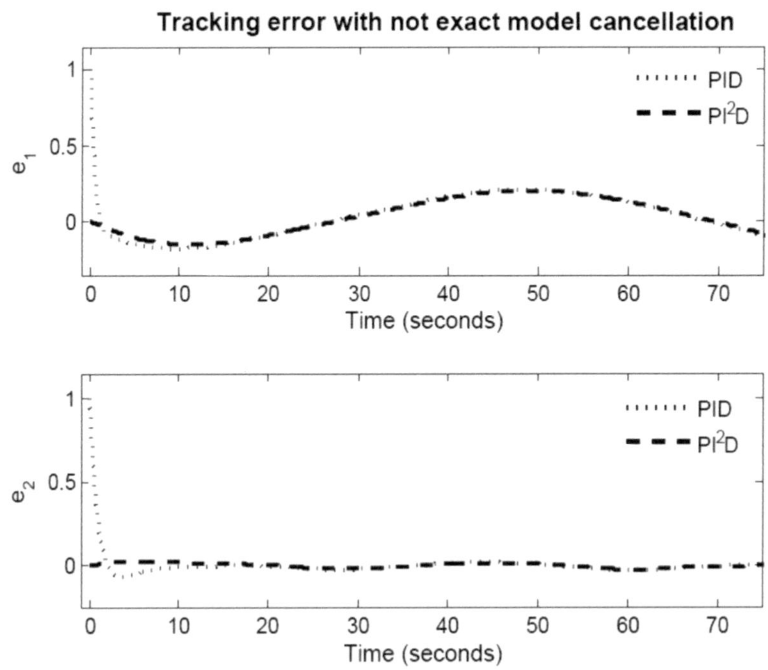

Fig. 15. Robot tracking error for PI²D and PID controllers without perfect cancellation.

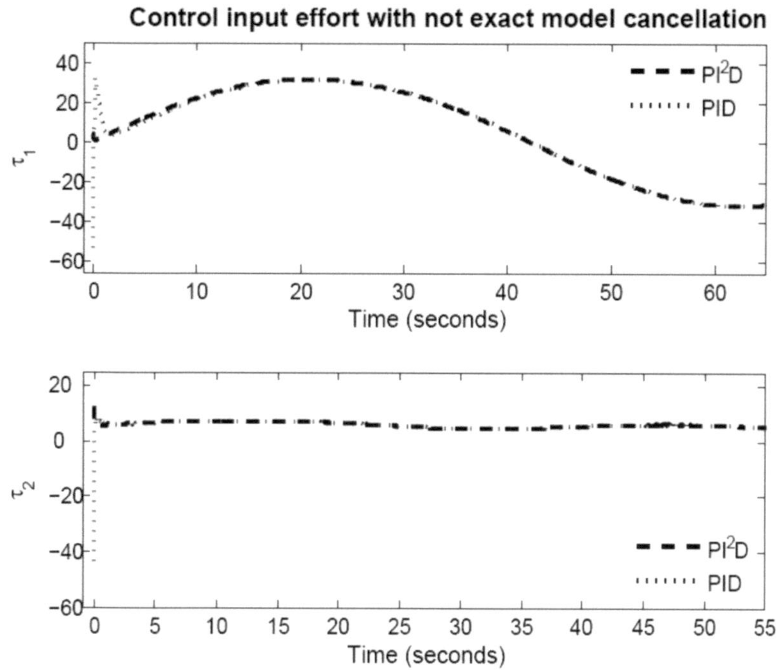

Fig. 16. Robot input torque for PI²D and PID controllers without perfect cancellation.

4. Conclusions

The proposed controller represents a version of the classical PID controller, where an extra feedback signal and integral term is added. The proposed PI²D controller shows better performance and convergence properties than the PID. The stability analysis yields easy and direct control gain tuning guidelines, which guarantee asymptotic convergence of the closed loop system.

As future work the proposed controller will be implemented at a real robot system, it is expected that the experimental results confirm the simulated ones, besides that it is well know that an integral action renders robustness against signal noise, by filtering it.

5. Acknowledgment

The second author acknowledges support from CONACyT via project 133527.

6. References

Astrom, K.J. and T. Hagglund (1995). PID Controllers: Theory, Design and Tuning. Instrument Society of America, USA. NC, USA.

Belanger, P.W. and W.L. Luyben (1997). Design of low-frequency compensators for improvement of plantwide regulatory perfromance. Ind. Eng. Chem. Res. 36, 5339–5347.

Kelly, R. (1998). Global positioning of robot manipulators via pd control plus a class of nonlinear integral action. IEEE Trans. Automat. Contr. 43, 934–938.

Khalil, H.K. (2002). Nonlinear Systems. Prentice Hall. New Jersey.

Monroy-Loperena, R., I. Cervantes, A. Morales and J. Alvarez-Ramirez, (1999). Robustness and parametrization of the proportional plus double integral compensator. Ind. Eng. Chem. Res. 38, 2013–2020.

Ortega, R., A. Loria and R. Kelly (1995). A semiglobally stable output feedback PI^2D regulator of robot manipulators. IEEE Trans. Automat. Contr. 40, 1432–1436.

Mark W. Spong and M. Vidyasagar (1989), Robot Dynamics and Control, Wiley, New York, 1989.

Loria, A., E. Lefeber, and H. Nijmeijer, Global asymptotic stability of robot manipulators with linear PID and PI^2D control, SACTA, Vol. 3, No. 2, 2000, pp. 138-149.

5

From Basic to Advanced PI Controllers: A Complexity vs. Performance Comparison

Aldo Balestrino, Andrea Caiti, Vincenzo Calabró,
Emanuele Crisostomi and Alberto Landi
Department of Energy and Systems Engineering,
University of Pisa
Italy

1. Introduction

The history of PIDs dates back to the beginning of the twentieth century when preliminary works of [Sperry (1922)] and [Minorski (1922)] provided mathematical results for the control of the ship motion and of automatically steered bodies in general. In particular, Minorski was the first to introduce three-term controllers with Proportional-Integral-Derivative (PID) actions. The success of PID was fast and solid, as nowadays they still represent the most popular choice in most industrial process control applications. The main reasons for their success are:

- **Reduced number of parameters:** A process control engineer only has to tune a small number of parameters to make the PID work effectively.
- **Well-established tuning rules:** There are predefined and well-known methods for deciding the values of the PID regulators. The most popular methods were given by [Ziegler and Nichols (1942)], [Astrom and Hagglund (1984)], and more recently [Zhuang and Atherton (1993)] and [Luyben and Eskinat (1994)]. Different tuning methods are due to different control objectives (e.g. reference following, disturbance rejection) and different plants (e.g. first-order model, second-order model).
- **Good performances:** The main reason for PID success is of course that good control performances are usually obtained; thus, the control engineer might not be interested in developing more complicated and less intuitive control schemes to improve something that is already working fine.

The ideal equation of a PID controller is

$$u_{PID}(t) = k_p\left(r(t) - y(t)\right) + k_p\left\{\int_0^t \left[\frac{1}{T_i}\left(r(\tau) - y(\tau)\right)\right] d\tau + T_d\frac{d\left(r(t) - y(t)\right)}{dt}\right\}, \quad (1)$$

where k_p is the proportional gain, k_p/T_i is the integral gain (sometimes denoted as k_i) and k_pT_d is the derivative gain (sometimes denoted as k_d). According to conventional notation $r(t)$, $y(t)$ and $u(t)$ denote respectively the reference, output and input signals. The error signal $r(t) - y(t)$ is sometimes denoted as $e(t)$. The classic feedback structure involving a PID controller is shown in Figure 1. Equation (1) is sometimes considered an ideal equation for PIDs, as

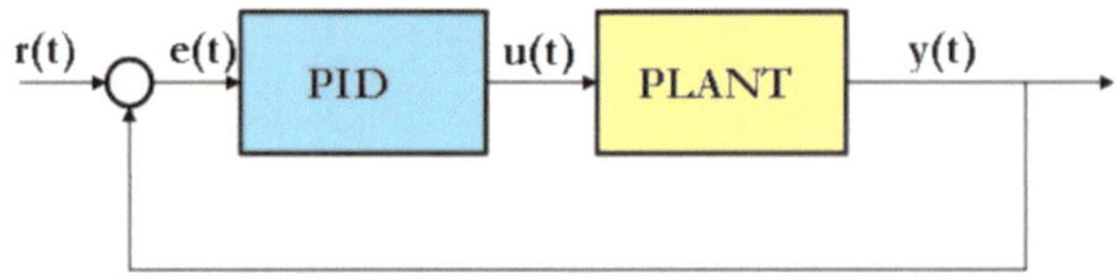

Fig. 1. Classic Feedback PID Control Scheme

usually a few tricks are required to avoid some typical well-known problems associated with ideal PIDs. For instance, a more realistic equation for PIDs is

$$u_{PID}(t) = k_p\left(b \cdot r(t) - y(t)\right) + k_p\left\{\int_0^t \left[\frac{1}{T_i}\left(r(\tau) - y(\tau)\right)\right] d\tau - T_d\frac{dy(t)}{dt}\right\}, \tag{2}$$

where the two main differences with the ideal equation (1) are the introduction of a set-point weighting factor b [Rasmussen (2009)] and the absence of the reference in the derivative term. The term b is used to mitigate kicking phenomena, that usually occur when there is an abrupt change in the reference signal $r(t)$. In the same circumstance, the derivative term of (1) is even larger, and a possible precaution is to relate the derivative only to the output signal. Assuming that only piecewise constant reference signals should be followed, the derivative term of (2) is equal to that of (1) after the transient stage.

Even the Equation (2) of a realistic PID is not always enough to obtain good control performances. One of the main drawbacks is that the PID's parameters are fixed, therefore if the plant's parameters vary in time then the fixed PID can not always represent the optimal solution. A simple way to deal with this problem is to use a relay auto-tuning procedure, i.e. the behaviour of the plant is continuously monitored, and the PID's parameters are automatically tuned in reaction to plant's changes.

A second drawback of conventional PIDs is that usually they are tuned either to have good tracking performances or to have good disturbance rejection. Of course, in practice both properties are desirable at the same time, thus requiring the design of suitable filters that separate the high-frequency components of noise from the low-frequency components of the reference signal. Alternatively, it is possible to design two PIDs which are optimal with respect to reference tracking and disturbance rejection respectively, and a smart device that switches between the two PIDs according to the most important objective in the particular moment.

A last problem of PIDs is that an anti-windup scheme must be adopted, especially in the common case that the inputs to the actuators are bounded by physical constraints, see for instance the tutorial [Peng et al. (1996)].

A consequence of the previous remarks is that some modifications to the basic structure of the PIDs might be required to achieve better control performances. A classic improvement is obtained by considering PID-like regulators whose parameters are not fixed, but are allowed to be time variant to maintain good control performances in different operating conditions. Moreover, it is clear that time varying parameters introduce more degrees of freedom in the control design, and in principle more flexibility in shaping the control response. Time variant PIDs can be designed according to several approaches well known in the literature: (a) fuzzy PIDs [Tang et al. (2001)]; in this case typically several fixed conventional PID controllers are designed for different operating conditions, and are then interpolated according to a set of fuzzy logic rules. Alternatively, fuzzy rules are used only to improve one component of the PID, for instance the set-point weighting term as in [Visioli (2004)]. (b) nonlinear PIDs; see

for instance [Haj-Ali and Ying (2004)] for a comparison with fuzzy PIDs. (c) variable structure PIDs, see for instance [Scottedward Hodel and Hall (2001)] or [Visioli (1999)] where a classic PID is modified by adding a feedforward term.
In this spirit, this work compares a traditional PI with three parameter varying PIs, characterised by an increasing level of complexity. As a support to the evalution of the control performance of each of the four regulators, conventional control indices, such as the integral of the absolute value or the square of the error signal, and the integral of the absolute value of the input and its derivative, have been used, as further detailed in Section 4. The final objective is to establish some thumb rules that quantify how much (or when) it is convenient to complicate the original conventional PI, in terms of improved control performances.

The discussion in this paper is restricted to PIs rather than PIDs for two main reasons:

- Many industrial controllers are simple PIs.
- The design of the derivative action simply follows the design of the other components. Therefore the derivative component can be introduced without affecting the general results of this work. Simple rules to introduce the derivative component can be found in the recent reference [Leva and Maggio (2011)].

Furthermore, the discussion is here restricted to linear time invariant systems, as in practice most industrial plants can be represented in such a form, eventually after a linearisation step around the desired operating point. While PI are successful in most stabilization and control problems, from a theoretical perspective there is a class of linear systems which are not stabilizable via output PI feedback; however, such examples will not be trated in this chapter.
This paper is organised as follows: next section introduces the 3 PIs that will be thoroughly compared with a conventional PI in benchmark examples. Section 3 illustrates the tuning procedures for all PIs. Section 4 compares the PI regulators in a challenging noisy reference tracking example, while Section 5 is dedicated to a realistic example. In the last section final conclusions are given and future work is outlined.

2. PI-like regulators

This sections presents the four regulators that will be later compared in challenging control problems.

2.1 Conventional PI controller

The conventional PI is described by Equation 1, without the derivative term, i.e.

$$u_{PI}(t) = k_p\left(b \cdot r(t) - y(t)\right) + k_p\left\{\int_0^t \left[\frac{1}{T_i}\left(r(\tau) - y(\tau)\right)\right] d\tau\right\}, \tag{3}$$

where all parameters are constant.

2.2 Variable Integral PI

A simple modification of the standard PI was proposed in [Balestrino et al. (2009)] and is here reproposed as a term of comparison as it represents a good trade-off between performances and complexity. In particular, the Variable Integral PI (VIPI), provides a control action equal to

$$u_{VIPI}(t) = k_p \left\{ (b \cdot r(t) - y(t)) + \int_0^t \left[\frac{1}{T_i} \left(e(\tau) \cdot \exp^{-\frac{e(\tau)^2}{2\sigma^2}} \right) \right] d\tau \right\}, \tag{4}$$

where σ is a further tuning parameter. The main difference with Equation (3) is that the integral gain is not simply k_p/T_i, but $k_p/T_i \cdot \exp^{-\frac{e(\tau)^2}{2\sigma^2}}$, and thus it is not fixed anymore, but depends on the instant error. The motivation of Equation (4) is that the integral gain

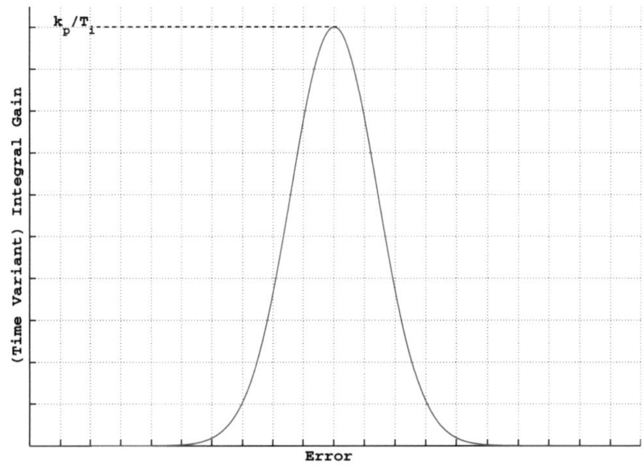

Fig. 2. The value of the integral gain depends on the error

is required to obtain a zero steady-state error, and therefore a precise and accurate tracking of the reference signal. However, in the transient stage when the error is large, it might be useless or at least misleading to perform accurate and refined control actions using the integral component. Moreover, this use of the integral action might also lead to windup problems. Accordingly, Equation (4) encourages the use of the proportional action to get close to the reference signal (as in presence of large errors the integral gain is close to zero), while the integral action recovers the nominal value k_p/T_i when the error is close to zero. For this purpose, the tuning parameter σ^2 plays the role of choosing when the integral action has to play an important role, as it corresponds to the variance of the Gaussian distribution centered around zero error, as in Figure 2. A consequence of VIPI is that the control effort is always inferior to that of a conventional PI as the integral action can never be greater than the nominal one.

2.3 API controller

The acronym API denotes an Adaptive PI controller, whose parameters are not fixed but change, i.e. adapt, in reaction to different operating conditions (e.g. plant parameter changes, ageing phenomena, input uncertainties). Adaptive controllers have a long history, but their use in practical applications has been limited by their usually high demanding computational requirements; indeed, computations like matrix inversions can not be easily embedded in real-time applications, as they might cause runtime faults and consequently even lead to plant damages.

More recently, adaptive techniques have been applied to PI controllers, as their simple structure is very attractive, by including additional adaptive terms to extend and robustify such controllers. One example is given by [Fisher (2009)], where the authors compare three different controllers: a classic PI, an Adaptive PI and a P-FI which is a Proportional+Fuzzy Integral term controller. In this paper, we use the second controller (API) as a term of comparison in our examples, because it is characterised by accurate and robust tracking performances. The main property of adaptive controllers is that parameters are not fixed, but vary in time searching for an optimal configuration. In [Fisher (2009)] the controller parameters update law is described by

$$\dot{k}_p = -\gamma_p k_p + \beta_p e^2 \tag{5}$$

$$\dot{k}_i = -\gamma_i k_i + \beta_i e \int_0^t e(\tau) d\tau \tag{6}$$

with positive constant parameters $\gamma_p, \gamma_i, \beta_p, \beta_i$; the resulting control law is as usual

$$u_{API}(t) = k_p e(t) + k_i \int_0^t e(\tau) d\tau \tag{7}$$

The rationale of this adaptive PI control is that the updating law is composed by a

$$\text{dissipative term} \begin{cases} -\gamma_p k_p \\ -\gamma_i k_i \end{cases} \tag{8}$$

and an

$$\text{anti-dissipative term} \begin{cases} \beta_p e^2 \\ \beta_i e \int_0^t e(\tau) d\tau \end{cases}. \tag{9}$$

The *dissipative term* is used to decrease the value of the corresponding gain, once that the *anti-dissipative terms* becomes small. For instance, a large error will cause an increase of the proportional gain through the anti-dissipative term; thus the error will decrease, and when close to zero ($e \approx 0$), the proportional gain decreases exponentially with decay rate γ_p.

2.4 FAPI controller

Similarly to many other recent approaches, we also propose here a Fuzzy variant of the Adaptive PI (FAPI). Fuzzy approximation property has been widely and successfully used in robotics and control theory, to handle model uncertainties and external unpredictable disturbances. A large number of controllers use the Wang universal approximation theorem [Wang (1997)], to design nonlinear integral terms to improve performance indices and address robustness issues. However, in many cases, as shown in [Fisher (2009)], the involved additional computational efforts do not match significative performance improvements, thus not making fuzzy techniques particularly attractive.

Here we present a different novel approach to fuzzy controllers, where the simplicity of the conventional PI regulator, the interesting idea of the VIPI integral action and the robustness properties of adaptive PI controllers, are all combined together into a single Fuzzy-Adaptive PI Control (FAPI).

Next section is dedicated to recall the basic ideas of Fuzzy Logic Theory that, in the following section, will be used to implement the FAPI controller, which is one of the main contributions of this paper.

2.4.1 Fuzzy logic theory background

A fuzzy set A on a domain X is a set defined by the membership function $\mu_A(x)$ which is a mapping from the domain X into the unit interval:

$$\mu_A(\cdot) : X \rightarrow [0,1]. \tag{10}$$

There are several ways to define a fuzzy set, in particular we define it here using the analytic description of its membership function $\mu_A(x) = f(x)$. For instance (see Fig. 3), the *triangular* membership function can be described as:

$$\mu(x;a,b,c) = \max\left(0, \min\left(\frac{x-a}{b-a}, 1, \frac{c-x}{c-b}\right)\right) \tag{11}$$

where a,b and c are parameters that is related to the coordinates of the triangle's vertices, whereas a Gaussian membership function can be described as

$$\mu(x;\eta,\sigma) = \exp\left[-\left(\frac{x-\eta}{\sigma}\right)^2\right]. \tag{12}$$

A static or dynamic system which makes use of fuzzy sets and the corresponding mathematical framework is called a *fuzzy system*. In order to derivate the FAPI controller

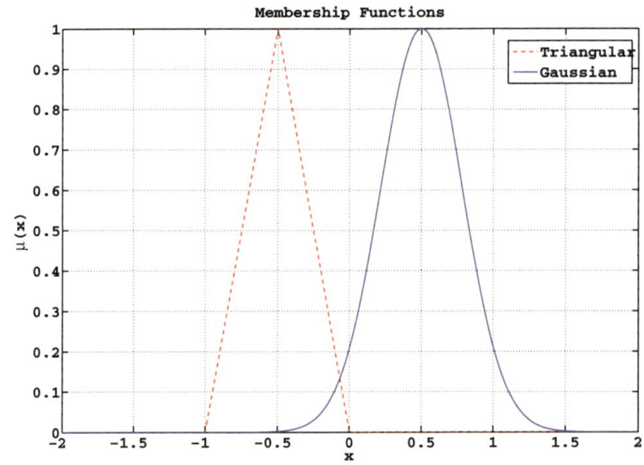

Fig. 3. Example of Membership Functions: Triangular $(a = -1, b = -0.5, c = 0)$ and Gaussian $(\eta = 0.5, \sigma = 0.4)$

updating law, it is necessary to define the intersection of fuzzy sets (connective *AND*) , obtained by considering a function $t : [0,1] \times [0,1] \rightarrow [0,1]$ that transforms the membership functions of fuzzy sets A and B into the membership function of the intersection of A and B, that is:

$$t\left[\mu_A(x), \mu_B(x)\right] = \mu_{A\cap B}(x). \tag{13}$$

A function t can be qualified as an intersection function, if it satisfies at least the following four requirements:

$$\begin{array}{cc} t(0,0) = 0, \quad t(a,1) = t(1,a) = a & \text{boundary condition} \\ t(a,b) = t(b,a) & \text{commutativity} \\ t(a,b) \leq t(a',b'), \quad \forall a \leq a',\ b \leq b' & \text{monotonicity} \\ t(t(a,b),c) = t(a,t(b,c)) & \text{associativity} \end{array} . \tag{14}$$

In the following analysis, the probabilistic connective *AND* will be used:

$$\mu_{A \cap B}(x) = \mu_A(x)\mu_B(x). \tag{15}$$

The most common fuzzy systems are defined by means of if-then rules: *rule-based fuzzy systems*. In the rule-based fuzzy systems, the relationships between variables are represented in the following general form:

if *antecedent proposition* **then** *consequent proposition*.

A *fuzzy proposition* is a statement like "x is big" where "big" is a *linguistic label*, defined by a fuzzy set on the universe of discourse of variable x. In the *linguistic fuzzy model* developed by [Zadeh (1978)] and [Mamdani (1977)], both the antecedent and the consequent are fuzzy propositions:

$$\mathcal{R}_i : \textbf{if } x \text{ is } A_i \textbf{ then } y \text{ is } B_i\,, \quad i = 1, ..., L, \tag{16}$$

where L is the number of propositions (rules). Here x is the input (antecedent) *linguistic variable*, and A_i are the antecedent *linguistic terms* (labels). Similarly, y is the output (consequent) linguistic variable and B_i are the consequent linguistic terms. The linguistic terms A_i,B_i are always fuzzy sets. After fuzzy theory gained popularity, many control problems have been recasted into control of Takagi-Sugeno-Kang (TSK) models:

$$\mathcal{R}_i : \textbf{if } x \textit{ is } A_i \textbf{ then } y = f_i(x)\,, \quad i = 1, ..., L \tag{17}$$

which is a particular case of the general fuzzy model (16), obtained when the consequent fuzzy sets B_i are functions of the variable x. In systems and control theory, TSK models are frequently used to model nonlinear systems over a fuzzy space. The resulting TSK model can efficiently clone the nonlinear system or alternatively, approximate it over a defined domain. For such a nonlinear systems representation, stability and synthesis of controllers and observers can be expressed in terms of Linear Matrix Inequalities, which in turn can be solved adopting convex optimization techniques as shown in [Tanaka (2001)]. It is important to mention that the output of a fuzzy system can be obtained using different *defuzzification* methods. In the remainder of this chapter we will use the following TSK model:

$$\mathcal{R}_i : \textbf{if } x_1 \textit{ is } A_{i1} \textit{ and ... } x_n \textit{ is } A_{in} \textbf{ then } y = f_i(x)\,, \quad i = 1, ..., L \tag{18}$$

where we consider that each rule has an antecedent proposition obtained by intersecting n fuzzy sets. The output can be evaluated by considering the Center of Gravity defuzzification method

$$y = \sum_{i=1}^{L} \alpha_i f_i(x) \tag{19}$$

where

$$\alpha(t) = (\alpha_1(t), ..., \alpha_L(t)), \qquad \alpha_i(t) = \frac{\beta_i(t)}{\sum_{i=1}^{L} \beta_i(t)},$$

$$\beta_i(t) = \prod_{j=1}^{n} \mu_{Aij}(x). \tag{20}$$

2.4.2 FAPI parameters update law

According to the discussion on fuzzy sets and rules introduced in the previous section, we introduce now the controller parameters update laws:

$$\text{IF error is SMALL, then } \dot{k}_p = -\beta_p k_p \tag{21}$$

$$\text{IF error is MEDIUM, then } \dot{k}_p = -\gamma_p(k_p - k_p^*) \tag{22}$$

$$\text{IF error is LARGE, then } \dot{k}_p = \alpha_p k_p^* e^2 \tag{23}$$

for the proportional gain k_p, while for the integral action we have

$$\text{IF error is LARGE, then } \dot{k}_i = -\beta_i k_i \tag{24}$$

$$\text{IF error is MEDIUM, then } \dot{k}_i = -\gamma_i(k_i - k_i^*) \tag{25}$$

$$\text{IF error is SMALL, then } \dot{k}_i = \alpha_i k_i^* e \int_0^t e(\tau)d\tau. \tag{26}$$

The main difference with respect to the API regulator is the presence of the two terms k_p^* and k_i^* that are the gains of a reference model regulator K^*. In order to compute the corresponding time-varying gain, we will consider a single Gaussian membership function $\mu_S(e)$ defined over the error domain to identify the fuzzy set SMALL (S), and also the fuzzy sets MEDIUM (M) and LARGE (L) as follows:

$$\mu_S = e^{-\left(\frac{x}{\sigma}\right)^2}, \qquad \mu_L(e) = 1 - \mu_S(e), \qquad \mu_M(e) = \mu_{S\cap L}(e) = \mu_S(e)\cdot\mu_L(e) \tag{27}$$

The philosophy of shaping the control effort on the basis of the error value is analogous to that of the previously introduced VIPI. The resulting k_p gain law is obtained as

$$\dot{k}_p = \frac{1}{1+\mu_M(e)}\left(\alpha_p\mu_L(e)k_p^*e^2 - \beta_p\mu_S(e)k_p - \gamma_p\mu_M(e)(k_p - k_p^*)\right) \tag{28}$$

while the integral gain k_i law is

$$\dot{k}_i = \frac{1}{1+\mu_M(e)}\left(\alpha_i\mu_S(e)k_i^*e\int_0^t e(\tau)d\tau - \beta_i\mu_L(e)k_i - \gamma_i\mu_M(e)(k_i - k_i^*).\right) \tag{29}$$

Each updating law it is composed of three terms:

$$\text{dissipative term}\begin{cases} -\beta_p\mu_S(e)k_p \\ -\beta_i\mu_L(e)k_i \end{cases} \tag{30}$$

used to decrease the (absolute) value of the gains,

$$\text{anti-dissipative term}\begin{cases} \alpha_p\mu_L(e)k_p^*e^2 \\ \alpha_i\mu_S(e)k_i^*e\int_0^t e(\tau)d\tau \end{cases} \tag{31}$$

used to increase the gain values analogously to the API control philosophy, and

$$\text{model reference tracking term}\begin{cases} -\gamma_p\mu_M(e)(k_p - k_p^*) \\ -\gamma_i\mu_M(e)(k_i - k_i^*) \end{cases} \tag{32}$$

used to force the adapting law to generate controller gains sufficiently close to the ideal controller K^*. In the end, the control law is as usual

$$u_{FAPI}(t) = k_p e(t) + k_i \int_0^t e(\tau)d\tau \tag{33}$$

In practice, when the error is large, the parameter update laws make the proportional gain increase due to its anti-dissipative term, while the integral action progressively disappears. This leads to a fast response (high proportional gain). On the other hand, when the error is small, the proportional gain is subject to the dissipative term and gets negligible values, while the integral component grows. This will result in a disturbance rejection behaviour. In any moment, good performances are guaranteed by the third term that makes the PI close to the model reference controller K^*.

Remark: Both the API controller developed in [Fisher (2009)] and the FAPI controller shown here are not symmetrical with respect to the error signal as their update rules are a function of the error, and thus depend on its sign. As a consequence, they can behave differently if the reference signal is larger or smaller than the actual output of the plant.

3. Tuning methods

3.1 Tuning of the conventional PI

In this paper we tune the conventional PI using Zhuang-Atherton optimal parameters [Zhuang and Atherton (1993)]. In particular we use the values of Table 1 of [Zhuang and Atherton (1993)], which correspond to PI tuning formulae for set-point changes in the case of first-order plus dead time plant model, optimised in order to minimise the Integral of the Square Error (ISE) signal. The set-point weighting factor is usually not used (i.e. $b = 1$), as in the examples a time-varying reference signal is used.

3.2 Tuning of the VIPI

Tuning of the VIPI is a two-step procedure:

1. Conventional tuning is first performed, and values of k_p and T_i are found according to the procedure outlined in Section 3.1.
2. The further parameter σ is computed to decide at which point the integral action should come into action. Namely, the integral action must already be active when the error is equal to the steady-state error obtained using only the proportional action.

Example :

Let us consider a plant described by the transfer function

$$G(s) = \frac{4}{s^2 + 4s + 4} \tag{34}$$

and let us design a classic PI characterised by $k_p = 6.122$, $T_i = 0.606$ and $b = 1$. Then the step response of the VIPI for different values of $\sigma = 0.1, 0.15, 0.25, 0.5, 1, 5$ are shown in Figure 4. As can be appreciated in Figure 4, the step response is contained between the one obtained using a single proportional controller, which is recovered from Equation (4) when σ tends to zero, and that of the conventional PI, which is recovered from Equation (4) when σ has large values (in practice they coincide already for $\sigma = 5$).

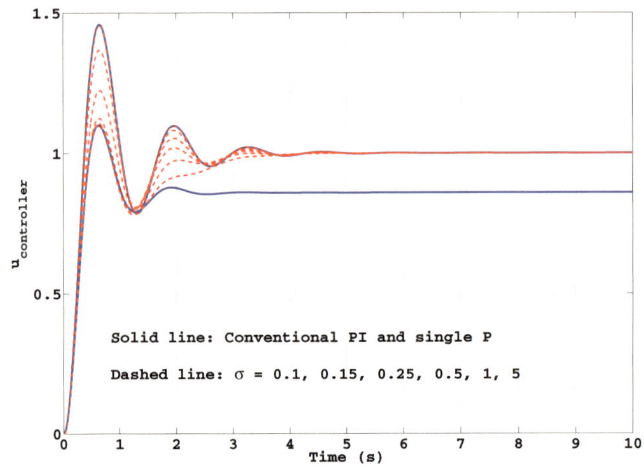

Fig. 4. Different step responses as a function of the free parameter σ of the VISI. The step response is contained between the one obtained using a single proportional controller (i.e. $\sigma \to 0$) and high values of the parameter. In this case, the step response when $\sigma = 5$ already coincides with the one obtained with the nominal PI.

3.3 Tuning of the API

Tuning of adaptive controllers is simpler than other PIs as the inner adaptive capacity allows the API to recover good performances against non optimal initial tunings. However, APIs are characterised by more degrees of freedom, e.g. parameters in the updating rules. For the purpose of the example shown in the following sections, the adaptive PI control parameters γ and β have been optimally tuned (using genetic algorithms) in order to get a good trade-off between tracking and disturbance rejection. Particular care is required to handle the anti-dissipative terms, which might yield to instability problems when a fault occurs. In fact, the anti-dissipative term should be neglected only when the error is close to zero.

3.4 Tuning of the FAPI

The FAPI controller parameters α, β, γ must be tuned, after a desired target controller K^* is chosen. In this case, we use a conventional PI tuned according to Zhuang-Atherton rules (see Section 3.1) as a reference model. Then, the parameters can be tuned keeping in mind that each parameter directly affects a different controller property:

- α: Adapting
- β: Low Gain Trend
- γ: K^* Model Reference Tracking.

Therefore, parameters are chosen in function of whether the priority objective is fast response to variations, or no overshoots or adherence to the ideal model controller. Particular care should be used in tuning α, that should be small in presence of significative system delays.

4. Comparison of the four PIs

As a preliminary comparison the step-responses of the four controllers are compared. Then, in the following sections, a more challenging example and a realistic scenario are simulated to further establish the differences among the proposed PI regulators. The step response of the four controllers is shown in Figure 5, in the case of the system plant (34). The shown comparison is performed after a transient time given to the adaptive controllers to adapt their parameters, and after Zhuang-Atherton tuning procedure for the other two controllers [Zhuang and Atherton (1993)]. The control performances of the four regulators are also

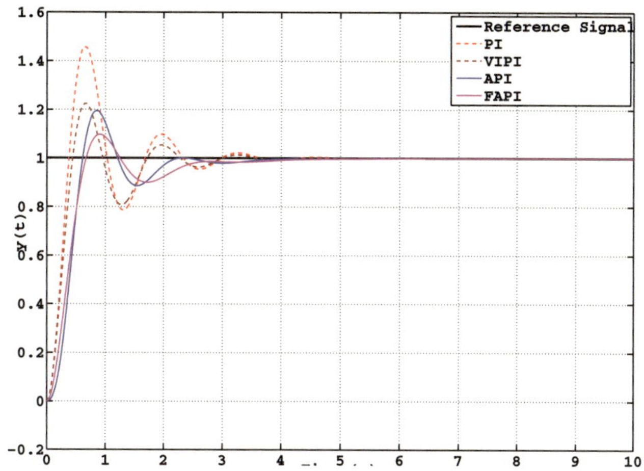

Fig. 5. Comparison of the four PI controllers in terms of the step response.

compared in Table 1 to further distinguish and classify the proposed regulators, where the following well known control indices were used

- **IAE:** Integral of the Absolute value of the Error, $IAE = \int_0^t |e(\tau)|\,d\tau$
- **ISE:** Integral of the Square Error, $ISE = \int_0^t (e(\tau))^2\,d\tau$
- **IAU:** Integral of the Absolute value of the input u, $IAU = \int_0^t |u(\tau)|\,d\tau$
- **IADU:** Integral of the Absolute value of the Derivative of the input u, $IADU = \int_0^t \left|\frac{du(\tau)}{d\tau}\right| d\tau$

	IAE	ISE	IAU	IADU
PI	0.58	0.26	16.75	22.20
VIPI	0.46	0.21	16.10	18.42
API	0.54	0.32	15.59	11.72
FAPI	0.50	0.27	15.47	8.69

Table 1. Comparison of the four controllers in terms of the Step Response. The best values of the indices have been highlighted in grey. The FAPI requires the least control effort, while the VIPI has the best overall control performances.

4.1 A more challenging example

The performances of the four controllers are again compared in a more challenging scenario where the plant transfer equation is the same (i.e. Equation (34)), but the reference signal is composed of a periodic sinusoidal component and of a pulse wave, plus a filtered Gaussian random signal $n(t)$ added to simulate sensor noise (i.e. $e(t) = r(t) - y(t) - n(t)$). As a consequence, this simulation is tailored on purpose to compare the robustness and disturbance rejection performances of the four controllers. The ability of the four controllers to track the reference signal despite the sensor noise is shown in Figure 6. Again, the comparison

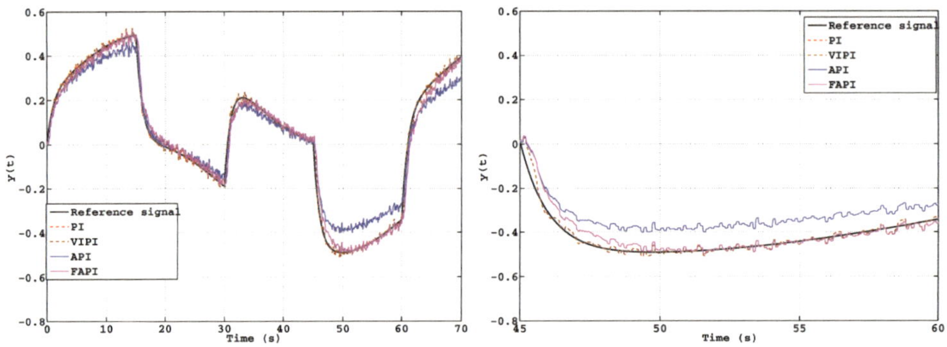

Fig. 6. Comparison of the four PI controllers in presence of a varying reference signal and sensor noise. This simulation aims at comparing the disturbance rejection abilities of the four controllers. On the left a long time interval, and a zoom is shown on the right. The API exhibits the worst tracking capabilities.

has been performed after some time that was required by the adaptive controllers to reach a steady-state behaviour. As illustrated in Figure 6, the conventional PI and the modified VIPI apparently have the best performance in terms of tracking, however, as better shown in Figure 7, the adaptive controllers, and especially the FAPI, are characterised by a less demanding input signal. This is particularly important because the input signal is usually required to vary slowly in time, to avoid actuators' stress.

Remark: In this example, the plant is required to follow small variations of the reference signal, therefore the error is usually small and the integral action of the VIPI is constantly set to the nominal value. As a consequence, the PI and the VIPI provide (almost) identical results.

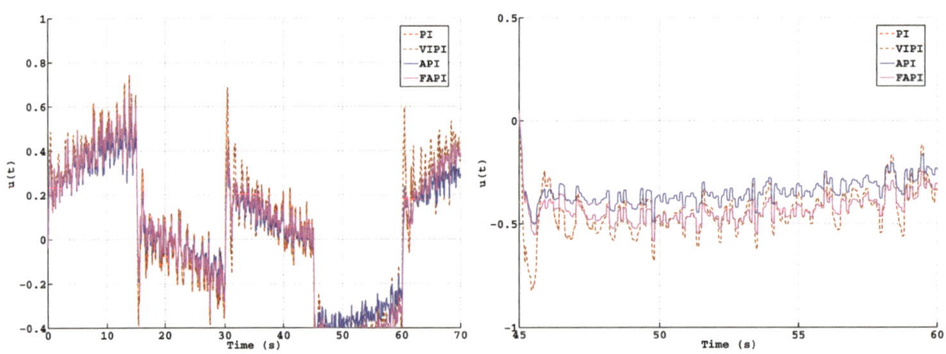

Fig. 7. Comparison of the four PI controllers in presence of a varying reference signal and sensor noise. This simulation shows the control effort of the four controllers. Clearly the FAPI is the most convenient one, as actuators are less stressed. On the left a long time interval, while on the right a shorter time interval is shown.

4.2 A realistic example: Ship course control

Let us consider a 3DoF model of a low-speed marine vessel [Fossen (2002)]:

$$M\dot{\nu} + C(\nu)\nu + D\nu = \tau + J^T(\eta)\tau_d \quad (35)$$

$$\dot{\eta} = J(\eta)\nu \quad (36)$$

where

- M represents the generalized mass-inertia matrix, including the added-masses contribution
- $C(\nu)$ contains the Coriolis-centripetal effects
- D represent the linear approximation of hydrodynamic drag
- τ is the generalized force-torque applied to the 3DoF model expressed in the body-fixed reference frame
- τ_d is an external disturbance expressed in the navigation referenceframe
- $\nu = [u, v, r]^T \in \mathbb{R}^3$ is the state variable related to the surge, sway and yaw rate speed
- $\eta = [p_n, p_e, \psi] \in \mathbb{R}^3$ represents the position and the orientation of the vessel with respect to the navigation frame
- $J(\eta)$ is the Jacobian matrix which relates body-fixed reference frame to navigation reference frame:

$$J(\eta) = \begin{bmatrix} \cos\psi & -\sin\psi & 0 \\ \sin\psi & \cos\psi & 0 \\ 0 & 0 & 1 \end{bmatrix} \quad (37)$$

Let us assume that the vessel is moving at constant speed u_0 , and $\sqrt{u_0^2 + v^2} \approx u_0$, then the previous 3DoF model can be decoupled into longitudinal and manoeuvring subsystems. Here we will analyse the manoeuvring subsystem in order to obtain a course control for a vessel equipped with a single rudder. For low surge speed, in addition the Eq. (35) can be approximated by:

$$\bar{M}\dot{\bar{v}} + N(u_0)\bar{v} = b\delta \quad (38)$$

where $\bar{v} = [v, r]^T$, $b = -[Y_\delta, N_\delta]^T \in \mathbb{R}^2$ and

$$\bar{M} = \begin{bmatrix} m - Y_{\dot{v}} & mx_g - Y_{\dot{r}} \\ mx_g - Y_{\dot{r}} & I_z - N_{\dot{r}} \end{bmatrix}, \quad N(u_0) = \begin{bmatrix} -Y_v & mu_0 - Y_r \\ -N_v & mx_g u_0 - N_r \end{bmatrix} \quad (39)$$

where the parameters Y_δ, N_δ are used to model the force and the torque generated by the rudder, $Y_{\dot{v}}, Y_{\dot{r}}, N_{\dot{r}}$ are parameters related to the added-masses, m, x_g, I_z are parameter of the rigid-body (mass, center of gravity and moment of inertia, respectively), Y_v, Y_r, N_v, N_r are coefficients related to the drag effects and δ is the rudder deflection. The equivalent state-space model of (38) can be found by observing that:

$$\dot{\bar{v}} = -\bar{M}^{-1}N(u_0)\bar{v} + \bar{M}^{-1}b\delta = A\bar{v} + B\delta \quad (40)$$

Considering the the parameters of the CyberShip II experimentally estimated in Fossen (2004), choosing a constant speed of $u_0 = 1.5m/s \approx 3knots$ and defining the output $y = r =$

$C_r \bar{v}$, $C_r = [0,\ 1] \in \mathbb{R}^2$, the following second linear time invariant system, also referred as Nomoto $2nd$ order model is obtained:

$$G_r(s) = C_r\,(sI - A)^{-1}\,B = \frac{r(s)}{\delta(s)} = \frac{-0.09185s - 0.002137}{s^2 + 0.8165s + 0.04882} \tag{41}$$

Since the course angle derivative is related to the yaw-rate as $\dot{\psi} = r$, we can finally derive the course model for the CyberShip II as:

$$G_\psi(s) = \frac{\psi(s)}{\delta(s)} = \frac{1}{s} G_r(s) = \frac{-0.09185s - 0.002137}{s^3 + 0.8165s^2 + 0.04882s} \tag{42}$$

The controller parameters used in the course-control problem are summarised in Table 2.

	ZA	VIPI	API	FAPI
K_p^*	7.7220	7.7220	-	7.7220
$K_i^* = K_p^*/T_i^*$	0.0978	0.0978	-	0.0978
σ	-	0.5	-	0.25
β_p	-	-	1.1612	0.0087
β_i	-	-	1.1343	0.1206
γ_p	-	-	0.0151	0.1142
γ_i	-	-	0.1363	0.1671
α_p	-	-	-	0.0011
α_i	-	-	-	0.7126

Table 2. Course Control Problem: controller parameters used in the simulation.

Note that we are not handling actuator saturations and limitations of the input rate. However, in order to use efficiently those controllers with such limitations the adoption of anti-windup systems and reference filters is strongly recommended. In practice, the use of a frequency-shaped reference signal causes a smoother and less demanding control action which is expected to satisfy the actuator limitations.
The four controllers are compared in the challenging scenario described in Figure 8. In this simulation we assume that the reference signal is a desired course angle (i.e. not a step reference, as it is not realistic in this context as previously remarked). Disturbance is modeled with two components: a filtered Gaussian noise, of the order of $2 - 3°$; and an aperiodic square pulse which refers to unpredictable external disturbance (e.g. wave current, wind gust). It is possible to note from Figure 8 that the API controller not always provide a satisfactory tracking of the reference signal. On the other hand, the other controllers have similar good performances, but the FAPI is characterised by a reduced control effort.

5. Conclusion

This chapter gives a comparison between a conventional PI regulator tuned according to Zhuang-Atherton rules with three less conventional controllers: a variable integral component PI (VIPI), an adaptive PI (API) and a fuzzy adaptive PI (FAPI). The VIPI is characterised by one time variant parameter, i.e. the integral one, and only one more degree of freedom (the parameter σ). Both the API and the FAPI have two time variant parameters and more degrees of freedom, as for instance the dissipative and anti-dissipative coefficients that regulate the parameters' update laws.

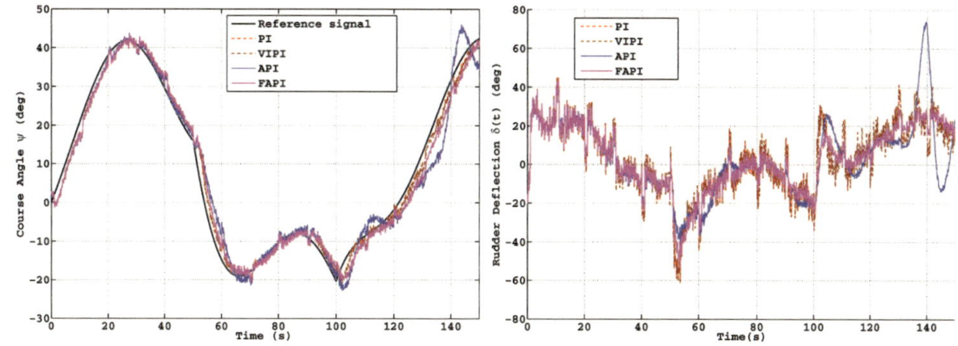

Fig. 8. Comparison of the four PI controllers in response to a course angle reference signal (on the left). Realistic disturbances are taken into consideration. On the right, the control effort of the four controllers.

Simulations show that the VIPI generally outperforms the simple PI both in terms of the control effort, which is always inferior, and in terms of settling time. The VIPI is very convenient, as it only contains one more parameter than the conventional PI, and better performances are usually achieved without requiring a complex tuning procedure for the extra parameter. On the other hand, the adaptive controllers require a more laborious tuning procedure (as more parameters are involved), and not always the control performance is so satisfactory, especially for the API, at least for the proposed examples. However, the FAPI, although provides similar control results to the PI and the VIPI, is characterised by a reduced small effort, both in terms of the absolute value and its derivative; for this reason it is particularly suitable in particular control applications: for instance when control components with moving parts are involved (e.g. valves) frequent fluctuations of the control action should be avoided to skip the high expenses of valve wear and maintenance programs.
Ongoing and future work will follow several directions:

- Robustness performances will be further investigated, so to account for time variant process plants. In some industrial applications, the plant coefficients change according to different factors (e.g. temperature, age, wear and tear of the machines).
- The controllers can be further compared on their ability to prevent wind-up phenomena.
- The proposed framework can be easily extended to decentralised Multiple Input Multiple Output (MIMO) control problems.
- The FAPI controller exhibits the best performance in terms of control effort, and for this reason it will be used in a real application in underwater robotics.

6. References

E. Sperry, *Automatic steering*, Society of Naval Architects and Marine Engineers, 1922.

N. Minorski, *Directional stability of automatically steered bodies*, Journal of American Society of Naval Engineers, 1922.

J. Ziegler and N. Nichols, *Optimum settings for automatic controller*, Trans. ASME, vol. 75, pp.827–833, 1942.

K. Astrom and E. Hagglund, *Adaptive tuning of simple regulators with specifications on phase and amplitude margins*, Automatica, vol. 20, pp. 645–651, 1984.

M. Zhuang and D.P. Atherton, *Automatic tuning of optimum PID controllers*, IEE Proceedings D on Control Theory and Applications, vol. 140, pp. 216–224, 1993.

W. Luyben and E. Eskinat, *Nonlinear auto-tune identification*, International Journal of Control, vol. 59, pp.595–626, 1994.

H. Rasmussen, *Automatic tuning of pid-regulators*, Textbook, Department of Control Engineering, Aalborg University, Denmark, 2009.

Y. Peng, D. Vrancic and R. Hanus, *Anti-windup, bumpless, and CT techniques for PID controllers*, IEEE Control systems magazine, vol. 16, pp.48–56, 1996.

K. Tang, K. Man, G. Chen and S. Kwong, *An optimal fuzzy PID controller*, IEEE Transactions on Industrial Electronics, vol. 48, pp. 757–765, 2001.

A. Visioli, *A new design for a PID plus feedforward controller*, Journal of Process Control, vol. 14, pp. 457–463, 2004.

A. Haj-Ali and H. Ying, *Structural analysis of fuzzy controllers with nonlinear input fuzzy sets in relation to nonlinear PID control with variable gain*, Automatica, vol. 40, pp. 1551–1559, 2004.

A. Scottedward Hodel and C.E. Hall, *Variable-Structure PID control to prevent integrator windup*, IEEE Transactions on Industrial Electronics, vol. 48, no. 2, 2001.

A. Visioli, *Fuzzy logic based set-point weight tuning of PID controllers*, IEEE Transactions on Systems, Man, and Cybernetics - Part A, vol. 29, no. 6, pp. 587–592, 1999.

A. Leva and M. Maggio, *A systematic way to extend ideal PID tuning rules to the real structure*, Journal of Process Control, vol. 21, pp. 130–136, 2011.

A. Balestrino, V. Biagini, P. Bolognesi and E. Crisostomi, *Advanced variable structure PI controllers*, IEEE Conference on Emerging Technologies and Factory Automation (ETFA), 2009.

A.D. Fisher, J.H. VanZwieten and T.S. VanZwieten, *Adaptive Control of Small Outboard-Powered Boats for Survey Applications*, OCEANS 2009, MTS/IEEE Biloxi - Marine Technology for Our Future: Global and Local Challenges, 2009.

L.X. Wang, *A Course in Fuzzy Systems and Control*, Prentice Hall,1997.

L.A. Zadeh, *Fuzzy Sets as basis for a theory of possibility.*, Fuzzy Sets and System 1, 1978.

E.H. Mamdani, *Application of Fuzzy Logic to Approximate Reasoning Using Linguistic Synthesis*, IEEE Transaction on Computers, 1977.

K. Tanaka and H.O. Wang, *Fuzzy Control Systems Design and Analysis: A Linear Matrix Inequality Approach*, John Wiley and Sons, 2001.

T.I. Fossen, *Marine Control Systems: Guidance, Navigation and Control of Ships, Rigs and Underwater Vehicles*, Marine Cybernetics, 2002.

T. I. Fossen, *Modeling, Identification, and Adaptive Maneuvering of CyberShip II: A complete design with experiments*, Proc. of the IFAC CAMS'04, Ancona, Italy.

6

Adaptive Gain PID Control for Mechanical Systems

Ricardo Guerra, Salvador González and Roberto Reyes
Universidad Autónoma de Baja California
México

1. Introduction

The design and use of PID controllers is a part of what has been denominated Classical Control, which as the name implies, has been studied for many years (DiStefano et al. 1996), however it continues to be a source for research (Alvarez et al. 2008), (Ang et al. 2008), (Su et al. 2010).
The structure of the controller contains a differential term to aid in the reduction of system friction and an integral term to attenuate steady state error. The drawbacks of this control scheme, particularly for nonlinear mechanical systems, include the difficulty in selecting adecuate controller gains, a process usually refered to as tuning. The difficulty usually lies in the fact that if the controller gains are set too small, the control objective may never be reached, whereas the selection of excesively large controller gains may result in system instability.
Many approaches have been proposed to properly tune PID gains (Ang et al. 2008), (Chang & Jung 2009), (Su et al. 2010), others have tried to improve upon the performance of the PID controller by including modern control techniques such as neural networks, fuzzy logic or variable structure control (Guerra et al. 2005).
Among these, variable structure control, specifically sliding mode control, has shown to possess certain desirable properties, such as disturbance rejection and finite time convergence; however it also presents unwanted behaviors mainly high frequency switching, a phenomenon refered to as chattering, which is undesirable in mechanical systems because it can cause accelerated wear of the mechanical components as well as activate unmodeled dynamics. One solution presented is to include an adaptive gain in the high frequency term so that the desirable properties may be exploited, and the undesirable effects minimized, achieving an enhanced performance (Guerra et al. 2005).

2. Background

The control of mechanical systems is subject to many difficulties, as evidenced by the research devoted to such aspects of mechanical systems as dead zone (Zhang & Gen 2009), and friction (Canudas de Wit et al. 1995).
Consider a first order mechanical system given by (Canudas de Wit et al. 1995)

$$\ddot{x} = \frac{u - f(\dot{x})}{m} \tag{1}$$

where x is the position variable, m is the mass, u is the control input and the function $f(\dot{x})$ denotes the nonlinear friction force. The PID control law is given by (Canudas de Wit et al. 1995):

$$u = K_p e + K_i \int_0^t e(\tau)d\tau + K_d \dot{e} \tag{2}$$

where K_p, K_d and K_i, are the proportional, derivative and integral gains, respectively and the error term is given by $e = x - x_d$, where x_d is the constant desired value.

Mechanical systems under integral control action have been known to present limit cycles, due in part to the complex nature of the friction force. This results in the system never reaching the desired position (Canudas de Wit et al. 1995).

The authors in (Guerra et al. 2005) present an approach considering a PD controller which is modified by the inclusion of a neural networks chattering controller that allows the high frequency swithching when the system is away from the desired position, but tends to vanish once the desired position is reached. In this chapter we will build upon that result and apply a similar stragegy to a PID controller.

3. Controller design

Consider the system (1) with unit mass and friction force given by (Makkar et al. 2005):

$$f(\dot{x}) = \gamma_1 [\tanh(\gamma_2 \dot{x}) - \tanh(\gamma_3 \dot{x})] + \gamma_4 \tanh(\gamma_5 \dot{x}) + \gamma_6 \dot{x} \tag{3}$$

The objective is for the error e to reach zero, *i.e.,*:

$$\lim_{t \to \infty} e(t) = 0 \tag{4}$$

where

$$e = x - x_d \tag{5}$$

to achieve this, the controller (2) is modified to:

$$u = -K_p e - K_i \zeta - [2\varepsilon + \delta K_d] \dot{e} \tag{6}$$

where

$$\dot{\zeta} = e \tag{7}$$

$$\dot{\delta} = -\alpha \ln(\delta + 1) + K_r \frac{[\delta + 1]}{\ln(\delta + 1) + 1} e^2 \tag{8}$$

where $\varepsilon > 0$, $\alpha > 0$ and $K_r > 0$ are constant parameters. The term ζ is used for simplicity in place of the term $\int_0^t e(\tau)d\tau$. It should be noted that for an intnitial condition $\delta(t_0) = \delta_0 \geq 0$, $\delta(t) \geq 0$, for all $t \geq t_0$ (Hench, 1999). In addtion, the adaptive gain can be considered to be bound by $\delta \leq \delta_M$ by taking into account that a practical controller is subject to saturation.

4. Closed loop system

To analyze the stability of the closed loop system, the following variable change is introduced:

$$\omega = \varepsilon\zeta + e \tag{9}$$

which is used to form the vector $\xi = [\omega \; e \; \dot{e}]^T$. Using equations (1), (3), (5), (6) and (7) the dynamic of the closed loop system is given by:

$$\dot{\xi} = A(\delta)\xi + B(\dot{e}) \tag{10}$$

where

$$A(\delta) = \begin{bmatrix} 0 & \varepsilon & 1 \\ 0 & 0 & 1 \\ -\varepsilon^{-1}K_i & -\left[K_p - \varepsilon^{-1}K_i\right] & -[2\varepsilon + \delta K_d] \end{bmatrix} \tag{11}$$

$$B(\dot{e}) = [0 \;\; 0 \;\; -f(\dot{e})]^T \tag{12}$$

The state δ contained in $A(\delta)_{3,3}$ is governed by the dynamic adaptation law (8). By setting (8) and (10) to zero, it can be seen that the origin of the state space ($\xi = 0$, $\delta = 0$) is the unique equillibrium for the system which, when applied to equation (9), implies $\zeta = 0$.

5. Stability analysis

Consider the candidate Lyapunov function:

$$V(\xi,\delta) = \xi^T P_c \xi + (\delta+1)\ln(\delta+1) \tag{13}$$

where

$$P_c = \frac{1}{2}\left(P + P^T\right) \tag{14}$$

$$P = \begin{bmatrix} \beta\varepsilon^{-1}K_i & 0 & 0 \\ 0 & \beta\left[K_p - \varepsilon^{-1}K_i\right] & 0 \\ 0 & 2\beta\varepsilon & \beta \end{bmatrix} \tag{15}$$

It should be noted that $V > 0$ implies that $P_c > 0$, which by applying Sylvester's Theorem (Kelly et al. 2005) requires that $\beta > 0$, the complete analysis to ensure positivity of matrix P_c is presented in the next section. To simplify stability analysis, the equality $\xi^T P_c \xi = \xi^T P \xi$ is considered so that expression (13) can be restated as

$$V(\xi,\delta) = \xi^T P \xi + (\delta+1)\ln(\delta+1) \tag{16}$$

The time derivative of (16) along the closed loop system (8) and (10) yields:

$$\dot{V} = -\xi^T Q(\delta)\xi - R(\xi) - \alpha\ln(\delta+1)[\ln(\delta+1)+1] \tag{17}$$

where

$$R(\xi) = -B(\dot{e})^T P\xi - \xi^T P B(\dot{e}) = 2\,\beta\,\varepsilon\, e\, f(\dot{e}) + 2\beta\dot{e}f(\dot{e}) \tag{18}$$

$$W(\delta) = -\left[PA(\delta) + A(\delta)^T P\right] \tag{19}$$

$$\begin{aligned} Q(\delta) &= \frac{1}{2}\left[W(\delta) + W(\delta)^T\right] - \hat{e}_2 K_r(\delta+1)\hat{e}_2^T \\ &= \begin{bmatrix} 0 & 0 & 0 \\ 0 & 2\beta\varepsilon\left[K_p - \varepsilon^{-1}K_i\right] - K_r(\delta+1) & \beta\varepsilon\left[2\varepsilon + \delta K_d\right] \\ 0 & \beta\varepsilon\left[2\varepsilon + \delta K_d\right] & 2\beta\left[\varepsilon + \delta K_d\right] \end{bmatrix} \end{aligned} \tag{20}$$

where $\hat{e}_2 = [0 \;\; 1 \;\; 0]^T$. Regarding equation (3) used in (18), every term in the expression can be bound by $b\tanh(c) \leq |b||c| \forall b, c \in \Re$. It can be stated that equation (3) satisfies:

$$-ef(\dot{e}) \leq K_\gamma |e||\dot{e}| \tag{21}$$

where

$$K_\gamma = \gamma_1|\gamma_2 - \gamma_3| + \gamma_4\gamma_5 + \gamma_6 \tag{22}$$

it should be remembered that all the parameters γ_ι for $\iota = 1 \ldots 6$ are positive constants. Regarding the term $\dot{e}f(\dot{e})$ in equation (18), it can be seen that this term is positive for $\gamma_2 \geq \gamma_3 > 0$ by using the properties of hiperbolic functions in equation (3) and considering $\dot{e} \to \Theta \geq 0$ (first quadrant) we find that:

$$\tanh\left([\gamma_2 - \gamma_3]\Theta\right)\left[1 - \tanh\left(\gamma_2\Theta\right)\tanh\left(\gamma_3\Theta\right)\right] \geq 0 \tag{23}$$

given that Θ, γ_2 and γ_3 are considered to be positive, the second term will always be non negative, whereas the first will be non negative if $\gamma_2 \geq \gamma_3 > 0$ (as was previously stated). These considerations apply also when $\dot{e} \to \Theta \leq 0$ (third quadrant). By applying (21), (22) and (23) in (18), along with the previously stated $\delta \leq \delta_M$ equation (17) can be bounded by:

$$\dot{V} \leq -\begin{bmatrix} |e| \\ |\dot{e}| \end{bmatrix}^T Q_c \begin{bmatrix} |e| \\ |\dot{e}| \end{bmatrix} - 2\beta\dot{e}f(\dot{e}) - \alpha\ln(\delta+1)\left[\ln(\delta+1)+1\right] \tag{24}$$

where

$$Q_c = \begin{bmatrix} 2\beta\varepsilon\left(K_p - \varepsilon^{-1}K_i\right) - K_r\left(\delta_M+1\right) & \beta\varepsilon\left(2\varepsilon + \delta_M K_d - K_\gamma\right) \\ \beta\varepsilon\left(2\varepsilon + \delta_M K_d - K_\gamma\right) & 2\beta\varepsilon \end{bmatrix} \tag{25}$$

In the follwing section, a process for tuning the controller gains will be introduced, this will also be useful in provinding sufficient conditions to guarantee the positiviy of matrices P_c and Q_c.

6. Controller tuning

In order to establish bounds on the controller gains, we first analyze the matrix P_c defined in expression (14). To find the roots of this symmetric matrix, we apply Sylvester's Theorem (Kelly et al. 2005), which generates a cubic polynomial of the form $\varepsilon^3 - 3b\varepsilon + 2a < 0$ with $a = \frac{K_i}{2}$ and $b = \frac{K_p}{3}$ which is satisfied for $b^3 > a^2$. Using $\exp(\bullet)$ to denote the exponential function, we define the terms $v_{1,2} = -a \pm ic = r\exp\left[\mp i\left(\theta - \pi\right)\right]$, $c = \sqrt{b^3 - a^2}$, $r = b^{\frac{3}{2}}$, $\theta = \arctan\left(\frac{c}{a}\right)$, $\vartheta_{1,2} = 2a + v_{1,2} = r\exp\left(\pm i\theta\right)$ and using Euler's formula the roots are:

$$\varepsilon_1 = -(p_1 + p_2) = -2r^{\frac{1}{3}} \cos\left(\frac{\theta}{3}\right) \tag{26}$$

$$\varepsilon_2 = \sqrt{y_1 y_2} = 2r^{\frac{1}{3}} \sin\left(\frac{\pi - 2\theta}{6}\right) \tag{27}$$

$$\varepsilon_1 = q_1 + q_2 = 2r^{\frac{1}{3}} \cos\left(\frac{\pi - \theta}{3}\right) \tag{28}$$

considering that $p_\iota = \vartheta_\iota^{\frac{1}{3}}$, $q_\iota = \upsilon_\iota^{\frac{1}{3}}$, and $y_\iota = p_\iota - q_\iota$ for $\iota = 1,2$. Given that $c \in \Re^+$, $\theta \in (0, \frac{\pi}{2}]$. Taking then $\varepsilon_1 < 0$, and $0 < \varepsilon_2 \leq \varepsilon_3$. The polynomial $\varepsilon^3 - 3b\varepsilon + 2a = (\varepsilon - \varepsilon_1)(\varepsilon - \varepsilon_2)(\varepsilon - \varepsilon_3) < 0$ is satisfied for all $\varepsilon_2 < \varepsilon < \varepsilon_3$. We propose the definition $b^3 = (\sigma^2 + 1)\, a^2$ with $\sigma \gg 0$, in other words, the proportional gain in equation (6) is tuned as

$$K_p = \left[\frac{27K_i^2}{4}\left(\sigma^2 + 1\right)\right]^{\frac{1}{3}} \tag{29}$$

Returning to Q_c defined in expression (25), this matrix can be defined as positive by applying Sylvester's Theorem (Kelly et al. 2005) and tuning the derivitave gain in (6) as

$$K_d = \frac{K_\gamma - 2\varepsilon}{\delta_M} \tag{30}$$

the numerator in this equation must be positive, specifically, the constant bound from equation (22) must satisfy $K_\gamma > 2\varepsilon$, so from equations (26)-(30) the positivity of matrices P_c and Q_c is restricted to

$$\max\left\{\frac{K_i}{K_p}, \frac{2\beta K_i + K_r(\delta_M + 1)}{2\beta K_p}, \varepsilon_2\right\} < \varepsilon < min\left\{\frac{K_\gamma}{2}, \varepsilon_3\right\} \tag{31}$$

By establishing conditions to satisfy (31), which include the values of K_r and K_i selected to generate a valid range for ε, we can conclude that expression (13) is positive definite and that expression (17) is locally negative semi-definite, consequently the system (8) and (10) has a stable equilibrium at the origin. Moreover, by restricting $\eta = \left[\xi^T \delta\right]^T$ by the bounds $\eta_{min} \leq \eta \leq \eta_{max}$ and applying LaSalle's Principle (Kelly et al. 2005) to expression (24) a closed set can be defined as:

$$\Omega = \{\eta \in \Re^4 : \dot{V}(\eta) = 0\} = \{\omega \in \Re, [e\ \dot{e}\ \delta]^T = 0\} \tag{32}$$

Solving (32) along (8) and (10) it can be seen that

$$\lim_{t \to \infty} \omega(t) = 0 \tag{33}$$

and by invoking the variable change (9) that

$$\lim_{t \to \infty} \zeta(t) = 0 \tag{34}$$

therefore the origin of the system defined by (8) and (10) is locally asymptotically stable.

7. Simulation results

In order to test the performance of the proposed controller simulations were carried out using the friction model (3) with the parameters set to $\gamma_1 = 1.25$, $\gamma_2 = 100$, $\gamma_3 = 10$, $\gamma_4 = \gamma_5 = 1.$, $\gamma_6 = 0.1$, $\alpha = 10$, $\beta = 1$, $\delta_M = 1$, $\sigma = 100$, $K_i = K_r = 10$ and the mass is considered to be unitary.
Using the mentioned values in equations (22), (29) and (30) we obtain $K_p = 188.9945$, $K_\gamma = 122.6$ and applying the obtained values to equation (31) we arrive at $\max\{0.53, 0.106, 0.52\} < \varepsilon < \min\{61.3, 13.7\}$ such that the value chosen was $\varepsilon = 6.808$ and hence $K_d = 108.9848$.
Figure 1 shows the performance of the position regulation. It should be noted that there is a very small overshot and that no limit cycles are present. The asymptotic stability can be easily seen in Figure 2 where the error is presented, it is clear that the error is still decreasing, achieving an accuracy within a micrometer after 200 seconds.

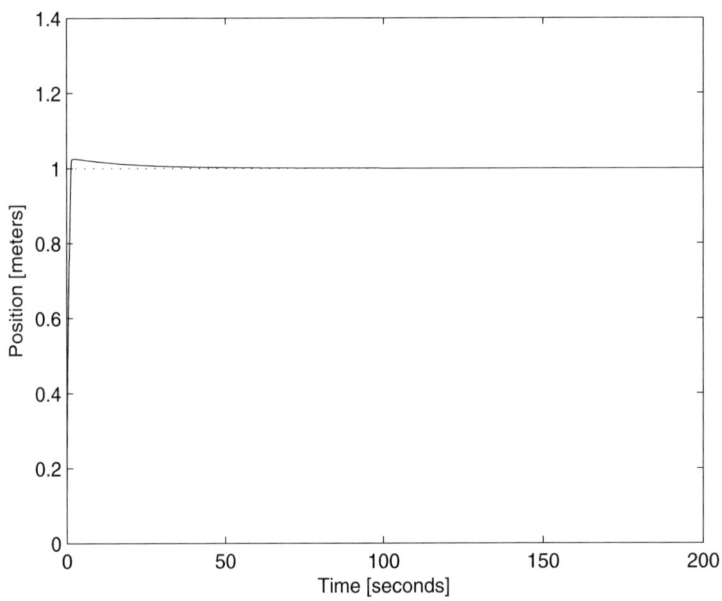

Fig. 1. Controller Performance: Achieved Position.

Figure 3 shows the evolution of the adaptive gain δ, it is clear that as the error approaches zero, so too does the value of the adaptive gain, and consequently so does the value of the control variable, shown in Figure 4.
The control variable initially presents a large value which then decreases. It can be inferred from the asymptotic stability that the control variable decreses asymptotically with time as shown in Figures 4 and 5. Figure 5 shows the control variable in more detail. During the first ten seconds a small oscillation can be seen but it is eliminated after approximately 3 seconds.
Figures 6 and 7 show that the term $\zeta = \int_0^t e(\tau)d\tau$ also approaches zero. It can be clearly seen, especially in Figure 7 that ζ asymptotically approaches zero.

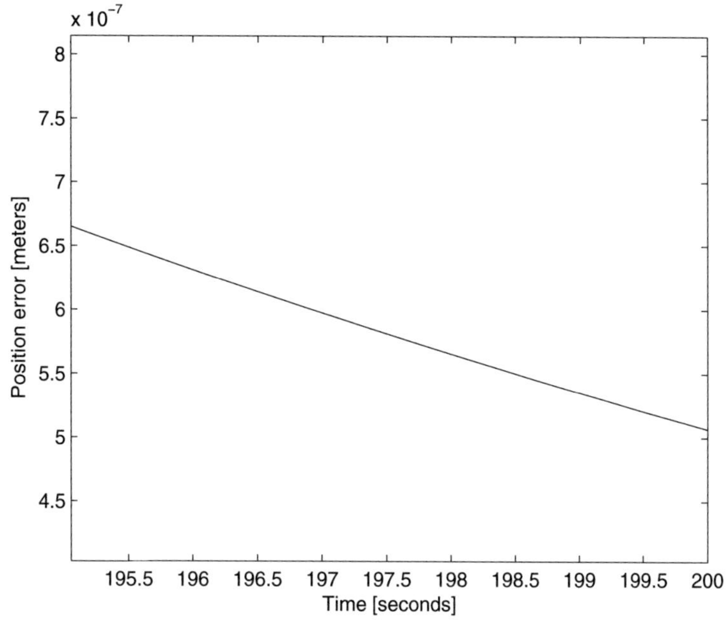

Fig. 2. Controller Performance: Position Error.

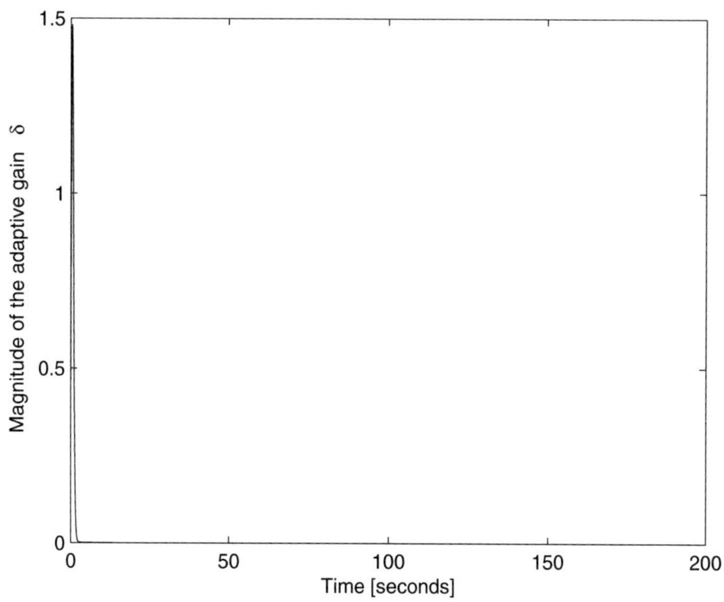

Fig. 3. Controller Performance: Adaptive Gain.

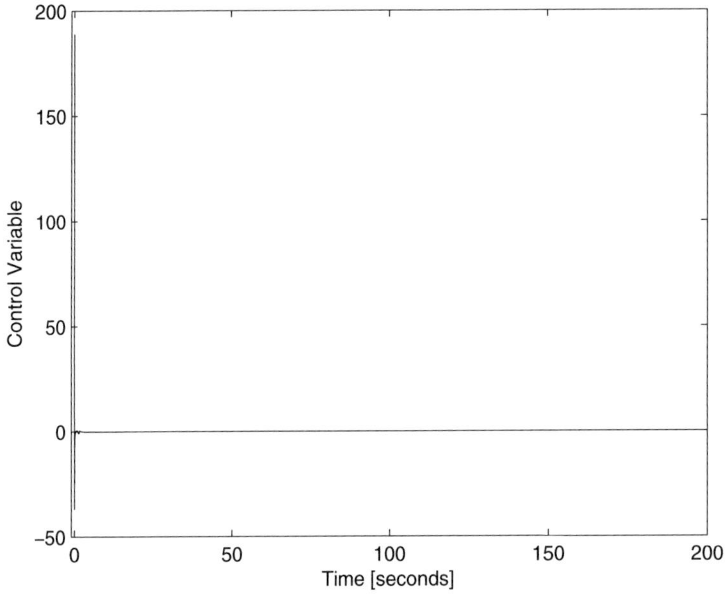

Fig. 4. Controller Performance: Control Variable.

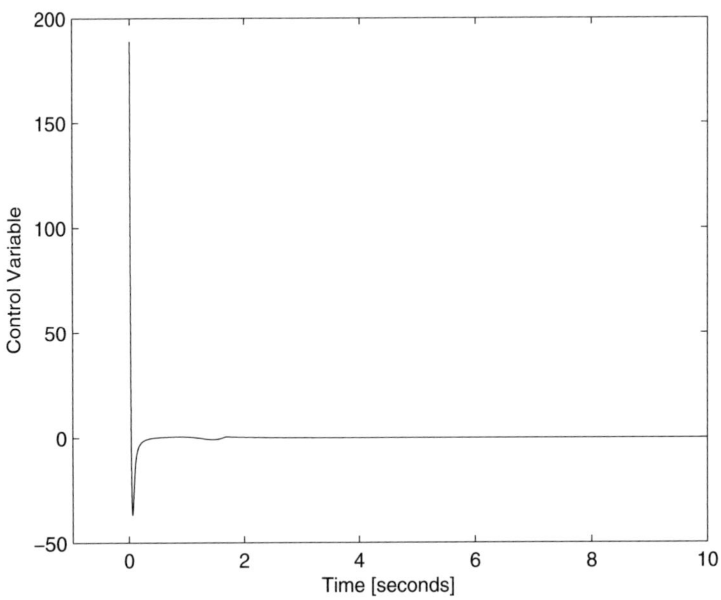

Fig. 5. Controller Performance: Control Variable (detail).

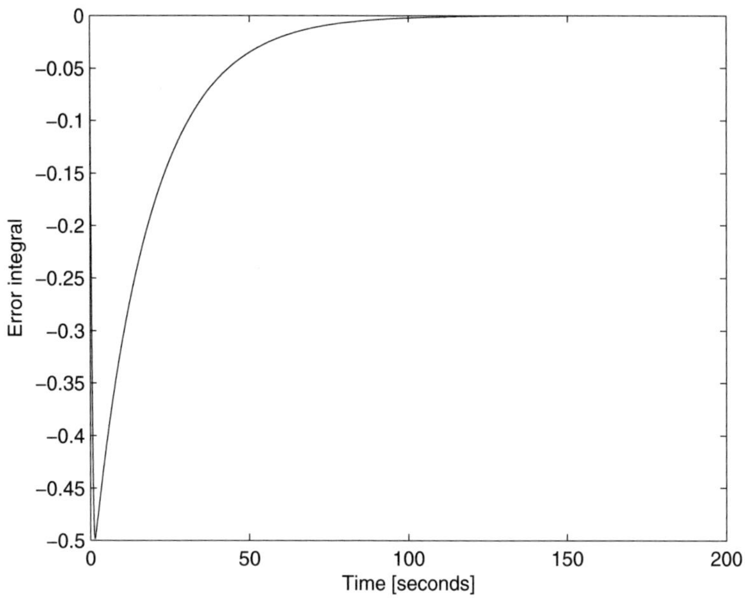

Fig. 6. Controller Performance: Error Integral.

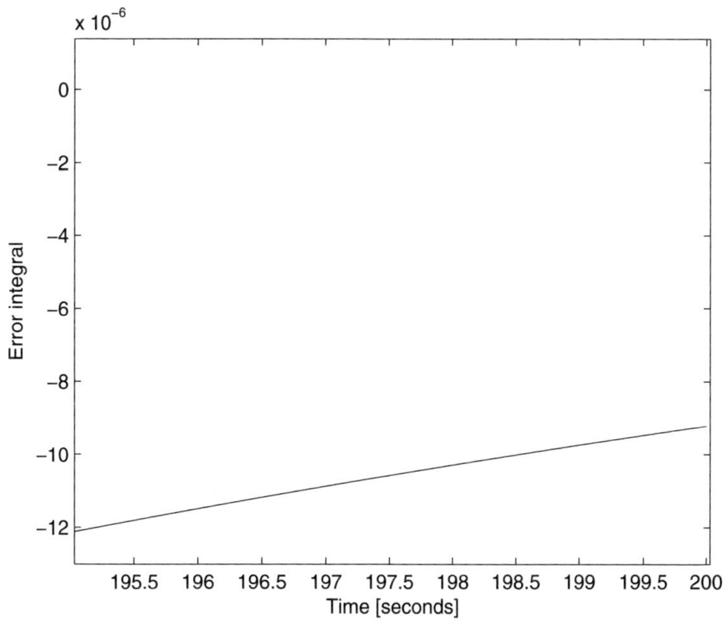

Fig. 7. Controller Performance: Error Integral (detail).

Figure 8 shows how with increasing time, the value of the adaptive gain draws even closer to zero. The same can be said of the error in Figure 9 and of ζ in Figure 10.

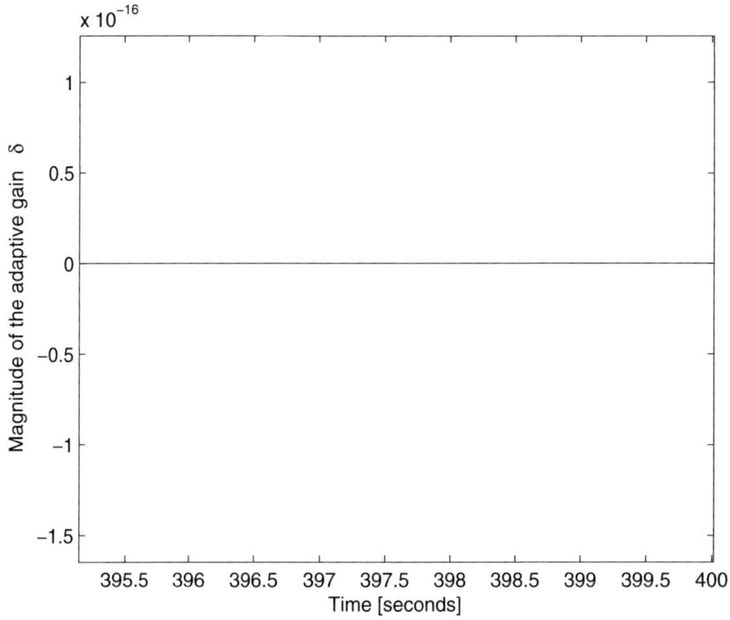

Fig. 8. Controller Performance: Adaptive Gain (detail).

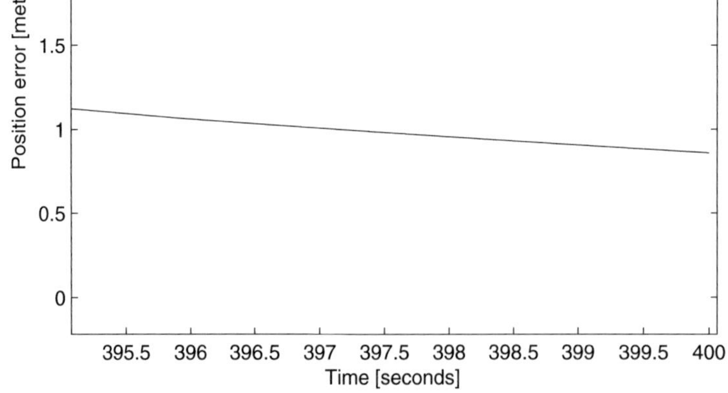

Fig. 9. Controller Performance: Position Error (detail).

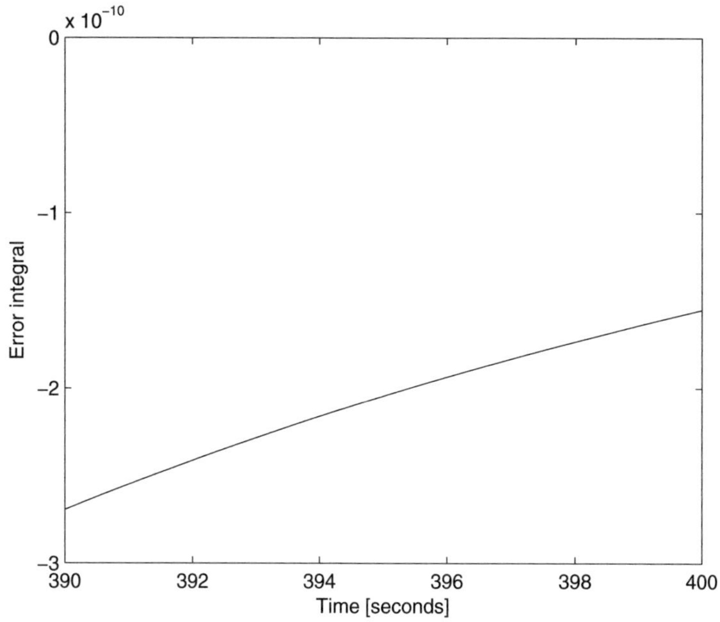

Fig. 10. Controller Performance: Error Integral (detail).

8. Conclusions

An extension to the traditional PID controller has been presented that incorporates an adaptive gain. The adaptive gain PID controller presented is demonstrated to asymptotically stabilize the system, this is shown in the simulations where the position error converges to zero.

In the presented analysis, considerations using known bounds of the system (such as friction coefficients) are used to show the stability of the system as well as to tune the controller gains K_p and K_d.

9. References

Alvarez, J.; Santibañez, V. & Campa, R. (2008). Stability of Robot Manipulators Under Saturated PID Compensation. *IEEE Transactions on Control Systems Technology*, Vol. 16, No. 6, Nov 2008, 1333 – 1341, ISSN 1063-6536

Ang, K. H.,; Chong, G. & Li, Y. (2005). PID Control System Analysis, Design, and Technology. *IEEE Transactions on Control Systems Technology*, Vol. 13, No. 4, Jul 2005, 559 – 576, ISSN 1063-6536

Canudas de Wit, C. ; Olsson, H. ; Astrom, K.J. & Lischinsky, P. (1995). A new model for control of systems with friction. *IEEE Transactions on Automatic Control*, Vol. 40, No. 3, Mar 1995, 419 – 425, ISSN 0018-9286

Chang, P. H. & Jung J.H. (2009). A Systematic Method for Gain Selection of Robust PID Control for Nonlinear Plants of Second-Order Controller Canonical Form. *IEEE Transactions on Automatic Control*, Vol. 17, No. 2, Mar 2009, 473 – 483, ISSN 1063-6536

Distefano, J. J.; Stuberud, A. R & Williams, I. J.(1990). *Feedback and Control Systems, 2nd Edition*, McGraw Hill, ISBN: 0-13228024-8, Upper Saddle River, New Jersey.

Guerra, R.; Acho, L. & Aguilar L.(2005). Chattering Attenuation Using Linear-in-the-Parameter Neural Nets in Variable Structure Control of Robot Manipulators with friction . *Proceedings of the International Conference on Fuzzy Systems and Genetic Algorithms 2005*, pp. 65 – 75, Tijuana, Mexico, October 2005.

Hench, J. J. (1999). On a class of adaptive suboptimal Riccati-based controllers. *Proceedings of the : American Control Conference, 1999.*, pp. 53 – 55, ISBN: 0-7803-4990-3 , San Diego, CA, June 1999.

Kelly, R.; Santibáñez, V. & Loría, A. (1996). *Control of Robot Manipulators in Joint Space*, Springer, ISBN: 978-1-85233-994-4, Germany.

Makkar, C.; Dixon, W.E.; Sawyer, W.G. & Hu, G. (2005). A new continuously differentiable friction model for control systems design. *Proceedings of the 2005 IEEE/ASME International Conference on Advanced Intelligent Mechatronics*, pp. 600 – 605, ISBN: 0-7803-9047-4, Monterey, CA, July 2005.

Su, Y.; Müller P. C. & Zheng, C. (2010). Global Asymptotic Saturated PID Control for Robot Manipulators. *IEEE Transactions on Control Systems Technology*, Vol. 18, No. 6, Nov 2010, 1280 – 1288, ISSN 1063-6536

Zhang, T. & Ge, S. S. (2009). Adaptive Neural Network Tracking Control of MIMO Nonlinear Systems With Unknown Dead Zones and Control Directions. *IEEE Transactions on Neural Networks*, Vol. 20, No.3, Mar 2009, 483 – 497, ISBN 1045-9227

7

PI/PID Control for Nonlinear Systems via Singular Perturbation Technique

Valery D. Yurkevich
Novosibirsk State Technical University
Russia

1. Introduction

The problem of output regulation for nonlinear time-varying control systems under uncertainties is one of particular interest for real-time control system design. There is a broad set of practical problems in the control of aircraft, robotics, mechatronics, chemical industry, electrical and electro-mechanical systems where control systems are designed to provide the following objectives: (i) robust zero steady-state error of the reference input realization; (ii) desired output performance specifications such as overshoot, settling time, and system type of reference model for desired output behavior; (iii) insensitivity of the output transient behavior with respect to unknown external disturbances and varying parameters of the system.

In spite of considerable advances in the recent control theory, it is common knowledge that PI and PID controllers are most widely and successfully used in industrial applications (Morari & Zafiriou, 1999). A great attention of numerous researchers during the last few decades was devoted to turning rules (Åström & Hägglund, 1995; O'Dwyer, 2003; Ziegel & Nichols, 1942), identification and adaptation schemes (Li et al., 2006) in order to fetch out the best PI and PID controllers in accordance with the assigned design objectives. The most recent results have concern with the problem of PI and PID controller design for linear systems. However, various design technics of integral controllers for nonlinear systems were discussed as well (Huang & Rugh, 1990; Isidori & Byrnes, 1990; Khalil, 2000; Mahmoud & Khalil, 1996). The main disadvantage of existence design procedures of PI or PID controllers is that the desired transient performances in the closed-loop system can not be guaranteed in the presence of nonlinear plant parameter variations and unknown external disturbances. The lack of clarity with regard to selection of sampling period and parameters of discrete-time counterparts for PI or PID controllers is the other disadvantage of the current state of this question.

The output regulation problem under uncertainties can be successfully solved via such advanced technics as control systems with sliding motions (Utkin, 1992; Young & Özgüner, 1999), control systems with high gain in feedback (Meerov, 1965; Young et al., 1977). A set of examples can be found from mechanical applications and robotics where acceleration feedback control is successfully used (Krutko, 1988; 1991; 1995; Lun et al., 1980; Luo et al., 1985; Studenny & Belanger, 1984; 1986). The generalized approach to nonlinear control system design based on control law with output derivatives and high gain in feedback, where integral action can be incorporated in the controller, is developed as well and one is used effectively under uncertainties (Błachuta et al., 1997; 1999; Czyba & Błachuta, 2003; Yurkevich, 1995; 2004). The distinctive feature of such advanced technics of control system design is the presence of two-time-scale motions in the closed-loop system. Therefore, a singular perturbation method (Kokotović et al., 1976; 1999; Kokotović & Khalil, 1986; Naidu & Calise, 2001; Naidu, 2002; Saksena et al., 1984; Tikhonov, 1948; 1952) should be used for analysis of closed-loop system properties in such systems.

The goal of the chapter is to give an overview in tutorial manner of the newest unified design methodology of PI and PID controllers for continuous-time or discrete-time nonlinear control systems which guarantees desired transient performances in the presence of plant parameter variations and unknown external disturbances. The chapter presents the up-to-date coverage of fundamental issues and recent research developments in singular perturbation technique of nonlinear control system design. The discussed control law structures are an extension of PI/PID control scheme. The proposed design methodology allows to provide effective control of nonlinear systems on the assumption of uncertainty, where a distinctive feature of the designed control systems is that two-time-scale motions are artificially forced in the closed-loop system. Stability conditions imposed on the fast and slow modes, and a sufficiently large mode separation rate, can ensure that the full-order closed-loop system achieves desired properties: the output transient performances are as desired, and they are insensitive to parameter variations and external disturbances. PI/PID control design methodology for continuous-time control systems, as well as corresponding discrete-time counterpart, is discussed in the paper. The method of singular perturbations is used to analyze the closed-loop system properties throughout the chapter.

The chapter is organized as follows. First, some preliminary results concern with properties of singularly perturbed systems are discussed. Second, the application of the discussed design methodology for a simple model of continuous-time single-input single-output nonlinear system is presented and main steps of the design method are explained. The relationship of the presented design methodology with problem of PI and PID controllers design for nonlinear systems is explained. Third, the discrete-time counterpart of the discussed design methodology for sampled-data control systems design is highlighted. Numerical examples with simulation results are included as well.

The main impact of the chapter is the presentation of the unified approach to continuous as well as digital control system design that allows to guarantee the desired output transient performances in the presence of plant parameter variations and unknown external disturbances. The discussed design methodology may be used for a broad class of nonlinear time-varying systems on the assumption of incomplete information about varying parameters of the plant model and unknown external disturbances. The advantage of the discussed singular perturbation technique for closed-loop system analysis is that analytical expressions for parameters of PI, PID, or PID controller with additional lowpass filtering can be found for nonlinear systems, where controller parameters depend explicitly on the specifications of the desired output behavior.

2. Singularly perturbed systems

2.1 Continuous-time singularly perturbed systems

The singularly perturbed dynamical control systems arise in various applications mainly due to two reasons. The first one is that fast dynamics of actuators or sensors leads to the plant model in the form of singularly perturbed system (Kokotović et al., 1976; Naidu & Calise, 2001; Naidu, 2002; Saksena et al., 1984). The second one is that the singularly perturbed dynamical systems can also appear as the result of a high gain in feedback (Meerov, 1965; Young et al., 1977). In accordance with the second one, a distinctive feature of the discussed control systems in this chapter is that two-time-scale motions are artificially forced in the closed-loop control system due to an application of a fast dynamical control law or high gain parameters in feedback.

The main notions of singularly perturbed systems can be considered based on the following continuous-time system:

$$\dot{X} = f(X, Z), \tag{1}$$

$$\mu\dot{Z} = g(X, Z), \tag{2}$$

where μ is a small positive parameter, $X \in \mathbb{R}^n$, $Z \in \mathbb{R}^m$, and f and g are continuously differentiable functions of X and Z. The system (1)–(2) is called the standard singularly perturbed system (Khalil , 2002; Kokotović et al., 1976; 1999; Kokotović & Khalil, 1986).

From (1)–(2) we can get the fast motion subsystem (FMS) given by

$$\mu\frac{dZ}{dt} = g(X, Z) \tag{3}$$

as $\mu \to 0$ where $X(t)$ is the frozen variable. Assume that

$$\det\left\{\frac{\partial g(X, Z)}{\partial Z}\right\} \neq 0 \tag{4}$$

for all $Z \in \Omega_Z$ where Ω_Z is the specified bounded set $\Omega_Z \subset \mathbb{R}^m$.

From (4) it follows that the function $\bar{Z} = \psi(X)$ exists such that $g(X(t), \bar{Z}(t)) = 0 \ \forall\ t$ holds where $\bar{Z}$ is an isolated equilibrium point of (3). Assume that the equilibrium point $\bar{Z}$ is unique and one is stable (exponentially stable).

After the fast damping of transients in the FMS (3), the state space vector of the system (1)–(2) belong to slow-motion manifold (SMM) given by

$$M_{smm} = \{(X, Z) : g(X, Z) = 0\}.$$

By taking $\mu = 0$, from (1)–(2), the slow motion subsystem (SMS) (or a so-called reduced system) follows in the form

$$\dot{X} = f(X, \psi(X)).$$

2.2 Discrete-time singularly perturbed systems

Let us consider the system of difference equations given by

$$X_{k+1} = \{I_n + \mu A_{11}\}X_k + \mu A_{12}Y_k, \tag{5}$$

$$Y_{k+1} = A_{21}X_k + A_{22}Y_k, \tag{6}$$

where μ is the small positive parameter, $X \in \mathbb{R}^n$, $Y \in \mathbb{R}^m$, and the A_{ij} are matrices with appropriate dimensions.

If μ is sufficiently small, then from (5)–(6) the FMS equation

$$Y_{k+1} = A_{21}X_k + A_{22}Y_k \tag{7}$$

results, where $X_{k+1} - X_k \approx 0$ (that is $X_k \approx$ const) during the transients in the system (7). Assume that the FMS (7) is stable. Then the steady-state of the FMS is given by

$$Y_k = \{I_m - A_{22}\}^{-1}A_{21}X_k. \tag{8}$$

Substitution of (8) into (5) yields the SMS

$$X_{k+1} = \{I_n + \mu[A_{11} + A_{12}(I_m - A_{22})^{-1}A_{21}]\}X_k.$$

The main qualitative property of the singularly perturbed systems is that: if the equilibrium point of the FMS is stable (exponentially stable), then there exists $\mu^\star > 0$ such that for all $\mu \in (0, \mu^\star)$, the trajectories of the singularly perturbed system approximate to the trajectories of the SMS (Hoppensteadt, 1966; Klimushchev & Krasovskii, 1962; Litkouhi & Khalil, 1985; Tikhonov, 1948; 1952). This property is important both from a theoretical viewpoint and for practical applications in control system analysis and design, in particular, that will be used throughout the discussed below design methodology for continuous-time or sampled-data nonlinear control systems.

3. PI controller of the 1-st order nonlinear system

3.1 Control problem statement

Consider a nonlinear system of the form

$$\frac{dx}{dt} = f(x, w) + g(x, w)u, \tag{9}$$

where t denotes time, $t \in [0, \infty)$, $y = x$ is the measurable output of the system (9), $x \in \mathbb{R}^1$, u is the control, $u \in \Omega_u \subset \mathbb{R}^1$, w is the vector of unknown bounded external disturbances or varying parameters, $w \in \Omega_w \subset \mathbb{R}^l$, $\|w(t)\| \leq w_{max} < \infty$, and $w_{max} > 0$.
We assume that dw/dt is bounded for all its components,

$$\|dw/dt\| \leq \bar{w}_{max} < \infty,$$

and that the conditions

$$0 < g_{min} \leq g(x, w) \leq g_{max} < \infty, \quad |f(x, w)| \leq f_{max} < \infty \tag{10}$$

are satisfied for all $(x, w) \in \Omega_{x,w}$, where $f(x, w)$, $g(x, w)$ are unknown continuous bounded functions of $x(t)$, $w(t)$ on the bounded set $\Omega_{x,w}$ and $\bar{w}_{max} > 0$, $g_{min} > 0$, $g_{max} > 0$, $f_{mux} > 0$.
Note, $g(x, w)$ is the so called a high-frequency gain of the system (9).
A control system is being designed so that

$$\lim_{t \to \infty} e(t) = 0, \tag{11}$$

where $e(t)$ is an error of the reference input realization, $e(t) := r(t) - y(t)$, $r(t)$ is the reference input, and $y = x$. Moreover, the output transients should have the desired performance indices. These transients should not depend on the external disturbances and varying parameters of the system (9).
Throughout the chapter a controller is designed in such a way that the closed-loop system is required to be close to some given reference model, despite the effects of varying parameters and unknown external disturbances $w(t)$ in the plant model. So, the destiny of the controller is to provide an appropriate reference input-controlled output map of the closed-loop system as shown in Fig. 1, where the reference model is selected based on the required output transient performance indices.

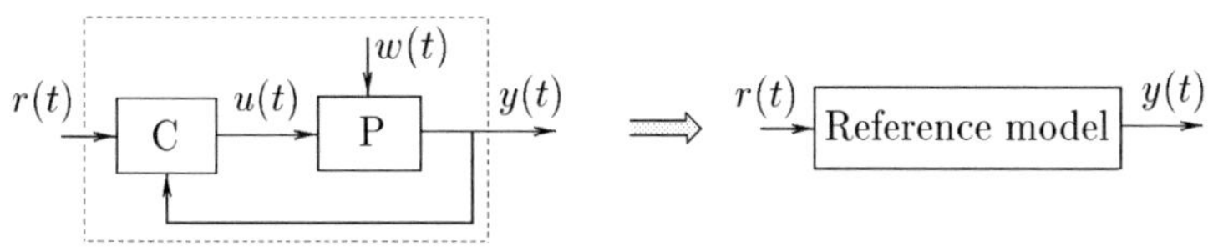

Fig. 1. Block diagram of the closed-loop control system

3.2 Insensitivity condition

Let us consider the reference equation of the desired behavior for (9) in the form of the 1st order stable differential equation given by

$$\frac{dx}{dt} = \frac{1}{T}(r - x), \tag{12}$$

which corresponds to the desired transfer function

$$G^d(s) = \frac{1}{Ts + 1},$$

where $y = x = r$ at the equilibrium point for $r =$ const and the time constant T is selected in accordance with the desired settling time of output transients.
Let us denote $F(x,r) := (r - x)/T$ and rewrite (12) as

$$\frac{dx}{dt} = F(x,r), \tag{13}$$

where $F(x,r)$ is the desired value of $\dot{x}$ for (9), $\dot{x} := dx/dt$. Hence, the deviation of the actual behavior of (9) from the desired behavior prescribed by (12) can be defined as the difference

$$e_F := F(x,r) - \frac{dx}{dt}. \tag{14}$$

Accordingly, if the condition

$$e_F = 0 \tag{15}$$

holds, then the behavior of $x(t)$ with prescribed dynamics of (13) is fulfilled. The expression (15) is an insensitivity condition for the behavior of the output $x(t)$ with respect to the external disturbances and varying parameters of the system (9).

Substitution of (9), (13), and (14) into (15) yields

$$F(x,r) - f(x,w) - g(x,w)u = 0. \tag{16}$$

So, the requirement (11) has been reformulated as a problem of finding a solution of the equation $e_F(u) = 0$ when its varying parameters are unknown. From (16) we get $u = u^{id}$, where

$$u^{id} = [g(x,w)]^{-1}[F(x,r) - f(x,w)] \tag{17}$$

and $u^{id}(t)$ is the analytical solution of (16). The control function $u(t) = u^{id}(t)$ is called a solution of the nonlinear inverse dynamics (id) (Boychuk, 1966; Porter, 1970; Slotine & Li, 1991). It is clear that the control law in the form of (17) is useless in practice under uncertainties, as far as one may be used only if complete information is available about the disturbances, model parameters, and state of the system (9).

Note, the nonlinear inverse dynamics solution is used in such known control design methodologies as exact state linearization method, dynamic inversion, the computed torque control in robotics, etc (Qu et al., 1991; Slotine & Li, 1991).

3.3 PI controller

The subject of our consideration is the problem of control system design given that the functions $f(x,w)$, $g(x,w)$ are unknown and the vector $w(t)$ of bounded external disturbances or varying parameters is unavailable for measurement. In order to reach the discussed control goal and, as a result, to provide desired dynamical properties of $x(t)$ in the specified region of the state space of the uncertain nonlinear system (9), consider the following control law:

$$\mu \frac{du}{dt} = k_0 \left\{ \frac{1}{T}(r - x) - \frac{dx}{dt} \right\}, \tag{18}$$

where μ is a small positive parameter. The discussed control law (18) may be expressed in terms of transfer functions, that is the structure of the conventional PI controller

$$u(s) = \frac{k_0}{\mu T s}[r(s) - x(s)] - \frac{k_0}{\mu} x(s). \tag{19}$$

For purposes of numerical simulation or practical implementation, let us rewrite the control law (18) in the state-space form. Denote

$$b_1 = -\frac{k_0}{\mu}, \quad b_0 = -\frac{k_0}{\mu T}, \quad c_0 = \frac{k_0}{\mu T}.$$

Then, (18) can be rewritten as $u^{(1)} = b_1 x^{(1)} + b_0 x + c_0 r$. Hence, the following expression $u^{(1)} - b_1 x^{(1)} = b_0 x + c_0 r$ results. Denote $u_1^{(1)} = b_0 x + c_0 r$. Finally, we obtain the equations of the controller given by

$$\begin{aligned} \dot{u}_1 &= b_0 x + c_0 r, \\ u &= u_1 + b_1 x. \end{aligned} \tag{20}$$

The block diagram of PI controller (20) is shown in Fig. 2(a).

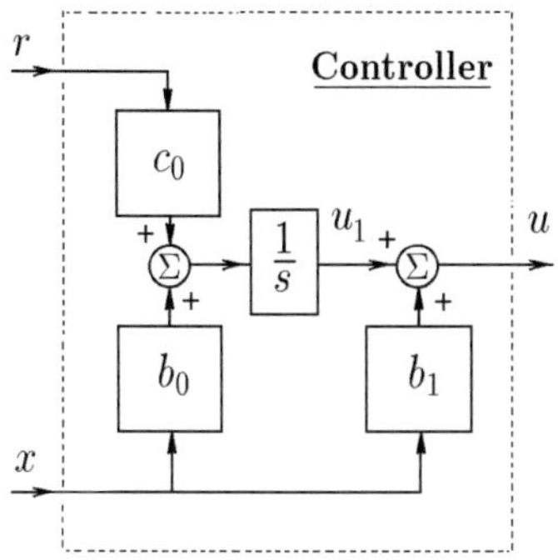

(a) PI controller (20)

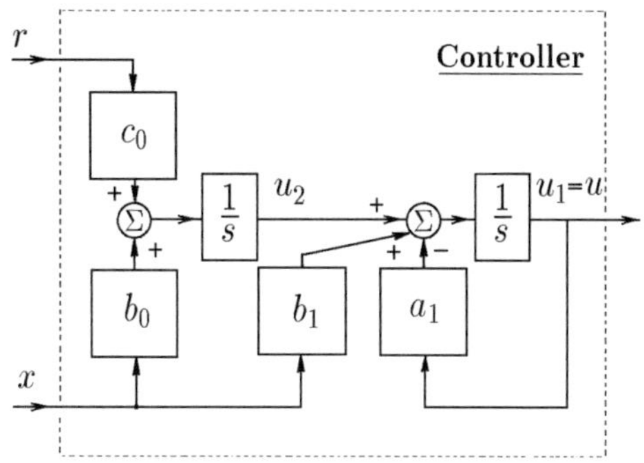

(b) PIF controller (42)

Fig. 2. Block diagrams of PI and PIF controllers

3.4 Two-time-scale motion analysis

In accordance with (9) and (18), the equations of the closed-loop system are given by

$$\frac{dx}{dt} = f(x,w) + g(x,w)u, \tag{21}$$

$$\mu\frac{du}{dt} = k_0\left\{\frac{1}{T}(r-x) - \frac{dx}{dt}\right\}. \tag{22}$$

Substitution of (21) into (22) yields the closed-loop system equations in the form

$$\frac{dx}{dt} = f(x,w) + g(x,w)u, \tag{23}$$

$$\mu\frac{du}{dt} = -k_0 g(x,w)u + k_0\left\{\frac{1}{T}(r-x) - f(x,w)\right\}. \tag{24}$$

Since μ is the small positive parameter, the closed-loop system equations (23)–(24) have the standard singular perturbation form given by (1)–(2). If $\mu \to 0$, then fast and slow modes are artificially forced in the system (23)–(24) where the time-scale separation between these modes depends on the parameter μ. Accordingly, the singular perturbation method (Kokotović et al., 1976; 1999; Kokotović & Khalil, 1986; Naidu & Calise, 2001; Naidu, 2002; Saksena et al., 1984; Tikhonov, 1948; 1952) may be used to analyze the closed-loop system properties.
From (23)–(24), we obtain the FMS given by

$$\mu\frac{du}{dt} + k_0 g(x,w)u = k_0\left[\frac{1}{T}(r-x) - f(x,w)\right], \tag{25}$$

where $x(t)$ and $w(t)$ are treated as the frozen variables during the transients in (25).
In accordance with the assumption (10), the gain k_0 can be selected such that the condition $g(x,w)k_0 > 0$ holds for all $(x,w) \in \Omega_{x,w}$, then the FMS is stable and, after the rapid decay of transients in (25), we have the steady state (more precisely, quasi-steady state) for the FMS (25), where $u(t) = u^{id}(t)$ and $u^{id}(t)$ is given by (17). Hence, if the steady state of the FMS (25) takes place, then the closed-loop system equations (23)–(24) imply that

$$\frac{dx}{dt} = \frac{1}{T}(r-x)$$

is the equation of the SMS, which is the same as the reference equation (12).
So, if a sufficient time-scale separation between the fast and slow modes in the closed-loop system and exponential convergence of FMS transients to equilibrium are provided, then after the damping of fast transients the desired output behavior prescribed by (12) is fulfilled despite that $f(x,w)$ and $g(x,w)$ are unknown complex functions of $x(t)$ and $w(t)$. Thus, the output transient performance indices are insensitive to parameter variations of the nonlinear system and external disturbances, by that the solution of the discussed control problem (11) is maintained.

3.5 Selection of PI controller parameters

The time constant T of the reference equation (12) is selected in accordance with the desired settling time of output transients. Take the gain $k_0 \approx g^{-1}(x,w)$. Then, in accordance with (25), the FMS characteristic polynomial is given by $\mu s + 1$. The time constant μ is selected as $\mu = T/\eta$ where η is treated as the degree of time-scale separation between the fast and slow modes in the closed-loop system, for example, $\eta \geq 10$.

3.6 Example 1

Consider the nonlinear system given by

$$\dot{x} = x^3 - (2 + x^2)u, \tag{26}$$

which is accompanied by the discussed PI controller (18). Substitution of (26) into (18) yields the singularly perturbed differential equations of the closed-loop system

$$\dot{x} = x^3 - (2 + x^2)u, \tag{27}$$

$$\mu\dot{u} = k_0[(r - x)/T - x^3 + (2 + x^2)u], \tag{28}$$

where fast and slow modes are forced as $\mu \to 0$. From (27)-(28), the FMS

$$\mu\dot{u} - k_0(2 + x^2)u = k_0[(r - x)/T - x^3] \tag{29}$$

follows, where x is treated as the frozen parameter during the transients in (29).
Take $k_0 = -0.5 < 0$, then the transients of (29) are exponentially stable and the unique exponentially stable isolated equilibrium point u_{id} of the FMS (29) is given by

$$u_{id} = (2 + x^2)^{-1}[(r - x)/T - x^3]. \tag{30}$$

Substitution of $\mu = 0$ into (27)-(28) yields the equation of the SMS which is the same as the reference equation (12).
Note, at the equilibrium point of the FMS (29), the state of the closed-loop system (27)-(28) belongs to the slow-motion manifold (SMM) given by

$$M_{smm} = \{(x,u) : (r - x)/T - x^3 + (2 + x^2)u = 0\}, \tag{31}$$

which is the attractive manifold when the FMS (29) is stable and the behavior of $x(t)$ on the SMM is described by (12).
The phase portrait of (27),(28) in case of $r(t) \equiv 1$ and the output response of (20),(26) are shown in Fig. 3, where the simulation has been done for $T = 1$, $\mu = 0.05$ s, $k_0 = -0.5$. It is

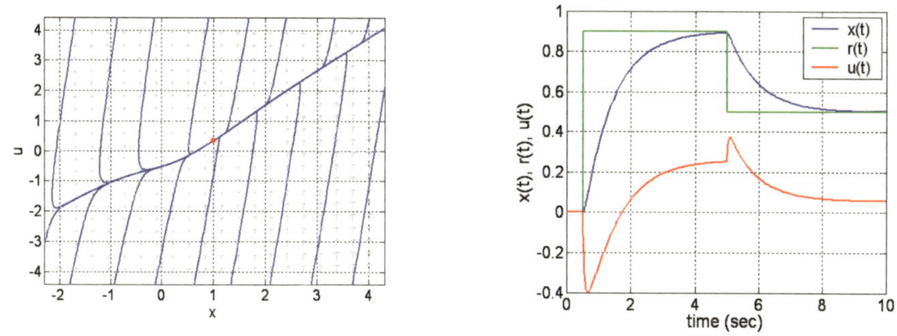

(a) The phase portrait of (27),(28) when $r(t) \equiv 1$

(b) The output response of (20),(26)

Fig. 3. The phase portrait and output response of the closed-loop system in Example 1

easy to see from Fig. 3(a), there is fast transition of the closed-loop system state trajectories on the SMM (31) where the motions along this manifold correspond to the SMS given by (12). Hence, after the damping of fast transients, the condition $x(t) \to r = \text{const}$ holds due to (12) for arbitrary initial conditions, that is the output stabilization of (26), where the desired settling time is defined by selection of the parameter T. The output response of the closed-loop system (20),(26) provided for initial conditions at origin reveals the transients behavior of the reference equation given by (12) as shown in Fig. 3(b).

4. PIF controller of the 1-st order nonlinear system

4.1 High-frequency sensor noise attenuation

Consider the nonlinear system (9) in presence of high-frequency sensor noise $n_s(t)$, that is

$$\frac{dx}{dt} = f(x,w) + g(x,w)u, \quad \hat{y} = x + n_s, \; y = x, \tag{32}$$

where the sensor output $\hat{y}(t)$ is corrupted by a zero-mean, high-frequency measurement noise $n_s(t)$. Hence, instead of (21)-(22), we get of the closed-loop system given by

$$\frac{dx}{dt} = f(x,w) + g(x,w)u, \quad \hat{y} = x + n_s, \tag{33}$$

$$\mu\frac{du}{dt} = k_0\left\{\frac{1}{T}(r - \hat{y}(t)) - \frac{d\hat{y}(t)}{dt}\right\}. \tag{34}$$

The main disadvantage of the sensor noise $n_s(t)$ in the closed-loop system is that it leads to high-frequency chattering in the control variable $u(t)$. At the same time, the effect of the high-frequency noise $n_s(t)$ on the behavior of the output variable $y(t)$ is much smaller since the system (32) rejects high frequencies.

From the closed-loop system equations given by (33)-(34), the FMS equation

$$\mu\dot{u} + k_0 g(x,w)u = k_0\left\{\frac{1}{T}(r - x) - f(x,w) - \frac{1}{T}n_s - \dot{n}_s\right\} \tag{35}$$

results, where $x(t)$ and $w(t)$ are treated as the frozen variables during the transients in (35). From (35), we obtain the transfer function $G_{un_s}(s) = u(s)/n_s(s)$, that is

$$G_{un_s}(s) = -\frac{k_0}{T}\frac{Ts+1}{\mu s + k_0 g},$$

where

$$\lim_{\omega\to\infty} |G_{un_s}(j\omega)| = k_0/\mu. \tag{36}$$

The transfer function $G_{un_s}(s)$ determines the sensitivity of the plant input $u(t)$ to the sensor noise signal $n_s(t)$ in the closed-loop system. In other words, $G_{un_s}(s)$ is an input sensitivity function with respect to noise $n_s(t)$ in the closed-loop system. The requirement on high-frequency sensor noise attenuation can be expressed by the following inequality:

$$|G_{un_s}(j\omega)| \leq \varepsilon_{un_s}(\omega), \quad \forall \quad \omega \geq \omega^{n_s}_{min}, \tag{37}$$

where $\varepsilon_{un_s}(\omega)$ is an upper bound on the amplitude of the input sensitivity function with respect to noise for high frequencies.

In order to provide a high-frequency measurement noise attenuation assigned by (37), we can consider, instead of (18), the control law given by

$$\mu^2\ddot{u} + d_1\mu\dot{u} = k_0\left\{\frac{1}{T}(r-\hat{y}) - \dot{\hat{y}}\right\}, \tag{38}$$

which can also be expressed in terms of transfer functions as

$$u(s) = \frac{k_0}{\mu(\mu s + d_1)}\left\{\frac{1}{Ts}[r(s) - \hat{y}(s)] - \hat{y}(s)\right\}.$$

that is, in compare with (19), the structure of PI controller with additional lowpass filtering (PIF controller).

The way for two-time-scale motion analysis in the closed-loop system is the same as it was shown above. Hence, from the closed-loop system equations given by (32) and (38), the FMS equation

$$\mu^2\ddot{u} + d_1\mu\dot{u} + k_0 g(x,w)u = k_0\left\{\frac{1}{T}(r-x) - f(x,w) - \frac{1}{T}n_s - \dot{n}_s\right\} \tag{39}$$

results, where $x(t)$ and $w(t)$ are treated as the frozen variables during the transients in (39). Accordingly, from (39), the transfer function

$$G_{un_s}(s) = -\frac{k_0}{T}\frac{Ts+1}{\mu^2 s^2 + d_1\mu s + k_0 g}$$

results, where

$$|G_{un_s}(j\omega)| = \frac{|k_0|}{T}\frac{\sqrt{(T\omega)^2+1}}{\sqrt{(k_0 g - \mu^2\omega^2)^2 + (d_1\mu\omega)^2}}. \tag{40}$$

Note, in contrast to (36), we have

$$\lim_{\omega\to\infty} |G_{un_s}(j\omega)| = 0.$$

So, the high-frequency measurement noise attenuation is provided in case of control law given by (38). The amplitude of high-frequency oscillations induced in behavior of the control variable $u(t)$ due to effect of the high-frequency harmonic measurement noise can be calculated by (40).

4.2 Selection of PIF controller parameters

The time constant T of the reference equation (12) is selected in accordance with the desired settling time of output transients. Take the gain $k_0 \approx g^{-1}(x, w)$ and parameter $d_1 = 2$. Then, in accordance with (39), the FMS characteristic polynomial is given by $(\mu s + 1)^2$. The time constant μ is selected as $\mu = T/\eta$ where η is treated as the degree of time-scale separation between the fast and slow modes in the closed-loop system, for example, $\eta \geq 10$.

4.3 Implementation of PIF controller

The discussed PIF controller (38) can be rewritten in the form given by

$$u^{(2)} + \frac{d_1}{\mu}u^{(1)} = -\frac{k_0}{\mu^2}x^{(1)} - \frac{k_0}{\mu^2 T}x + \frac{k_0}{\mu^2 T}r, \tag{41}$$

where $\hat{y}$ is replaced by x. Denote

$$a_1 = \frac{d_1}{\mu}, \quad b_1 = -\frac{k_0}{\mu^2}, \quad b_0 = -\frac{k_0}{\mu^2 T}, \quad c_0 = \frac{k_0}{\mu^2 T}.$$

From (41), we have $u^{(2)} + a_1 u^{(1)} = b_1 x^{(1)} + b_0 x + c_0 r$ and, thereafter, $u^{(2)} + a_1 u^{(1)} - b_1 x^{(1)} = b_0 x + c_0 r$. Denote $u_2^{(1)} = b_0 x + c_0 r$. Then we get $u^{(1)} + a_1 u - b_1 x = u_2$. Denote $u_1^{(1)} = u_2 - a_1 u + b_1 x$. Hence, $u = u_1$. Finally, the state space equations of the PIF controller are given by

$$\begin{aligned} \dot{u}_1 &= u_2 - a_1 u_1 + b_1 x, \\ \dot{u}_2 &= b_0 x + c_0 r, \\ u &= u_1. \end{aligned} \tag{42}$$

The block diagram of the PIF controller (42) is shown in Fig. 2(b).

5. PID controller of the 2-nd order nonlinear system

5.1 Control problem and insensitivity condition

Consider a nonlinear system of the 2-nd order given by

$$\ddot{x} = f(X, w) + g(X, w)u, \tag{43}$$

where x is the measurable output of the system (43), $y = x$, and $\dot{x}$ is the unmeasurable variable of the state $X = [x, \dot{x}]^T$. Assume that the inequalities

$$0 < g_{min} \leq g(X,w) \leq g_{max} < \infty, \quad |f(X,w)| \leq f_{max} < \infty \tag{44}$$

are satisfied for all $(X,w) \in \Omega_{X,w}$, where $f(X,w)$, $g(X,w)$ are unknown continuous bounded functions of $X(t)$, $w(t)$ on the bounded set $\Omega_{X,w}$.
The control objective is given by (11), where the desired settling time and overshoot have to be provided for $x(t)$ regardless the presence of the external disturbances and varying parameters $w(t)$ of the system (43).
Consider the reference equation of the desired behavior for (43) in the form of the 2nd order stable differential equation given by

$$T^2\ddot{x} + a_1^d T\dot{x} + x = b_1^d T\dot{r} + r.$$

Hence, we have

$$\ddot{x} = \frac{1}{T}[b_1^d\dot{r} - a_1^d\dot{x}] + \frac{1}{T^2}[r - x]. \tag{45}$$

Let us rewrite (45) in the form

$$\ddot{x} = F(X,R),$$

where $R = [r,\dot{r}]^T$ and the parameters T, a_1^d, and b_1^d are selected in accordance with the desired system type, settling time, and overshoot for $x(t)$. Denote

$$e_F := F(X,R) - \ddot{x}.$$

Hence, the behavior of $x(t)$ with prescribed dynamics of (45) is fulfilled in presence of the external disturbances and varying parameters of (43), if the insensitivity condition $e_F = 0$ holds. Similar to the above, the nonlinear inverse dynamics solution is given by

$$u^{id} = [g(X,w)]^{-1}[F(X,R) - f(X,w)]. \tag{46}$$

5.2 PID controller

Consider the control law in the form

$$\mu^2\ddot{u} + d_1\mu\dot{u} = k_0[F(X,R) - \ddot{x}], \tag{47}$$

where μ is a small positive parameter. In accordance with (45), the controller (47) can be represented as

$$\mu^2\ddot{u} + d_1\mu\dot{u} = k_0\left\{-\ddot{x} + \frac{1}{T}[b_1^d\dot{r} - a_1^d\dot{x}] + \frac{1}{T^2}[r - x]\right\}. \tag{48}$$

The discussed control law (48) can also be expressed in terms of transfer functions

$$u(s) = \frac{k_0}{\mu(\mu s + d_1)}\left\{\frac{1}{T}[b_1^d r(s) - a_1^d x(s)] + \frac{1}{T^2 s}[r(s) - x(s)] - sx(s)\right\}, \tag{49}$$

which corresponds to the PID controller and (49) is implemented without an ideal differentiation of $x(t)$ or $r(t)$ due to the presence of the term $k_0/[\mu(\mu s + d_1)]$. Note, PID controller with additional lowpass filtering (PIDF controller)

$$\mu^q u^{(q)} + d_{q-1}\mu^{q-1}u^{(q-1)} + \cdots + d_1 \mu u^{(1)} = k_0[F(X,R) - x^{(2)}] \tag{50}$$

can be considered as well, where $q > 2$.

5.3 Two-time-scale motion analysis

Consider the closed-loop system equations (43),(47), that are

$$\ddot{x} = f(X,w) + g(X,w)u, \tag{51}$$

$$\mu^2 \ddot{u} + d_1 \mu \dot{u} = k_0[F(X,R) - \ddot{x}]. \tag{52}$$

Substitution of (51) into (52) yields

$$\ddot{x} = f(X,w) + g(X,w)u, \tag{53}$$

$$\mu^2 \ddot{u} + d_1 \mu \dot{u} + k_0 g(x,w)u = k_0[F(X,R) - f(X,w)]. \tag{54}$$

Denote $u_1 = u$ and $u_2 = \mu u$. Hence, the system (53)–(54) can be represented as a standard singular perturbation system, that is

$$\begin{aligned}
\dot{x}_1 &= x_2, \\
\dot{x}_2 &= f(x_1, x_2, w) + g(x_1, x_2, w)u_1, \\
\mu \dot{u}_1 &= u_2, \\
\mu \dot{u}_2 &= -k_0 g(x,w)u_1 - d_1 u_2 + k_0[F(x_1, x_2, R) - f(x_1, x_2, w)].
\end{aligned}$$

From the above system, the fast-motion subsystem (FMS) equation

$$\mu^2 \ddot{u} + d_1 \mu \dot{u} + k_0 g(x,w)u = k_0[F(X,R) - f(X,w)] \tag{55}$$

follows, where $X(t)$ and $w(t)$ are frozen variables during the transients in (55).
By selection of μ, d_1, and k_0, we can provide the FMS stability as well as the desired degree of time-scale separation between fast and slow modes in the closed-loop system. Then, after the rapid decay of transients in (55) (or, by taking $\mu = 0$ in (55)), we obtain the steady state (more precisely, quasi-steady state) for the FMS (55), where $u(t) = u^{id}(t)$. Hence, from (53)–(54), we get the slow-motion subsystem (SMS) equation, which is the same as (45) in spite of unknown external disturbances and varying parameters of (43) and by that the desired behavior of $x(t)$ is provided.

5.4 Selection of PID controller parameters

The time constant T of the reference equation (45) is selected in accordance with the desired settling time of output transients. The parameter a_1^d is defined by permissible overshoot of the output step response. Take, for example $a_1^d = 2$. Take $b_1^d = 0$ if the reference model given by (45) is a system of type 1. Take $b_1^d = a_1^d$ if the reference model given by (45) is a system

of type 2. Take the gain $k_0 \approx g^{-1}(X, w)$, and parameter $d_1 = 2$. Then, in accordance with (55), the FMS characteristic polynomial is given by $(\mu s + 1)^2$. The time constant μ is selected as $\mu = T/\eta$ where η is the desired degree of time-scale separation between the fast and slow modes in the closed-loop system, for example, $\eta \geq 10$.
Note, in case of PIDF controller given by (50), the FMS characteristic polynomial has the form $(\mu s + 1)^q$ when the parameters $d_{q-1}, \ldots, d_2, d_1$ are selected as the coefficients of the binomial polynomial, that is

$$(s+1)^q = s^q + d_{q-1}s^{q-1} + \cdots + d_2 s^2 + d_1 s + 1.$$

The more detailed results and procedures for selection of controller parameters can be found in (Yurkevich, 2004).

5.5 Implementation of PID controller

The discussed control law (48) can be rewritten in the form given by

$$u^{(2)} + \frac{d_1}{\mu} u^{(1)} = -\frac{k_0}{\mu^2} x^{(2)} - \frac{k_0 a_1^d}{\mu^2 T} x^{(1)} - \frac{k_0}{\mu^2 T^2} x + \frac{k_0 b_1^d}{\mu^2 T} r^{(1)} + \frac{k_0}{\mu^2 T^2} r,$$

that is

$$u^{(2)} + a_1 u^{(1)} = b_2 x^{(2)} + b_1 x^{(1)} + b_0 x + c_1 r^{(1)} + c_0 r, \tag{56}$$

where

$$a_1 = \frac{d_1}{\mu}, \; b_2 = -\frac{k_0}{\mu^2}, \; b_1 = -\frac{k_0 a_1^d}{\mu^2 T}, \; b_0 = -\frac{k_0}{\mu^2 T^2}, \; c_1 = \frac{k_0 b_1^d}{\mu^2 T}, \; c_0 = \frac{k_0}{\mu^2 T^2}.$$

The block diagram representation of the discussed control law (56) can be obtained based on the following derivations:

$$u^{(2)} - b_2 x^{(2)} + a_1 u^{(1)} - b_1 x^{(1)} - c_1 r^{(1)} = \underbrace{b_0 x + c_0 r}_{=\dot{u}_2} \Longrightarrow u^{(1)} - b_2 x^{(1)} + a_1 u - b_1 x - c_1 r = u_2$$

$$\Longrightarrow u^{(1)} - b_2 x^{(1)} = \underbrace{u_2 - a_1 u + b_1 x + c_1 r}_{=\dot{u}_1} \Longrightarrow u = u_1 + b_2 x.$$

Hence, we obtain the equations of the controller given by

$$\begin{aligned} \dot{u}_1 &= u_2 - a_1 u + b_1 x + c_1 r, \\ \dot{u}_2 &= b_0 x + c_0 r, \\ u &= u_1 + b_2 x. \end{aligned} \tag{57}$$

From (57), the block diagram of the controller follows as shown in Fig. 4(a).

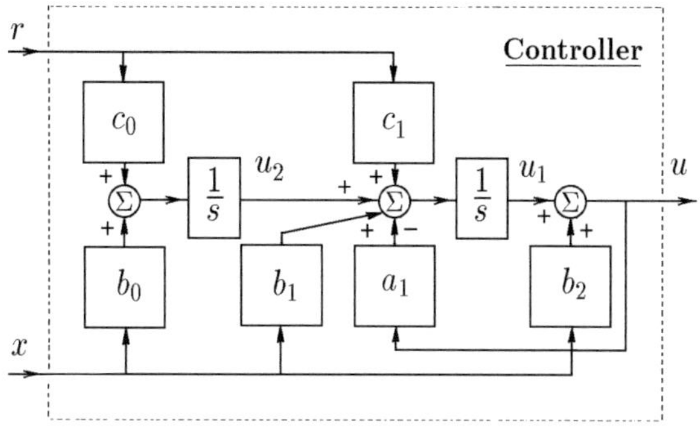

(a) Block diagram of PID controller (48) represented in the form (57)

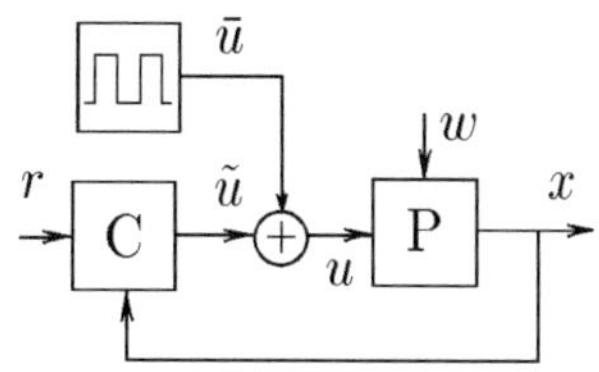

(b) Control system with additional pulse signal $\bar{u}(t)$

Fig. 4. Control system with PID controller

6. On-line tuning of controller parameters

Let us consider the closed-loop system with an additional pulse signal $\bar{u}(t)$ as shown in Fig. 4(b). Then, instead of (51)–(52), we get

$$\ddot{x} = f(X,w) + g(X,w)[\tilde{u} + \bar{u}],$$

$$\mu^2\tilde{u}^{(2)} + d_1\mu\tilde{u}^{(1)} = k_0[F(X,R) - x^{(2)}].$$

From the above system, the FMS equation

$$\mu^2\tilde{u}^{(2)} + d_1\mu\tilde{u}^{(1)} + k_0 g(x,w)\tilde{u} = k_0[F(X,R) - f(X,w) - g(x,w)\bar{u}] \tag{58}$$

results, where $X(t)$ and $w(t)$ are frozen variables during the transients in (58). In accordance with (58) and $u = \tilde{u} + \bar{u}$, the input sensitivity function with respect to pulse signal $\bar{u}(t)$ can be defined as the following transfer function $G_{u\bar{u}}(s) = u(s)/\bar{u}(s)$, that is

$$G_{u\bar{u}}(s) = \frac{\mu^2 s^2 + d_1\mu s}{\mu^2 s^2 + d_1\mu s + k_0 g},$$

or we may consider sensitivity function defined as $G_{\tilde{u}\bar{u}}(s) = \tilde{u}(s)/\bar{u}(s)$, that is

$$G_{\tilde{u}\bar{u}}(s) = -\frac{k_0 g}{\mu^2 s^2 + d_1\mu s + k_0 g}.$$

For example, if $d_1 = 2$ and $k_0 g = 1$, then the shape of the fast-motion transients excited by $\bar{u}(t)$ in behavior of $u(t)$ and $\tilde{u}(t)$ is easily predictable one. Therefore, on-line tuning of controller parameters can be provided based on direct observations of the fast-motion transients that are excited by the pulse signal $\bar{u}(t)$. In particular, if $d_1 = 2$ and the high-frequency gain $g(x,w)$ is unknown, then the gain k_0 can be manually adjusted such that to provide acceptable small oscillations of FMS transients excited by $\bar{u}(t)$.

6.1 Example 2

Consider a SISO nonlinear continuous-time system in the form

$$x^{(2)} = x^3 + |x^{(1)}| - (2 + x^2)u + w, \tag{59}$$

where the reference equation of the desired behavior for the output $x(t)$ is assigned by (45) and the control law structure is given by (48).
Take $T = 0.3$ s, $a_1^d = 2$, $\mu = 0.03$ s, $k_0 = -0.5$, , and $d_1 = 2$, where the control law (48) is represented in the form (57). The simulation results of the system (59) controlled by the algorithm (57) are displayed in Figs. 5–9, where the initial conditions are zero. The output response of the system (59) with controller (57) for a ramp reference input $r(t)$, in case where $b_1^d = 0$ (the reference model is a system of type 1) reveals the large value of a velocity error as shown in Fig 6. The velocity error can be significantly reduced by taking $b_1^d = a_1^d$ (the reference model is a system of type 2) as shown in Fig 8. Note, the high pulse in control variable, as shown in Fig 7(b), is caused by discrepancy between relative degree of the system (59) and relative degree of (45) when $b_1^d = a_1^d$. This high pulse can be eliminated by the use of a smooth reference input function $r(t)$ as shown in Fig. 9.

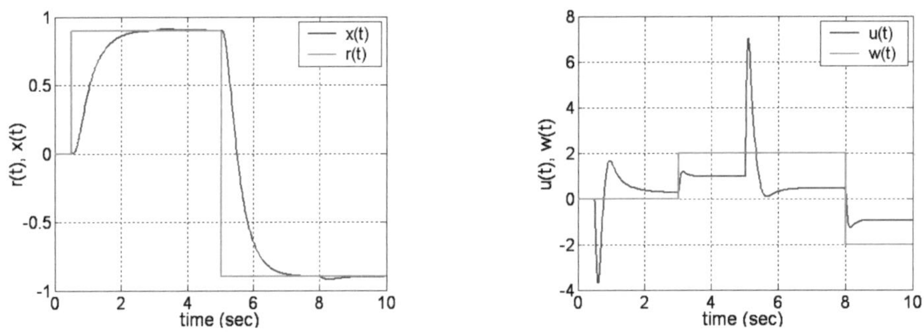

(a) Reference input $r(t)$ and output $x(t)$ (b) Control $u(t)$ and disturbance $w(t)$

Fig. 5. Output response of the system (59) with controller (57) for a step reference input $r(t)$ and a step disturbance $w(t)$, where $b_1^d = 0$ (the reference model is a system of type 1)

7. Sampled-data nonlinear system of the 1-st order

7.1 Control problem and insensitivity condition

In this section the discrete-time counterpart of the above singular perturbation design methodology is discussed. Let us consider the backward difference approximation of the nonlinear system (9) preceded by a zero-order hold (ZOH) with the sampling period T_s, that is

$$x_k = x_{k-1} + T_s[f(x_{k-1}, w_{k-1}) + g(x_{k-1}, w_{k-1})u_{k-1}], \tag{60}$$

where x_k, w_k, and u_k represent samples of $x(t)$, $w(t)$, and $u(t)$ at $t = kT_s$, respectively.
The objective is to design a control system having

$$\lim_{k \to \infty} e_k = 0. \tag{61}$$

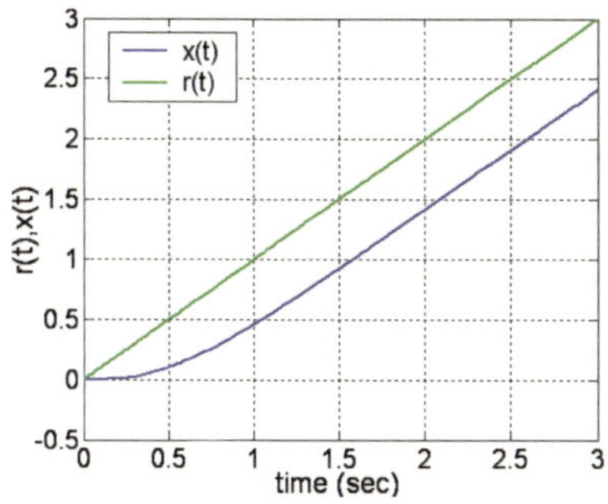

(a) Reference input $r(t)$ and output $x(t)$

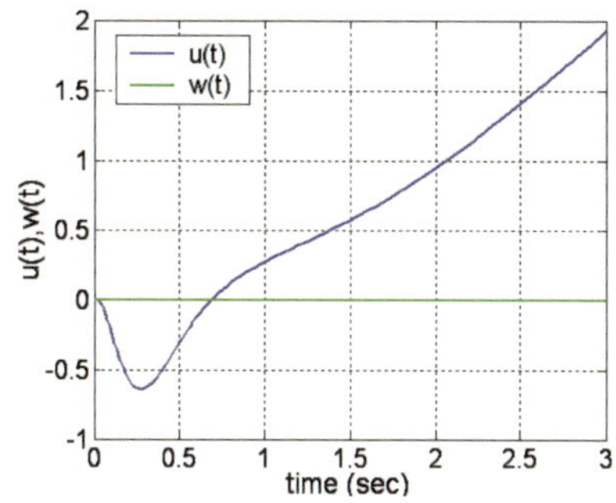

(b) Control $u(t)$ and disturbance $w(t)$

Fig. 6. Output response of the system (59) with controller (57) for a ramp reference input $r(t)$, where $b_1^d = 0$ and $w(t) = 0$ (the reference model is a system of type 1)

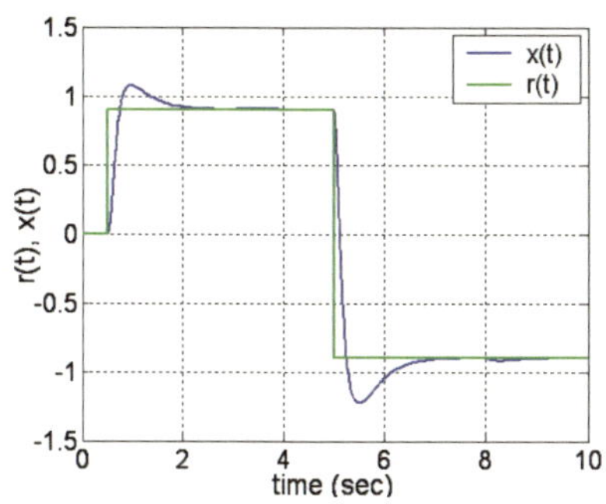

(a) Reference input $r(t)$ and output $x(t)$

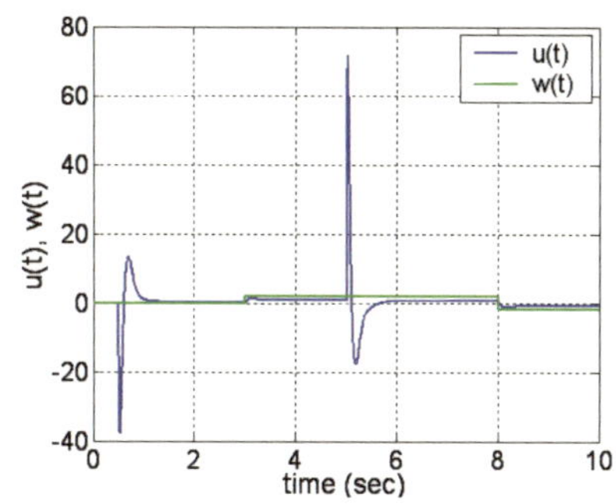

(b) Control $u(t)$ and disturbance $w(t)$

Fig. 7. Output response of the system (59) with controller (57) for a step reference input $r(t)$ and a step disturbance $w(t)$, where $b_1^d = a_1^d$ (the reference model is a system of type 2)

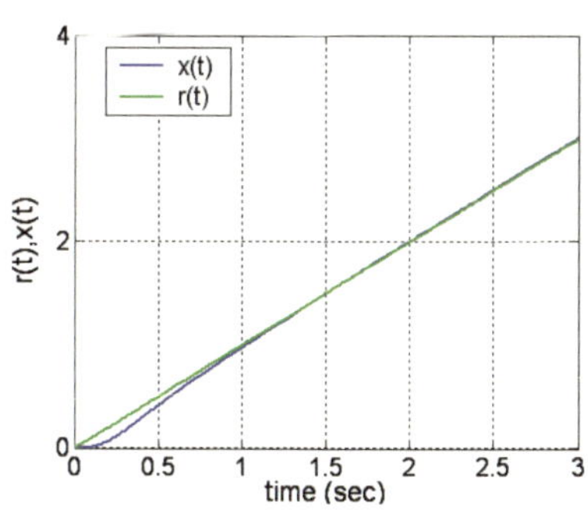

(a) Reference input $r(t)$ and output $x(t)$

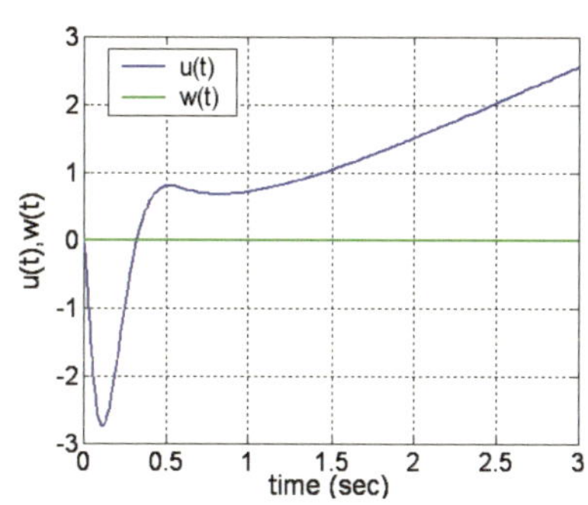

(b) Control $u(t)$ and disturbance $w(t)$

Fig. 8. Output response of the system (59) with controller (57) for a ramp reference input $r(t)$, where $b_1^d = a_1^d$ and $w(t) = 0$ (the reference model is a system of type 2)

Here $e_k := r_k - x_k$ is the error of the reference input realization, r_k being the samples of the reference input $r(t)$, where the control transients $e_k \to 0$ should meet the desired performance

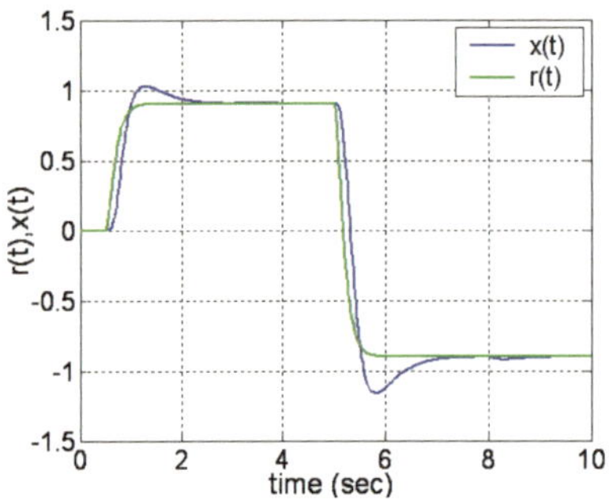

(a) Reference input $r(t)$ and output $x(t)$

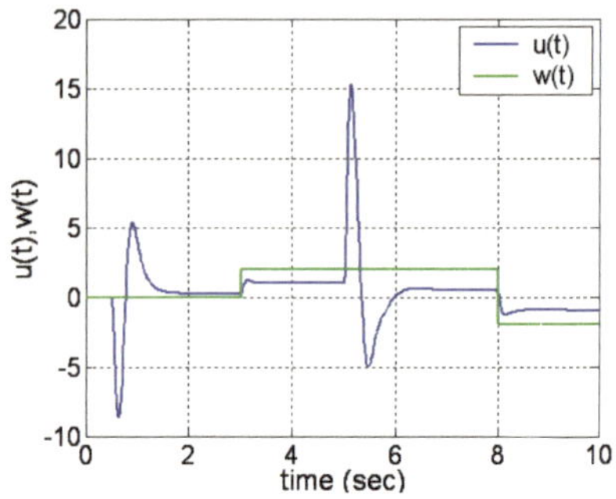

(b) Control $u(t)$ and disturbance $w(t)$

Fig. 9. Output response of the system (59) with controller (57) for a smooth reference input $r(t)$ and a step disturbance $w(t)$, where $b_1^d = a_1^d$ (the reference model is a system of type 2)

specifications given by (12).
By a $\mathcal{Z}$-transform of (12) preceded by a ZOH, the desired pulse transfer function

$$H_{xr}^d(z) = \frac{z-1}{z}\mathcal{Z}\left\{\mathcal{L}^{-1}\left[\frac{1/T}{s(s+1/T)}\right]\bigg|_{t=kT_s}\right\} = \frac{1-e^{-T_s/T}}{z-e^{-T_s/T}} \tag{62}$$

follows. Hence, from (62), the desired stable difference equation

$$x_k = x_{k-1} + T_s a(T_s)[r_{k-1} - x_{k-1}] \tag{63}$$

results, where

$$a(T_s) = \frac{1-e^{-T_s/T}}{T_s}, \quad \lim_{T_s\to 0} a(T_s) = \frac{1}{T},$$

and the output response of (63) corresponds to the assigned output transient performance indices.
Let us rewrite, for short, the desired difference equation (63) as

$$x_k = F(x_{k-1}, r_{k-1}), \tag{64}$$

where we have $r_k = x_k$ at the equilibrium of (64) for $r_k = \text{const}$, $\forall\, k$. Denote

$$e_k^F := F(x_{k-1}, r_{k-1}) - x_k, \tag{65}$$

where e_k^F is the realization error of the desired dynamics assigned by (64). Accordingly, if for all $k = 0, 1, \ldots$ the condition

$$e_k^F = 0 \tag{66}$$

holds, then the desired behavior of x_k with the prescribed dynamics of (64) is fulfilled. The expression (66) is the insensitivity condition for the output transient performance with respect to the external disturbances and varying parameters of the plant model given by (60). In other words, the control design problem (61) has been reformulated as the requirement (66).

The insensitivity condition given by (66) is the discrete-time counterpart of (15) which was introduced for the continuous-time system (9).

7.2 Discrete-time counterpart of PI controller

Let us consider the following control law:

$$u_k = u_{k-1} + \lambda_0[F(x_{k-1}, r_{k-1}) - x_k], \tag{67}$$

where $\lambda_0 = T_s^{-1}\tilde{\lambda}$ and the reference model of the desired output behavior is given by (63). In accordance with (63) and (65), the control law (67) can be rewritten as the difference equation

$$u_k = u_{k-1} + \tilde{\lambda}\left\{a(T_s)[r_{k-1} - x_{k-1}] - \frac{x_k - x_{k-1}}{T_s}\right\}. \tag{68}$$

The control law (68) is the discrete-time counterpart of the conventional continuous-time PI controller given by (18).

7.3 Two-time-scale motion analysis

Denote $f_{k-1} = f(x_{k-1}, w_{k-1})$ and $g_{k-1} = g(x_{k-1}, w_{k-1})$ in the expression (60). Hence, the closed-loop system equations have the following form:

$$x_k = x_{k-1} + T_s[f_{k-1} + g_{k-1}u_{k-1}], \tag{69}$$

$$u_k = u_{k-1} + \tilde{\lambda}\left\{a(T_s)[r_{k-1} - x_{k-1}] - \frac{x_k - x_{k-1}}{T_s}\right\}. \tag{70}$$

Substitution of (69) into (70) yields

$$x_k = x_{k-1} + T_s[f_{k-1} + g_{k-1}u_{k-1}], \tag{71}$$

$$u_k = [1 - \tilde{\lambda}g_{k-1}]u_{k-1} + \tilde{\lambda}\{a(T_s)[r_{k-1} - x_{k-1}] - f_{k-1}\}. \tag{72}$$

The sampling period T_s can be treated as a small parameter, then the closed-loop system equations (71)–(72) have the standard singular perturbation form given by (5)–(6). First, the stability and the rate of the transients of u_k in (71)–(72) depend on the controller parameter $\tilde{\lambda}$. Second, note that $x_k - x_{k-1} \to 0$ as $T_s \to 0$. Hence, we have a slow rate of the transients of x_k as $T_s \to 0$. Thus, if T_s is sufficiently small, the two-time-scale transients are artificially induced in the closed-loop system (71)–(72), where the FMS is governed by

$$u_k = [1 - \tilde{\lambda}g_{k-1}]u_{k-1} + \tilde{\lambda}\{a(T_s)[r_{k-1} - x_{k-1}] - f_{k-1}\} \tag{73}$$

and $x_k = x_{k-1}$, i.e., $x_k = \text{const}$ (hence, x_k is the frozen variable) during the transients in the FMS (73).

Let $g = g_k \; \forall \, k$. From (73), the FMS characteristic polynomial

$$z - 1 + \tilde{\lambda}g \tag{74}$$

results, where its root lies inside the unit disk (hence, the FMS is stable) if $0 < \tilde{\lambda} < 2/g$. To ensure stability and fastest transient processes of u_k, let us take the controller parameter

$\tilde{\lambda} = 1/g$, then the root of (74) is placed at the origin. Hence, the deadbeat response of the FMS (73) is provided. We may take $T_s \leq T/\eta$, where $\eta \geq 10$.
Third, assume that the FMS (73) is stable and consider its steady state (quasi-steady state), i.e.,

$$u_k - u_{k-1} = 0. \tag{75}$$

Then, from (73) and (75), we get $u_k = u_k^{id}$, where

$$u_k^{id} = g^{-1} \{a(T_s)[r_{k-1} - x_{k-1}] - f_{k-1}\}. \tag{76}$$

Substitution of (75) and (76) into (71) yields the SMS of (71)–(72), which is the same as the desired difference equation (63) in spite of unknown external disturbances and varying parameters of (60) and by that the desired behavior of x_k is provided.

8. Sampled-data nonlinear system of the 2-nd order

8.1 Approximate model

The above approach to approximate model derivation can also be used for nonlinear system of the 2-nd order, which is preceded by ZOH with high sampling rate. For instance, let us consider the nonlinear system given by (43)

$$x^{(2)} = f(X,w) + g(X,w)u, \quad y = x,$$

which is preceded by ZOH, where $y \in \mathbb{R}^1$ is the output, available for measurement; $u \in \mathbb{R}^1$ is the control; w is the external disturbance, unavailable for measurement; $X = \{x, x^{(1)}\}^T$ is the state vector.
We can obtain the state-space equations of (43) given by

$$\begin{aligned} \dot{x}_1 &= x_2, \\ \dot{x}_2 &= f(\cdot) + g(\cdot)u, \\ y &= x_1. \end{aligned}$$

Let us introduce the new time scale $t_0 = t/T_s$. We obtain

$$\begin{aligned} \frac{d}{dt_0}x_1 &= T_s x_2, \\ \frac{d}{dt_0}x_2 &= T_s\{f(\cdot) + g(\cdot)u\}, \\ y &= x_1, \end{aligned} \tag{77}$$

where $dX/dt_0 \to 0$ as $T_s \to 0$. From (77) it follows that

$$\frac{d^2y}{dt_0^2} = T_s^2\{f(\cdot) + g(\cdot)u\}. \tag{78}$$

Assume that the sampling period T_s is sufficiently small such that the conditions $X(t) = \text{const}$, $g(X, w) = \text{const}$ hold for $kT_s \leq t < (k+1)T_s$. Then, by taking the $\mathcal{Z}$-transform of (78), we get

$$y(z) = \frac{\mathcal{E}_2(z)}{2!\,(z-1)^2} T_s^2 \left\{ f(z) + \{gu\}(z) \right\}, \tag{79}$$

where $\mathcal{E}_2(z) = z + 1$. Denote $\mathcal{E}_2(z) = \epsilon_{2,1} z + \epsilon_{2,2}$ and $z^2 - a_{2,1} z - a_{2,2} = (z-1)^2$, where $\epsilon_{2,1} = \epsilon_{2,2} = 1$, $a_{2,1} = 2$, and $a_{2,2} = -1$. From (79) we get the difference equation

$$y_k = \sum_{j=1}^{2} a_{2,j} y_{k-j} + T_s^2 \sum_{j=1}^{2} \frac{\epsilon_{2,j}}{2!} \left\{ f_{k-j} + g_{k-j} u_{k-j} \right\} \tag{80}$$

given that the high sampling rate takes place, where $g_k = g(X(t), w(t))|_{t=kT_s}$, $f_k = f(X(t), w(t))|_{t=kT_s}$, and

$$y_k - y_{k-j} \to 0, \ \forall\, j = 1, 2 \text{ as } T_s \to 0. \tag{81}$$

8.2 Reference equation and insensitivity condition

Denote $e_k := r_k - y_k$ is the error of the reference input realization, where r_k being the reference input. Our objective is to design a control system having

$$\lim_{k \to \infty} e_k = 0. \tag{82}$$

Moreover, the control transients $e_k \to 0$ should have desired performance indices such as overshoot, settling time, and system type. These transients of y_k should not depend on the external disturbances and varying parameters of the nonlinear system (43).

Let us consider the continuous-time reference model for the desired behavior of the output $y(t) = x(t)$ in the form given by (45), which can be rewritten as

$$y(s) = G^d(s) r(s),$$

where the parameters of the 2nd-order stable continuous-time transfer function $G^d(s)$ are selected based on the required output transient performance indices and such that

$$G^d(s)\Big|_{s=0} = 1.$$

By a $\mathcal{Z}$-transform of $G^d(s)$ preceded by a ZOH, the desired pulse transfer function

$$H_{yr}^d(z) = \frac{z-1}{z} \mathcal{Z} \left\{ \mathcal{L}^{-1} \left[\frac{G_{yr}^d(s)}{s} \right] \Bigg|_{t=kT_s} \right\} = \frac{B^d(z)}{A^d(z)} \tag{83}$$

can be found, where

$$H_{yr}^d(z)\Big|_{z=1} = 1.$$

Hence, from (83), the desired stable difference equation

$$y_k = \sum_{j=1}^{2} a_j^d y_{k-j} + \sum_{j=1}^{2} b_j^d r_{k-j} \tag{84}$$

results, where

$$1 - \sum_{j=1}^{2} a_j^d = \sum_{j=1}^{2} b_j^d, \qquad \sum_{j=1}^{2} b_j^d \neq 0,$$

and the parameters of (84) correspond to the assigned output transient performance indices. Let us rewrite, for short, the desired difference equation (84) as

$$y_k = F(Y_k, R_k), \tag{85}$$

where $Y_k = \{y_{k-2}, y_{k-1}\}^T$, $R_k = \{r_{k-2}, r_{k-1}\}^T$, and $r_k = y_k$ at the equilibrium of (85) for $r_k = \text{const}$, $\forall\, k$. By definition, put $F_k = F(Y_k, R_k)$ and denote

$$e_k^F := F_k - y_k, \tag{86}$$

where e_k^F is the realization error of the desired dynamics assigned by (85). Accordingly, if for all $k = 0, 1, \ldots$ the condition

$$e_k^F = 0 \tag{87}$$

holds, then the desired behavior of y_k with the prescribed dynamics of (85) is fulfilled. The expression (87) is the insensitivity condition for the output transients with respect to the external disturbances and varying parameters of the plant model (80). In other words, the control design problem (82) has been reformulated as the requirement (87). The insensitivity condition (87) is the discrete-time counterpart of the condition $e_F = 0$ for the continuous-time system (43).

8.3 Discrete-time counterpart of PIDF controller

In order to fulfill (87), let us construct the control law as the difference equation

$$u_k = \sum_{j=1}^{q\geq 2} d_j u_{k-j} + \lambda_0 [F_k - y_k], \tag{88}$$

where

$$d_1 + d_2 + \cdots + d_q = 1, \quad \text{and} \quad \lambda_0 \neq 0. \tag{89}$$

From (89) it follows that the equilibrium of (88) corresponds to the insensitivity condition (87). In accordance with (84) and (86), the control law (88) can be rewritten as the difference equation

$$u_k = \sum_{j=1}^{q\geq 2} d_j u_{k-j} + \lambda_0 \left\{ -y_k + \sum_{j=1}^{2} a_j^d y_{k-j} + \sum_{j=1}^{2} b_j^d r_{k-j} \right\}. \tag{90}$$

The control law (90) is the discrete-time counterpart of the continuous-time PIDF controller (50). In particular, if $q = 2$, then (90) can be rewritten in the following state-space form:

$$\begin{aligned} \bar{u}_{1,k} &= \bar{u}_{2,k-1} + d_1 \bar{u}_{1,k-1} + \lambda_0 [a_1^d - d_1] y_{k-1} + \lambda_0 b_1^d r_{k-1}, \\ \bar{u}_{2,k} &= d_2 \bar{u}_{1,k-1} + \lambda_0 [a_2^d - d_2] y_{k-1} + \lambda_0 b_2^d r_{k-1}, \\ u_k &= \bar{u}_{1,k} - \lambda_0 y_k. \end{aligned} \tag{91}$$

Then, from (91), we get the block diagram of the controller as shown in Fig. 10.

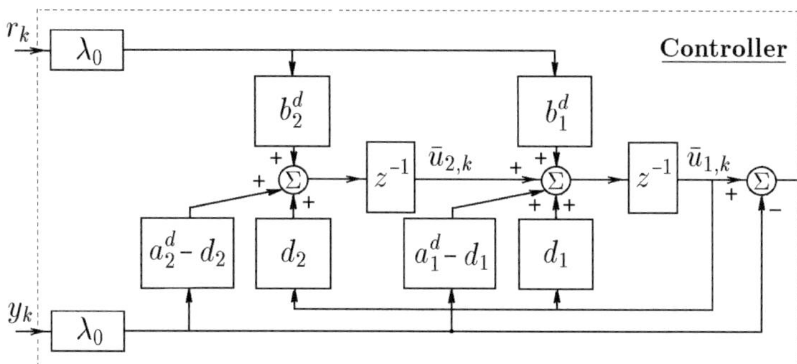

Fig. 10. Block diagram of the control law (90), where $q = 2$, represented in the form (91)

8.4 Two-time-scale motion analysis

The closed-loop system equations have the following form:

$$y_k = \sum_{j=1}^{2} a_{2,j} y_{k-j} + T_s^2 \sum_{j=1}^{2} \frac{\epsilon_{2,j}}{2!} \left[f_{k-j} + g_{k-j} u_{k-j} \right], \tag{92}$$

$$u_k = \sum_{j=1}^{q\geq 2} d_j u_{k-j} + \lambda_0 [F_k - y_k]. \tag{93}$$

Substitution of (92) into (93) yields

$$y_k = \sum_{j=1}^{2} a_{2,j} y_{k-j} + T_s^2 \sum_{j=1}^{2} \frac{\epsilon_{2,j}}{2!} \left[f_{k-j} + g_{k-j} u_{k-j} \right], \tag{94}$$

$$u_k = \sum_{j=n+1}^{q>2} d_j u_{k-j} + \sum_{j=1}^{2} [d_j - \lambda_0 T_s^2 \frac{\epsilon_{2,j}}{2!} g_{k-j}] u_{k-j} + \lambda_0 \left\{ F_k - \sum_{j=1}^{2} \left\{ a_{2,j} y_{k-j} - T_s^2 \frac{\epsilon_{2,j}}{2!} f_{k-j} \right\} \right\}. \tag{95}$$

First, note that the rate of the transients of u_k in (94)–(95) depends on the controller parameters $\lambda_0, d_1, \ldots, d_q$. At the same time, in accordance with (81), we have a slow rate of the transients of y_k, because the sampling period T_s is sufficiently small one. Therefore, by choosing the controller parameters it is possible to induce two-time scale transients in the closed-loop system (94)–(95), where the rate of the transients of y_k is much smaller than that of u_k. Then, as an asymptotic limit, from the closed-loop system equations (94)–(95) it follows that the FMS is governed by

$$u_k = \sum_{j=3}^{q>2} d_j u_{k-j} + \sum_{j=1}^{2} [d_j - \lambda_0 T_s^2 \frac{\epsilon_{2,j}}{2!} g_{k-j}] u_{k-j} + \lambda_0 \left\{ F_k - \sum_{j=1}^{2} \left\{ a_{2,j} y_{k-j} - T_s^2 \frac{\epsilon_{2,j}}{2!} f_{k-j} \right\} \right\}, \quad (96)$$

where $y_k - y_{k-j} \approx 0,\ \forall\, 1, \ldots, q$, i.e., $y_k = \text{const}$ during the transients in the system (96). Second, assume that the FMS (96) is exponentially stable (that means that the unique equilibrium point of (96) is exponentially stable), and $g_k - g_{k-j} \to 0,\ \forall\, j = 1, 2, \ldots, q$ as $T_s \to 0$. Then, consider steady state (or more exactly quasi-steady state) of (96), i.e.,

$$u_k - u_{k-j} = 0, \quad \forall\, j = 1, \ldots, q. \quad (97)$$

Then, from (89), (96), and (97) we get $u_k = u_k^{id}$, where

$$u_k^{id} = [T_s^2 g_k]^{-1} \left\{ F_k - \sum_{j=1}^{2} \left\{ a_{2,j} y_{k-j} + T_s^2 \frac{\epsilon_{2,j}}{2!} f_{k-j} \right\} \right\}. \quad (98)$$

The discrete-time control function u_k^{id} given by (98) corresponds to the insensitivity condition (87), that is, u_k^{id} is the discrete-time counterpart of the nonlinear inverse dynamics solution (46). Substitution of (97) into (94)–(95) yields the SMS of (94)–(95), which is the same as the desired difference equation (85) and by that the desired behavior of y_k is provided.

8.5 Selection of discrete-time controller parameters

Let, the sake of simplicity, $q = 2$, $\bar{g} = g_k = \text{const}\ \forall\, k$, and take

$$\lambda_0 = \{T_s^2 \bar{g}\}^{-1}, \quad d_j = \frac{\epsilon_{2,j}}{2!}, \ \forall\, i = 1, 2. \quad (99)$$

Then all roots of the characteristic polynomial of the FMS (96) are placed at the origin. Hence, the deadbeat response of the FMS (96) is provided. This, along with assumption that the sampling period T_s is sufficiently small, justifies two-time-scale separation between the fast and slow motions. So, if the degree of time-scale separation between fast and slow motions in the closed-loop system (94)–(95) is sufficiently large and the FMS transients are stable, then after the fast transients have vanished the behavior of y_k tends to the solution of the reference equation given by (85). Accordingly, the controlled output transient process meets the desired performance specifications. The deadbeat response of the FMS (96) has a finite settling time given by $t_{s,FMS} = 2T_s$ when $q = 2$. Then the relationship

$$T_s \leq \frac{t_{s,SMS}}{2\eta} \quad (100)$$

may be used to estimate the sampling period in accordance with the required degree of time-scale separation between the fast and slow modes in the closed-loop system. Here $t_{s,SMS}$ is the settling time of the SMS and η is the degree of time-scale separation, $\eta \geq 10$.
The advantage of the presented above method is that knowledge of the high-frequency gain g suffices for controller design; knowledge of external disturbances and other parameters of the system is not needed. Note that variation of the parameter g is possible within the domain where the FMS (96) is stable and the fast and slow motion separation is maintained.

8.6 Example 3

Let us consider the system (59). Assume that the specified region of $x(t)$ is given by $x(t) \in [-2,2]$. Hence, the range of high-frequency gain variations has the following bounds $g(x) \in [2,6]$. We have that $\mathcal{E}_2(z) = z + 1$. Let the desired output behavior is described by the reference equation (45) where $a_1^d = 2$. Therefore, from (45), the desired transfer function

$$G^d(s) = \frac{b_1^d Ts + 1}{T^2 s^2 + a_1^d Ts + 1} = \frac{b_1^d Ts + 1}{T^2 (s + \bar{\alpha})^2} \tag{101}$$

results, where $\bar{\alpha} = 1/T$. The pulse transfer function $H^d(z)$ of a series connection of a zero-order hold and the system of (101) is the function given by

$$H^d(z) = \frac{\bar{b}_1^d z + \bar{b}_2^d}{z^2 - \bar{a}_1^d z - \bar{a}_2^d}, \tag{102}$$

where $\bar{a}_1^d = 2d$, $\bar{a}_2^d = -d^2$, $\bar{b}_1^d = T^{-2}[1 - d + (b_1^d T - \bar{\alpha}) d T_s]$, and $\bar{b}_2^d = T^{-2} d[d - 1 + (\bar{\alpha} - b_1^d T) T_s]$. Take, for simplicity, $q = 2$. Hence, in accordance with (90) and (99), the discrete-time controller has been obtained

$$u_k = d_1 u_{k-1} + d_2 u_{k-2} + [T_s^2 \bar{g}]^{-1} \{-y_k + \bar{a}_1^d y_{k-1} + \bar{a}_2^d y_{k-2} + \bar{b}_1^d r_{k-1} + \bar{b}_2^d r_{k-2}\}, \tag{103}$$

where $d_1 = d_2 = 0.5$. The controller given by (103) is the discrete-time counterpart of PID controller (48). Let the sampling period T_s is so small that the degree of time-scale separation between fast and slow motions in the closed-loop system is large enough, then $g_k = g_{k-1} = g_{k-2}$, $\forall\, k$. From (96) and (99), the FMS characteristic equation

$$z^2 + 0.5\left[\frac{g}{\bar{g}} - 1\right] z + 0.5\left[\frac{g}{\bar{g}} - 1\right] = 0 \tag{104}$$

results, where the parameter g is treated as a constant value during the transients in the FMS. Take $\bar{g} = 4$, then it can be easily verified, that $\max\{|z_1|, |z_2|\} \leq 0.6404$ for all $g \in [2,6]$, where z_1 and z_2 are the roots of (104). Hence, the stability of the FMS is maintained for all $g \in [2,6]$. Let $T = 0.3$ s. and $\eta = 10$. Take $T_s = T/\eta = 0.03$ s. The simulation results for the output of the system (59) controlled by the algorithm (103) are displayed in Figs. 11–15, where the initial conditions are zero. Note, the simulation results shown in Figs. 11–15 approach ones shown in Figs. 5–9 when T_s becomes smaller.

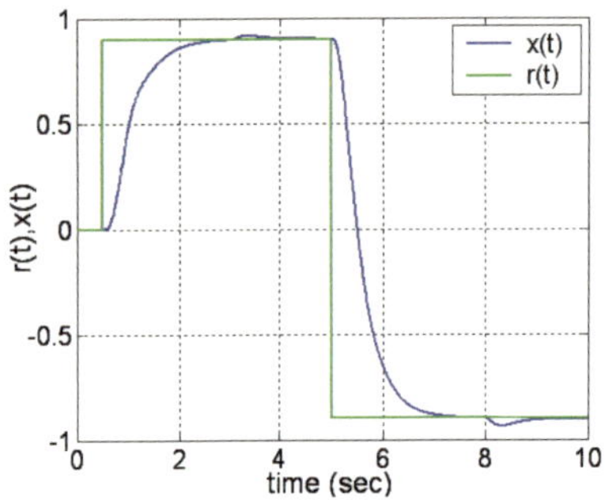

(a) Reference input $r(t)$ and output $x(t)$

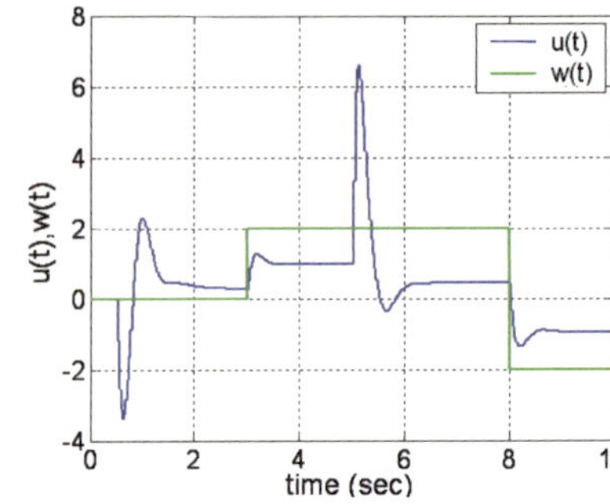

(b) Control $u(t)$ and disturbance $w(t)$

Fig. 11. Output response of the system (59) with controller (103) for a step reference input $r(t)$ and a step disturbance $w(t)$, where $b_1^d = 0$ (the reference model is a system of type 1)

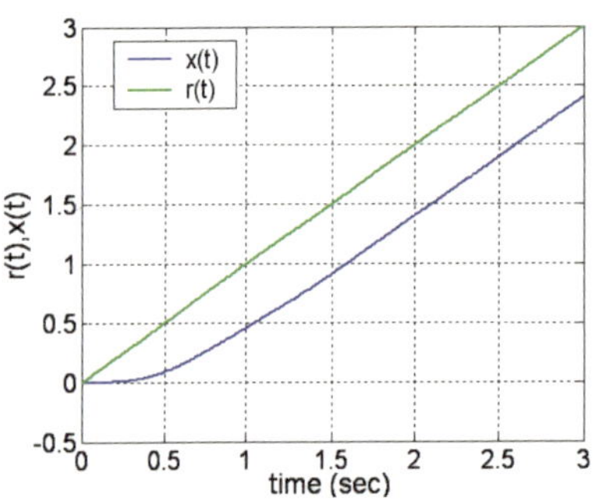

(a) Reference input $r(t)$ and output $x(t)$

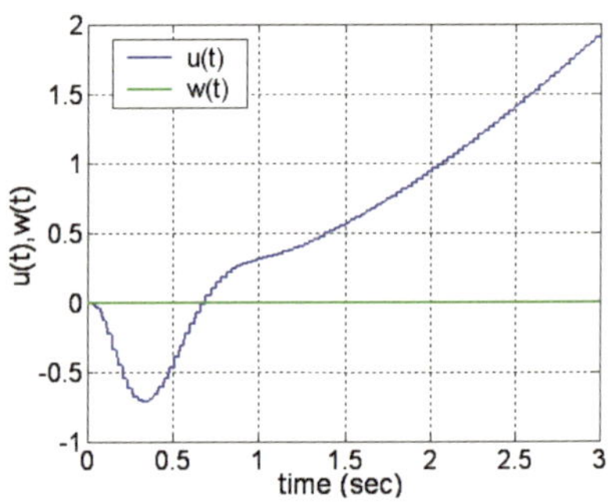

(b) Control $u(t)$ and disturbance $w(t)$

Fig. 12. Output response of the system (59) with controller (103) for a ramp reference input $r(t)$, where $b_1^d = 0$ and $w(t) = 0$ (the reference model is a system of type 1)

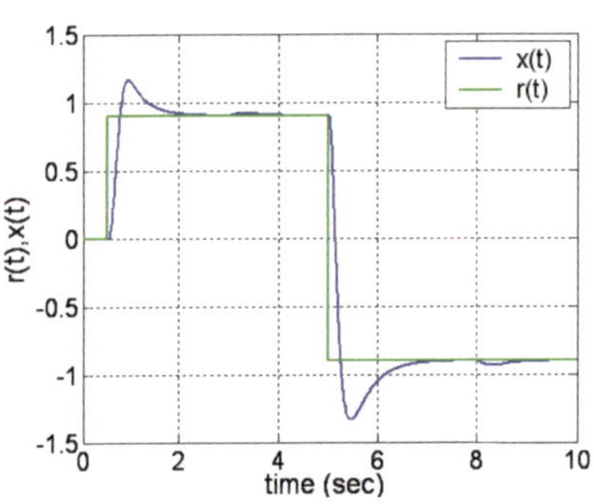

(a) Reference input $r(t)$ and output $x(t)$

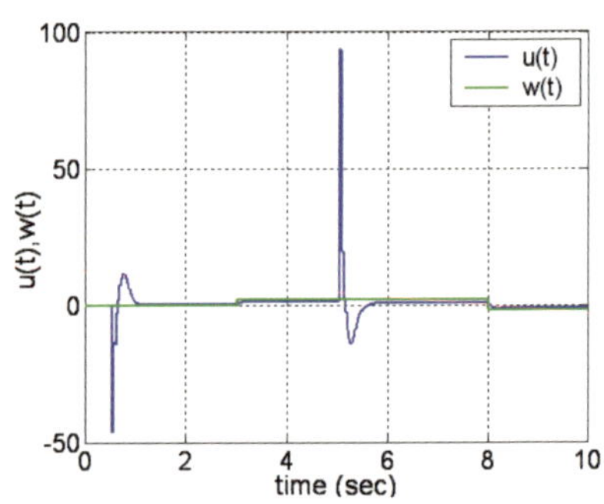

(b) Control $u(t)$ and disturbance $w(t)$

Fig. 13. Output response of the system (59) with controller (103) for a step reference input $r(t)$ and a step disturbance $w(t)$, where $b_1^d = a_1^d$ (the reference model is a system of type 2)

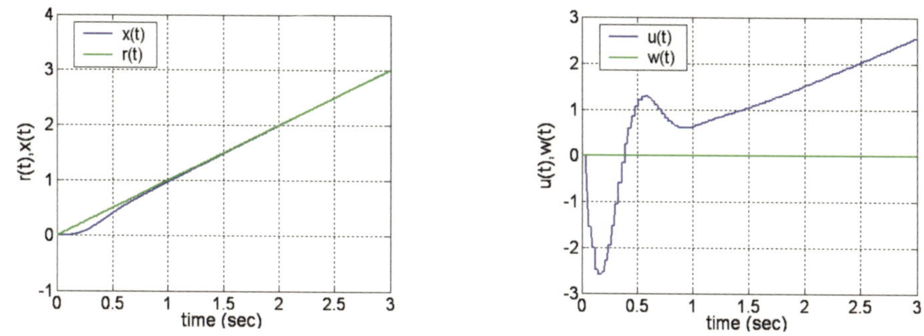

(a) Reference input $r(t)$ and output $x(t)$ (b) Control $u(t)$ and disturbance $w(t)$

Fig. 14. Output response of the system (59) with controller (103) for a ramp reference input $r(t)$, where $b_1^d = a_1^d$ and $w(t) = 0$ (the reference model is a system of type 2)

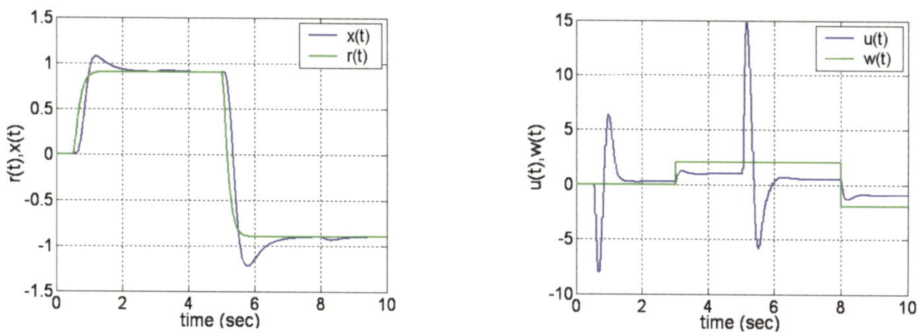

(a) Reference input $r(t)$ and output $x(t)$ (b) Control $u(t)$ and disturbance $w(t)$

Fig. 15. Output response of the system (59) with controller (103) for a smooth reference input $r(t)$ and a step disturbance $w(t)$, where $b_1^d = a_1^d$ (the reference model is a system of type 2)

9. Conclusion

In accordance with the presented above approach the fast motions occur in the closed-loop system such that after fast ending of the fast-motion transients, the behavior of the overall singularly perturbed closed-loop system approaches that of the SMS, which is the same as the reference model. The desired dynamics realization accuracy and an acceptable level of disturbance rejection can be provided by increase of time-scale separation degree between slow and fast motions in the closed-loop system. However, it should be emphasized that the time-scale separation degree is bounded above in practice due to the presence of unmodeled dynamics or time delay in feedback loop. So, the effect of unmodeled dynamics and time delay on FMS transients stability should be taken in to account in order to proper selection of controller parameters (Yurkevich, 2004). This effect puts the main restriction on the practical implementation of the discussed control design methodology via singular perturbation technique. The presented design methodology may be used for a broad class of nonlinear time-varying systems, where the main advantage is the unified approach to continuous as well as digital control system design that allows to guarantee the desired output transient performances in the presence of plant parameter variations and unknown external disturbances. The other advantage, caused by two-time-scale technique for closed-loop system analysis, is that analytical expressions for parameters of PI, PID, or PID controller with additional lowpass filtering for nonlinear systems can be found, where controller parameters depend explicitly on the specifications of the desired output behavior. The presented design methodology may be useful for real-time control system design under uncertainties and illustrative examples can be found in (Czyba & Błachuta, 2003; Khorasani et al., 2005).

10. References

Åström, K.J. & Hägglund, T. (1995). *PID controllers: Theory, Design, and Tuning*, Research Triangle Park, NC: Instrum. Soc. Amer., ISBN 1-55617-516-7.

Błachuta, M.J.; Yurkevich, V.D. & Wojciechowski, K. (1997). Design of analog and digital aircraft flight controllers based on dynamic contraction method, *Proc. of the AIAA Guidance, Navigation and Control Conference, 97'GN&C*, New Orleans, LA, USA, Part 3, pp. 1719–1729.

Błachuta, M.J.; Yurkevich, V.D. & Wojciechowski, K. (1999). Robust quasi NID aircraft 3D flight control under sensor noise, *Int. J. Kybernetika*, Vol. 35, No. 5, pp. 637–650, ISSN 0023-5954.

Boychuk, L. M. (1966). An inverse method of the structural synthesis of automatic control nonlinear systems *Automation*, Kiev: Naukova Dumka, No. 6, pp. 7–10.

Czyba, R. & Błachuta, M.J. (2003). Robust longitudinal flight control design: dynamic contraction method *Proc. of American Control Conf*, Denver, Colorado, USA, pp. 1020–1025, ISBN 0-7803-7896-2.

Hoppensteadt, F.C. (1966). Singular perturbations on the infinite time interval, *Trans. of the American Mathematical Society*, Vol. 123, pp. 521–535.

Huang, J. & Rugh, W.J. (1990). On a nonlinear multivariable servomechanism problem, *Automatica*, Vol. 26, pp. 963–972, ISSN 0005-1098.

Isidori, A. & Byrnes, C.I., (1990). Output regulation of nonlinear systems, *IEEE Trans. Automat. Contr.*, Vol. AC-35, No. 2, pp. 131–140, ISSN 0018-9286.

Khalil, H.K. (2000). Universal integral controllers for minimum-phase nonlinear systems, *IEEE Trans. Automat. Contr.*, Vol. AC-45 (No. 3), pp. 490–494, ISSN 0018-9286.

Khalil, H.K. (2002). *Nonlinear Systems*, 3rd ed., Upper Saddle River, N.J. : Prentice Hall, ISBN 0130673897.

Khorasani, K.; Gavriloiu, V. & Yurkevich, V. (2005). A novel dynamic control design scheme for flexible-link manipulators, *Proc. of IEEE Conference on Control Applications (CCA 2005)*, Toronto, Canada, 2005, pp. 595–600.

Klimushchev, A.I. & Krasovskii, N.N. (1962). Uniform asymptotic stability of systems of differential equations with a small parameter in the derivative terms, *J. Appl. Math. Mech.*, Vol. 25, pp. 1011–1025.

Kokotović, P.V.; O'Malley, R.E. & Sannuti, P. (1976). Singular perturbations and order reduction in control – an overview, *Automatica*, Vol. 12, pp. 123–132, ISSN 0005-1098.

Kokotović, P.V.; Khalil, H.K.; O'Reilly, J. & O'Malley, R.(1999). *Singular perturbation methods in control: analysis and design*, Academic Press, ISBN 9780898714449.

Kokotović, P.V. & Khalil, H.K. (1986). *Singular perturbations in systems and control*, IEEE Press, ISBN 087942205X.

Krutko, P.D. (1988). The principle of acceleration control in automated system design problems, *Sov. J. Comput. Syst. Sci. USA*, Vol. 26, No. 4, pp. 47–57, ISSN 0882-4002.

Krutko, P.D. (1991). Optimization of control systems with respect to local functionals characterizing the energy of motion *Sov. Phys.-Dokl. USA*, Vol. 36, No. 9, pp. 623–625, ISSN 0038-5689.

Krutko, P.D. (1995). Optimization of multidimensional dynamic systems using the criterion of minimum acceleration energy, *Sov. J. Comput. Syst. Sci. USA*, Vol. 33, No. 4, pp. 27–42, ISSN 0882-4002.

Li, Y.; Ang, K.H. & Chong, G.C.Y. (2006). PID control system analysis and design, *IEEE Contr. Syst. Mag.*, Vol. 26, No. 1, pp. 32–41, ISSN 1066-033X.

Litkouhi, B. & Khalil, H. (1985). Multirate and composite control of two-time-scale discrete-time systems, *IEEE Trans. Automat. Contr.*, Vol. AC-30, No. 7, pp. 645–651, ISSN 0018-9286.

Lun, J.Y.S.; Walker, M.W. & Paul, R.P.C. (1980). Resolved acceleration control of mechanical manipulator *IEEE Trans. Automat. Contr.*, Vol. AC-25, No. 3, pp. 468–474, ISSN 0018-9286.

Luo, G. & Saridis, G. (1985). *L-Q* design of PID controllers for robot arms, *IEEE Journal of Robotics and Automation*, Vol. RA-1, No. 3, pp. 152–159, ISSN 0882-4967.

Mahmoud, N.A. & Khalil, H. (1996). Asymptotic regulation of minimum-phase nonlinear systems using output feedback, *IEEE Trans. Automat. Contr.*, Vol. AC-41, No. 10, pp. 1402–1412, ISSN 0018-9286.

Meerov, M. V. (1965). *Structural synthesis of high-accuracy automatic control systems*, Pergamon Press international series of monographs on automation and automatic control, Vol. 6, Oxford, New York : Pergamon Press.

Morari, M. & Zafiriou, E. (1999). *Robust Process Control*, Englewood Cliffs, NJ. Prentice-Hall, ISBN 0137819560.

Naidu, D.S. & Calise, A.J. (2001). Singular perturbations and time scales in guidance and control of aerospace systems: a survey, *Journal of Guidance, Control, and Dynamics*, Vol. 24, No. 6, pp. 1057-1078, ISSN 0731-5090.

Naidu, D.S. (2002). Singular perturbations and time scales in control theory and applications: an overview, *Dynamics of Continuous, Discrete & Impulsive Systems (DCDIS)*, Series B: Applications & Algorithms, Vol. 9, No. 2, pp. 233-278.

O'Dwyer, A. (2003). *Handbook of PI and PID Tuning Rules*, London: Imperial College Press, ISBN 1860946224.

Porter, W.A. (1970). Diagonalization and inverses for nonlinear systems, *Int. J. of Control*, Vol. 11, No. 1, pp. 67–76, ISSN: 0020-7179.

Qu, Z.; Dorsey, J.F.; Zhang, X. & Dawson, D.M. (1991). Robust control of robots by the computed torque method, *Systems Control Lett.*, Vol. 16, pp.25–32, ISSN 0167-6911.

Saksena, V.R.; O'Reilly, J. & Kokotović, P.V. (1984). Singular perturbations and time-scale methods in control theory: survey 1976-1983, *Automatica*, Vol. 20, No. 3, pp. 273–293, ISSN 0005-1098.

Slotine, J.-J. E. & Li, W. (1991). *Applied nonlinear control*, Prentice Hall, ISBN 0-13-040890-5.

Studenny, J. & Belanger, P.R. (1984). Robot manipulator control by acceleration feedback, *Proc. of 23th IEEE Conf. on Decision and Control*, pp. 1070-1072, ISSN 0191-2216.

Studenny, J. & Belanger, P.R. (1986). Robot manipulator control by acceleration feedback: stability, design and performance issues, *Proc. of 25th IEEE Conf. on Decision and Control*, pp. 80–85, ISSN 0191-2216.

Tikhonov, A.N. (1948). On the dependence of the solutions of differential equations on a small parameter, *Mathematical Sb.*, Moscow, Vol. 22, pp. 193–204.

Tikhonov, A.N. (1952). Systems of differential equations containing a small parameter multiplying the derivative, *Mathematical Sb.*, Moscow, 1952, Vol. 31, No. 3, pp. 575–586.

Utkin, V.I. (1992). *Sliding Modes in Control and Optimization*, Springer-Verlag, ISBN-10: 0387535160.

Young, K.D.; Kokotović, P.V. & Utkin, V.I. (1977). A singular pertubation analysis of high-gain feedback systems, *IEEE Trans. Automat. Contr.*, Vol. AC-22, No. 3, pp. 931–938, ISSN 0018-9286.

Young, K.D. & Özgüner, Ü. (1999). *Variable structure systems, sliding mode, and nonlinear control*, Series: Lecture notes in control and information science, Vol. 247, ISBN 1852331976, London, New York: Springer.

Yurkevich, V.D. (1995). Decoupling of uncertain continuous systems: dynamic contraction method, *Proc. of 34th IEEE Conf. on Decision & Control*, Vol. 1, pp. 196–201, ISBN 0780326857, New Orleans. Louisiana.

Yurkevich, V.D. (2004). *Design of nonlinear control systems with the highest derivative in feedback*, World Scientific Publishing Co., ISBN 9812388990, Singapore.

Ziegel, J.G. & Nichols, N.B. (1942). Optimum settings for automatic controllers, *Trans. ASME*, Vol. 64, No. 8, pp. 759-768.

8

High-Speed and High-Precision Position Control Using a Nonlinear Compensator

Kazuhiro Tsuruta[1], Kazuya Sato[2] and Takashi Fujimoto[1]
[1]*Kyushu Sangyo University*
[2]*Saga University*
Japan

1. Introduction

To achieve high-speed, high-precision position control for semiconductor product machines and industrial robots, it is well known that PID control is widely applied (Kojima, T., 2004). Many PID control methods have been proposed for such a system so far. In general, proportional position control and proportional plus integral velocity control or integral plus proportional velocity control (P,PI/I-P), which is a type of proportional plus integral plus deferential control (PID), is applied in many industrial applications in Japan. However, the parameters of the control of P,PI/I-P must be changed to maintain a good motion performance when the characteristic of the control target change. As a method to compensate for a change of the control target, a disturbance observer is proposed (Ohishi, K. et al., 1999). Using this method, a load disturbance force is estimated based on a model set beforehand and adds an estimated disturbance force to a control input. However, it might cause the deterioration of the control performance when a control target model is different from real control target greatly. Furthermore, it is difficult to get a correct model because a lot of nonlinear elements such as the friction may change in control target in time and environment (Canudas-de-Wit, C. et al., 1995; Lischinsky, P. et al., 1999; Futami, S. et al., 1990; Otsuka, J. & Masuda, T., 1998; Iwasaki, M. et al., 2000; Tsuruta, K. et al., 2000, 2003). In addition, based on a control target model estimated in real time, the method to change the several control parameters is proposed to compensate it for a change of the control target (Kuwon, T. et al., 2006), but might cause the deterioration of the control response under the influence of an estimate error. Therefore, adaptive control (Suzuki, T., 2001; Sato, K. et al., 2005, 2006, 2007, 2008) and sliding mode control (Nonami, K. & Den, H., 1994; Utkin, V, I., 1977; Harashima, F. & Hashimoto, H., 1986; Fujimoto, T. et al., 2006) are given to solve these problems. However, that is hard to understand these methods to the engineer who got used to PID control; have a problem. In this chapter, we propose a new P,PI/I-P control method that includes a nonlinear compensator, that it is easy to understand for a PID control designer. The control objective is to get a high-speed and high-precision positioning response regardless of the case of control references or load characteristics change. The algorithm of the nonlinear compensator is based on sliding mode control with chattering compensation. The effectiveness of the proposed control method is evaluated with three kinds of single-axis slide system experimentally. The first experiment system is two slider tables comprised of an AC servo motor, a coupling and a ball-screw, the second one is a slide table using an AC linear motor and the third one is a slide table using synchronous piezoelectric device driver (Egashira, Y. et al., 2002; Kosaka, K. et al., 2006). By the first experiments, it is evaluated using single-axis slide system comprised of full closed feedback via point-to-point control response and tracking control response when load characteristics of the control target change. By the second experiments, it is evaluated using a linear motor driven slider system via tracking control at low-velocity, and the resolution of this system is 10nm. By the third experiments, it is evaluated a stepping motion and tracking motion using a synchronous piezoelectric device driver. Then, we derive control algorithm with nonlinear compensator and describe each experimental results.

2. Control method

In this section, we describe a P,PI/I-P+FF control method and propose the P,PI/I-P+FF control method with nonlinear compensator. The control objective is to design a control input to track a given position reference.

2.1 Conventional P,PI/I-P+FF control method

In general, P,PI/I-P control method is applied in many industrial applications. To achieve high-speed positioning response, velocity feed-forward compensation (FF) is usually applied. The FF compensation is effective in compensating it for a response delay of the positioning. As for P,PI/I-P method, positioning and velocity control are comprised of cascade control. When we do not make positioning, we can control velocity by inputting a direct velocity reference. Fig. 1 shows a block diagram of P,PI/I-P+FF control method. In this figure, a signal x_d is position reference, a signal x is table position, a signal $\dot{x}$ is table velocity, a signal e_1 is position error, a signal e_2 is velocity error and a signal u is control input which means torque command, respectively. Here, the differentiation uses backward difference equation. K_p is position loop gain, K_i is velocity integral gain, K_v is velocity loop gain, α is velocity feed-forward gain, β is the change fixed number to change velocity PI control method or velocity I-P control method. If β is 1, the velocity control is I-P control method, else if β is 0, the velocity control is PI control method. K_f is the torque conversion fixed number. The symbol s is Laplace transfer operator, s means differentiator and $1/s$ means integrator.

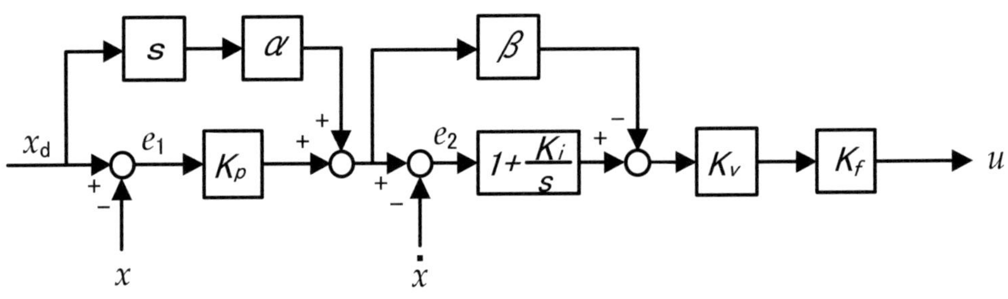

Fig. 1. Block diagram of P,PI/I-P+FF control method.

The control input u is given as follows

$$u = K_v K_f \left\{ \left(1 + \frac{K_i}{s}\right) e_2 - \beta\left(K_p e_1 + \alpha \dot{x}_d\right) \right\} \tag{1}$$

2.2 Proposed control method

In this paper, we will design a PID+FF controller with a nonlinear compensator for high accuracy and a fast response with small overshoot. Fig. 2 shows a block diagram of the P,PI/I-P+FF control method with proposed compensator. The algorithm of the nonlinear friction compensator will be given.

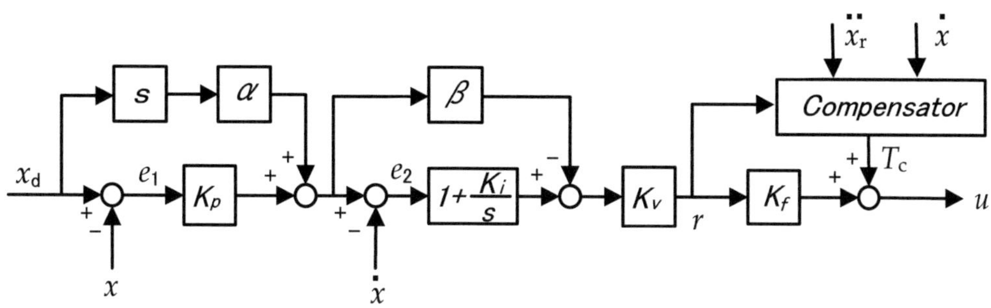

Fig. 2. Block diagram of proposed control method.

The dynamic equation of positioning table can be modelled as follows

$$J\ddot{x} + D\dot{x} + F = u \tag{2}$$

where J is the inertia, D is the viscous friction coefficient, F is the constant disturbance force, x is table position , $\dot{x}$ is table velocity, and u is the control input. Let the error value be

$$e_1 = x_d - x \tag{3}$$

$$e_2 = K_p e_1 - \dot{x} + \alpha \dot{x}_d \tag{4}$$

Taking the second time derivative of both sides for (4) and substituting it into (2), we have

$$\ddot{x} = K_p \dot{e}_1 + \alpha \ddot{x}_d - \dot{e}_2 \tag{5}$$

$$J(K_p \dot{e}_1 + \alpha \ddot{x}_d - \dot{e}_2) + D\dot{x} + F = u \tag{6}$$

$$J\dot{e}_2 = -u + J(K_p \dot{e}_1 + \alpha \ddot{x}_d) + D\dot{x} + F \tag{7}$$

Now, we can define the new signal as

$$r = K_v \left\{ \left(1 + \frac{K_i}{s}\right) e_2 - \beta\left(K_p e_1 + \alpha \dot{x}_d\right) \right\} \tag{8}$$

Taking the time derivative of both sides for (8) and multiplying J to both sides, we have

$$
\begin{aligned}
J\dot{r} &= JK_v\dot{e}_2 + JK_vK_ie_2 - J\beta K_pK_v\dot{e}_1 - JK_v\alpha\beta\ddot{x}_d \\
&= K_v\left(-u + JK_p\dot{e}_1 + J\alpha\ddot{x}_d + D\dot{x} + F\right) + JK_vK_ie_2 \\
&\quad - J\beta K_pK_v\dot{e}_1 - JK_v\alpha\beta\ddot{x}_d \\
&= -K_vu + K_vD\dot{x} + K_vF + JK_v\alpha\ddot{x}_d(1-\beta) \\
&\quad + JK_pK_v\dot{e}_1(1-\beta) + JK_vK_ie_2
\end{aligned} \tag{9}
$$

We define the augmented signal as

$$\ddot{x}_r = \alpha\ddot{x}_d(1-\beta) + K_p\dot{e}_1(1-\beta) + K_ie_2 \tag{10}$$

Then, (9) can be rewritten as

$$J\dot{r} = -K_vu + JK_v\ddot{x}_r + K_vD\dot{x} + K_vF \tag{11}$$

Now, we give the control input u as

$$u = K_fr + T_c \tag{12}$$

where T_c will be given later. Substituting (12) into (11), we have

$$J\dot{r} = -K_vK_fr - K_vT_c + JK_v\ddot{x}_r + K_vD\dot{x} + K_vF \tag{13}$$

The control input u must be determined that the closed-loop system becomes stable. Analysing the closed loop stability, we give the following positive definite function as

$$V = \frac{1}{2}Jr^2 \tag{14}$$

Taking the time derivative of both sides for (14) and substituting (13), we have

$$\dot{V} = Jr\dot{r} = -K_vK_fr^2 + K_vr\left(J\ddot{x}_r + D\dot{x} + F - T_c\right) \tag{15}$$

To achieve a negative $\dot{V}$, the following inequality must be satisfied

$$K_vr\left(J\ddot{x}_r + D\dot{x} + F - T_c\right) \le 0 \tag{16}$$

Then, if we design T_c as follows

$$T_c = \mathrm{sgn}(r)(J_{\max}|\ddot{x}_r| + D_{\max}|\dot{x}| + F_{\max}) \tag{17}$$

where J_{max}, D_{max} and F_{max} are maximum values which are predetermined and known, then inequality (16) is satisfied. This sgn function of r is established by a sliding mode control theory. Therefore, if the control input (12) and (17) is applied, then the closed loop system is stable in meaning of Lyapunov stability theory. However, chattering phenomena may occur, because (17) contains the sgn function of r. To avoid the chattering phenomena, we introduce an approximated function of the sign function as follows

$$u = r + \frac{r}{\delta + |r|}(J_{\max}|\ddot{x}_r| + D_{\max}|\dot{x}| + F_{\max}) \tag{18}$$

where δ is the chattering avoidance parameter. Consequently, if we select sufficiently large values of J_{max}, D_{max} and F_{max} in (18), then the time derivative of (14) is always negative and the control objective is accomplished. The P,PI/I-P+FF control method with nonlinear compensator was derived.

3. Experimental results

In this section, we evaluated positioning responses and tracking responses by three kinds of single-axis slide system experimentally. The first experimental system is two slider tables that consists of an AC servo motor, a coupling and a ball-screw, and the second one is a slide table using an AC linear motor and the third one is a slide table using synchronous piezoelectric device driver.

3.1 A table drive system using AC servo motor with a coupling and a ball-screw

The first experiment system is two slider tables comprised of an AC servo motor, a coupling and a ball-screw.

3.1.1 Experimental system

Fig. 3 shows the experimental setup which consists of the following parts. The control system was implemented using a Pentium IV PC with a D/A converter board and a counter board. The control input was calculated by the controller, and its value was translated into a voltage input for the current amplifier through the D/A board. The positions of the positioning table were measured by a position sensor with a resolution of 50 nm. The sensor's signal was provided as a full-closed feedback signal. The sampling period was 0.25 ms. The table with 5 kg weight was mounted on a driving rail. The total inertia of the moving part of the positioning table was approximately 1.128e-4 kgm^2. The table was supported by a rolling guide through the coupling that was connected with the motor, and the table was driven by an AC servo motor (SGMAS-02ACA21,Yaskawa Electric. Co., Ltd), a ball-screw lead of 20 mm (KR4620A+540L, THK Co., Ltd). The control parameters were set to K_p=75/s, K_v=377 rad/s, K_i=250 rad/s, α=0.55 or 0.60, β=1 and δ=5. The value of J_{max}, D_{max} and F_{max} were selected as five times of J, D, F of the slide table (there is not a weight) which measured beforehand, respectively.

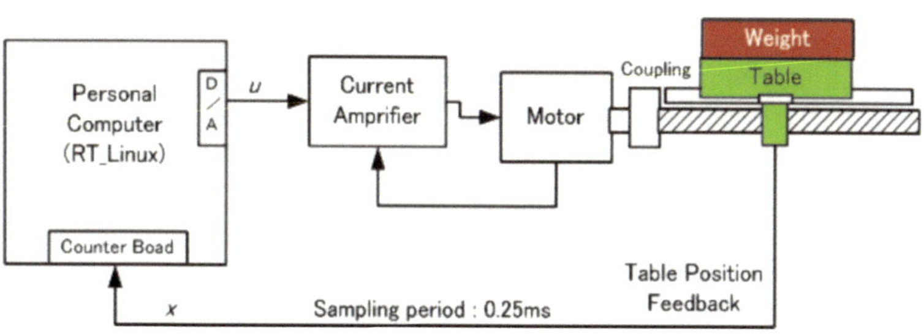

Fig. 3. Experimental system of single axis slider.

Next, the results of positioning responses and tracking responses are shown.

3.1.2 Experimental results

To evaluate our proposed method, we carried out three kinds of experiments. In the first type experiment, it is evaluated that the case of positioning responses when the acceleration/deceleration changes. In the second type of experiment, it is evaluated by the case of positioning responses when the load changes. In the third type of experiment, it is evaluated that the case of tracking responses when the load changes. Fig. 4 shows the experimental table positioning results with the 5 kg-weight in which the positioning reference acceleration/deceleration was changed. The acceleration/deceleration of the first (left side) positioning reference is ±1.0 G, the second is ±1.5 G, the third is ±2.0 G, and the fourth (right side) is ±3.0 G. In this figure, signal ① was the position reference x_d (right side vertical axis), signal ② was the position error (left side vertical axis) using the conventional control method in which the feed-forward gain was set to 0.55, signal ③ was the position error using the conventional control method in which feed-forward gain was set to 0.60, and signal ④ was the position error using the proposed control method in which feed-forward gain was set to 0.55. Fig. 5a shows an expanded graph at 1.0 G and Fig. 5b shows at 3.0 G. In the case of an acceleration/deceleration of 1.0 G, all responses showed approximately the same positioning control performance. However, in the case of an acceleration/deceleration 3.0 G, it is clearly found that there was undesired motion in the form of windup and overshoot using the conventional control methods. On the other hand, there was no windup or overshoot using the proposed control method.

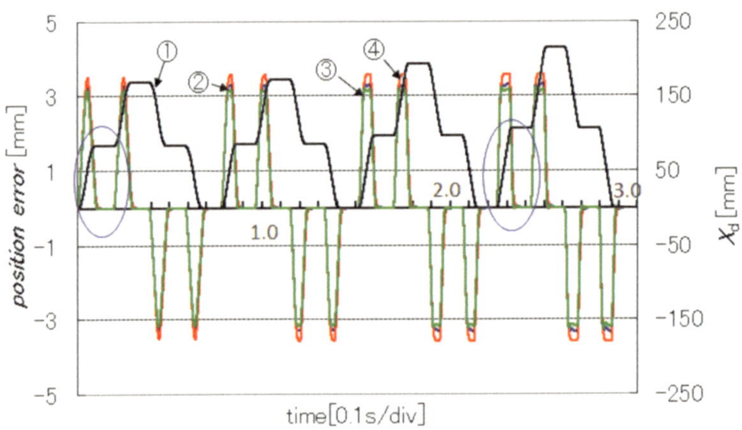

Fig. 4. Position reference and table error.

These results demonstrated the effectiveness of the proposed control method. Generally, it may be said that the acceleration 3.0 G in this experiment is very large because acceleration is used in less than 2.0 G at the ball screw drive table. Fig. 6 shows a torque reference with the conventional control method, and Fig. 7 shows a torque reference (signal ①) and compensated torque T_c (signal ②) with the proposed control method. The rate-torque of the motor is 0.637 Nm, and in both figures, the maximum torque is about 150 % of the rate-torque and is the same value in the conventional method and the proposed method. We found that if a nonlinear compensated torque T_c was very smoothly made, a chattering phenomenon would probably not occur, and we could get a smooth response without any vibration. Fig. 8 shows the response when δ changes in equation (18). In this figure, signal ① is the position reference with an acceleration/deceleration of 3.0 G, ②, ③, and ④ are the table position errors using the proposed control method, that is, ② with δ=500, ③ with δ=50, and ④ with δ=5. This figure shows the effect of the positioning response when δ changed. These experimental results suggested that we could get the disturbance attenuation performance using nonlinear compensator by changing δ.

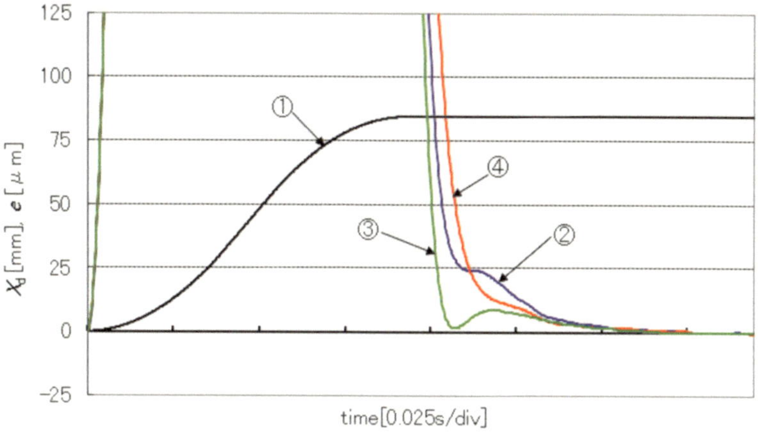

Fig. 5a. Experimental results of PTP control (acceleration=1.0 G).

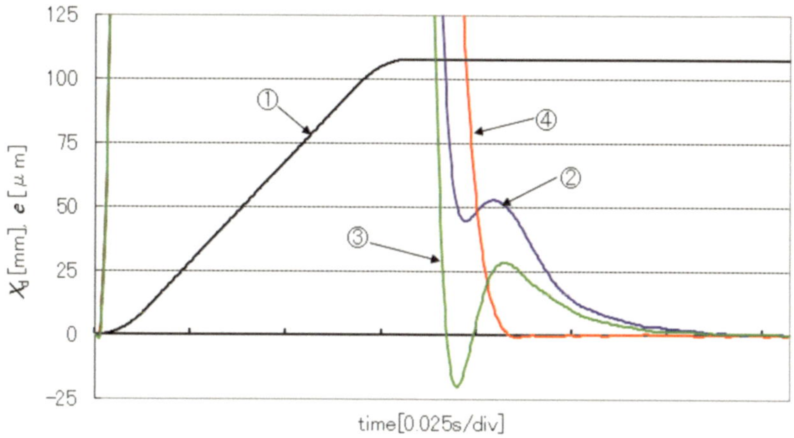

Fig. 5b. Experimental results of PTP control (acceleration=3.0 G).

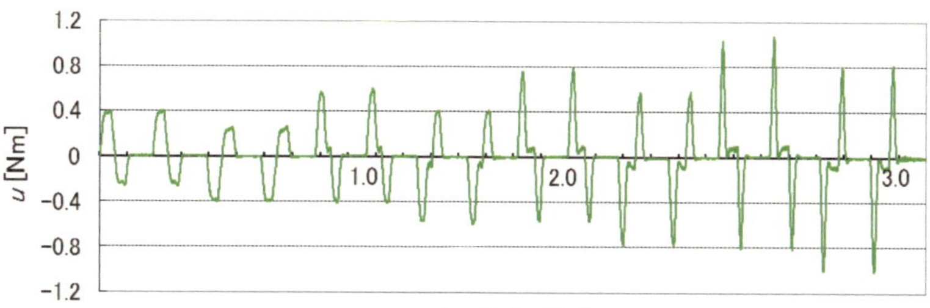

Fig. 6. Torque reference and compensation torque with the conventional control method.

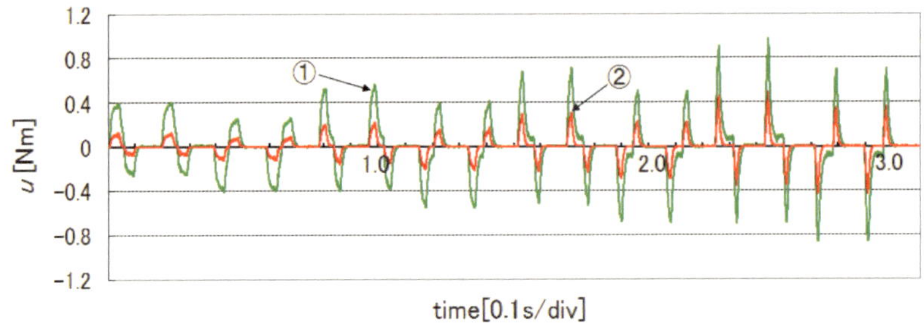

Fig. 7. Torque reference and compensation torque with the proposed control method.

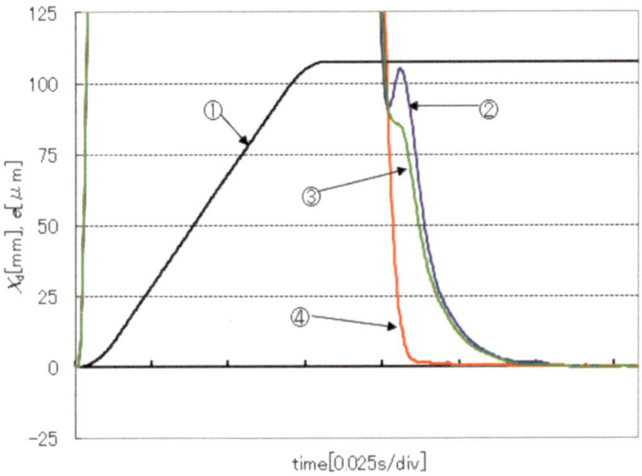

Fig. 8. The experimental results of PTP control (δ changed).

In the second type of experiment was a positioning response with load changing. Here, we describe the experimental result that verified the robustness of the proposed control method in the case of real-time load inertia change. To change the load inertia in real time, we prepared two sets of positioning tables, each consisting of a single-axis slider, a coupling, a motor, a servo amplifier, and a linear scale, as shown in Fig. 9. The D/A channel of the torque reference (the voltage) to output through the D/A board from the PC and the counter channel of the table position signal (the pulse) which is entered from the counter board were made to be able to be changed at the same time by the software. Therefore, the weight added or removed, can be imitated, and it is possible to perform the experiment based on the actual mobile status of the production machine. In this experiment, the trapezoid velocity accelerates from zero velocity to 0.4 m/s in 13.5 ms, moving to a max velocity of 0.4 m/s at the constant in 26.75 ms, decelerates to zero velocity in 13.5 ms in Fig. 10 and Fig. 11. The maximum velocity is 0.4m/s by this experiment, but, by the use of the high lead ball screw and the improvement of the frequency response of the counter, can put up the maximum velocity. The present position reference x_d used in the experiment is the value of this trapezoid velocity pattern integrated among at the time, and x_d is the same as the position reference in Fig. 4. The positioning response using the conventional method is shown in Fig. 10, and the positioning response using the proposed method is shown in Fig.

11. In these figures, d_x_d (left side vertical axis) is the position reference differential value which is the trapezoid velocity pattern, d_x (left side vertical axis) is the table velocity, e (right side vertical axis) is the table error of position, and u (right side vertical axis) is the torque reference. The dimension of u % means the ratio for the rating torque. In addition, at 0-200 ms, it is the response with the loop of channel_1 (weight=0 kg), and after 200 ms, it is the response with the loop of channel_2 (weight=5 kg). Incidentally, α=0.60 of the control parameter was the velocity feed-forward gain with the set value shown in section 3.1.1.

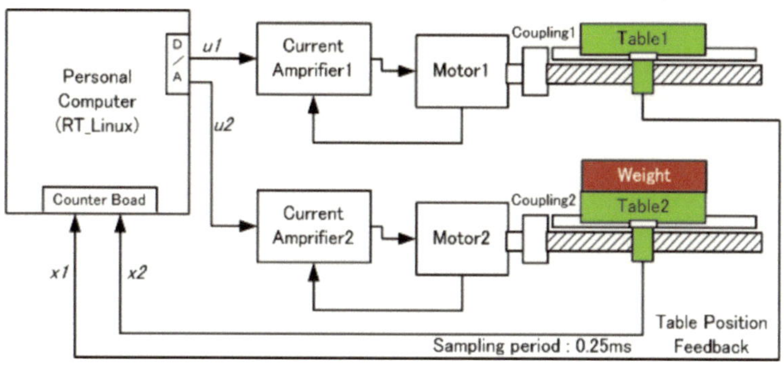

Fig. 9. Experimental system of the load changed.

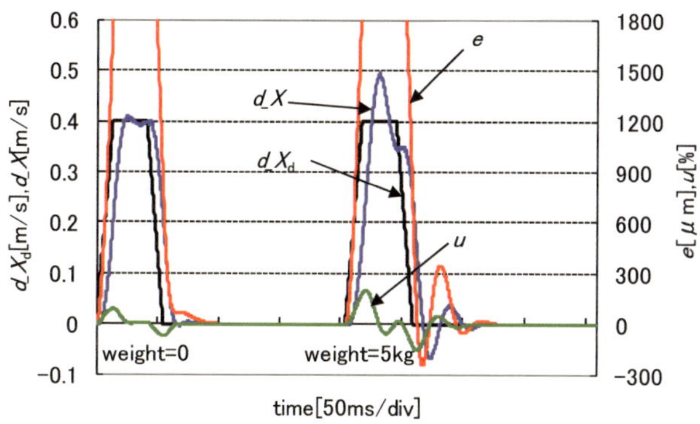

Fig. 10. Experimental results of PTP control using the conventional control method.

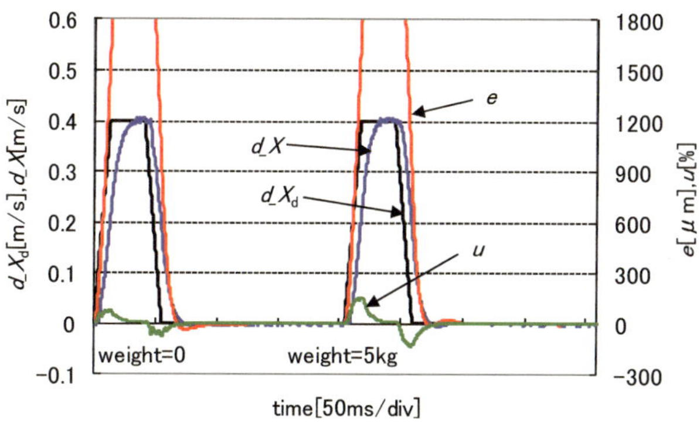

Fig. 11. Experimental results of PTP control using the proposed control method.

In the result using the conventional method shown in Fig. 10, a windup and a big overshoot occurred in the positioning. This is similar to the unstable phenomenon that occurs in the response of the velocity loop to the position loop when the stability is affected by the velocity loop-gain is becoming small. On the other hand, in the result using the proposed method shown in Fig. 11, there is no windup or overshoot when the weight is increased. Moreover, the torque reference is smoothly made and no vibration occurs. Therefore, as shown in both figures, the high-speed positioning responses following load changes were confirmed when the proposed control method was used.

In the third type of experiment we evaluated the tracking control characteristic when the trapezoid velocity was constant at 13 mm/s or 6.5 mm/s using the single-axis rolling guide slider, as described in section 3.1.1 for a 2-cycle period. There is no weight on the table at 1st period (0-3.6s), and there is 5 kg weight on the table at 2nd period (3.6-7.2s). The result with the 1st period when driving with the conventional control method is shown in Fig. 12 (left side), and the result with the 2nd period is shown in Fig. 12 (right side). Also, the result with the 1st period when driving with the proposed method is shown in Fig. 13 (left side), and the result with the 2nd period is shown in Fig. 13 (right side). In these figures, d_x_d is the position reference differential value, which is the trapezoid velocity pattern, d_x is the table velocity, and e is the table error of position. The control parameters were set to the same values as listed in section 3.1.1, and the velocity feed-forward gain was changed to α=1.0 to improve the tracking control from the set value when evaluating positioning response. In all cases of Figs. 12, 13, the maximum error occurred when the operation was influenced by the initial maximum static friction force, and a large error occurred when the velocity reversal was equivalent to the stroke end of the table.

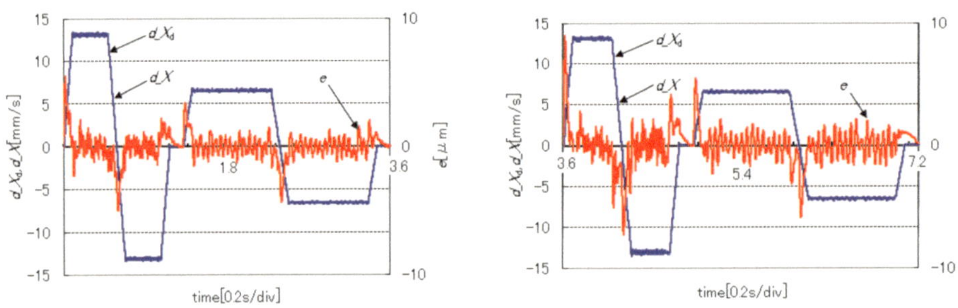

Fig. 12. The experimental results of tracking control using conventional control method. (left side: without weight, right side: with 5kg weight)

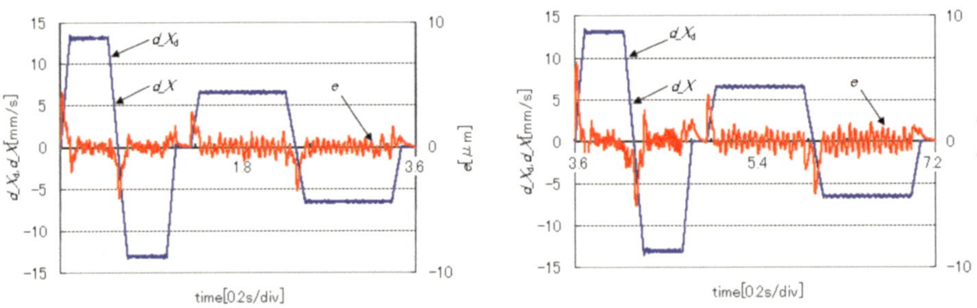

Fig. 13. The experimental results of tracking control using the proposed control method. (left side: without weight, right side: with 5kg weight)

Also, there was an error having to do with a ripple under constant velocity. The comparison results of tracking errors are shown in Table 1. It is obvious that the proposed method remains robust under controllability with or without the weight in the case of low-changing load conditions.

	tracking error[um]		
	start time	constant velocity	stroke end
conventional control without weight	+5.4	Max: +1.4 Min: −1.1	−4.8
conventional control with weight	+8.7	Max: +2.2 Min: −2.1	−7.0
proposed control without weight	+4.2	Max: +1.0 Min: −1.0	−4.0
proposed control with weight	+6.3	Max: +1.2 Min: −1.3	−4.9

Table 1. The comparison results of tracking errors.

3.2 A table drive system using AC linear motor

Next, we evaluated a tracking response in the low speed using a table drive system driven a linear motor, and the resolution of this system is 10 nm. After having investigated friction characteristics of this system because it was easy to receive a bad influence of the friction at the low-velocity movement, we inspected the effect of the proposed method.

3.2.1 Experimental system

Fig. 14 shows the photograph of single axis slider and the experimental system shown in Fig. 15. It consists of the following: (i) a one-axis stage mechanism consisting of an AC linear coreless motor which has no cogging force, (ii) a rolling guide mechanism, (iii) a position-sensor (1pulse=10nm), (iv) two current amplifiers, and (v) a personal computer with the controller, a D/A board and a counter board. In a practical application, high precision positioning at a low velocity is required, but in general, it is well known that the conventional control methods can not accomplish such a requirement. Moreover, the tracking error becomes large at the end of a stroke because of the effect of a friction force.

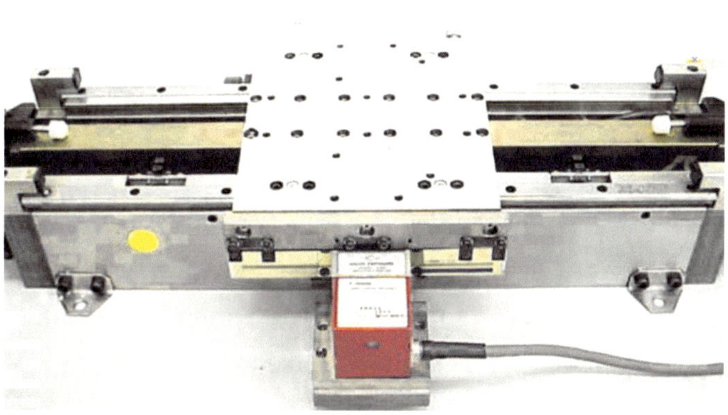

Fig. 14. The photograph of single axis slider.

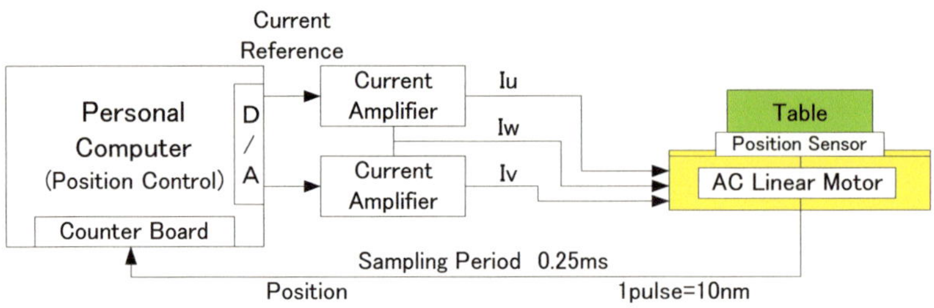

Fig. 15. The experimental control system.

In the previous researches, a friction force can be regarded as a static function of velocity in spite of its complicated phenomenon. Therefore, the servo characteristics of this experimental system were investigated. Experiments have shown that there is a deflection or relative movement in the pre-sliding region, indicating that the relationship between the deflection and the input force resembles a non-linear spring with a hysteretic behavior. In this experiment, general PID control is used. Thus, the present study focuses on the nonlinear behavior at the end of a stroke during changes in velocity as shown in Fig. 16 (left side). In this figure, the signals of ①, ④ and ⑦ are velocity references, the signals of ②, ⑤ and ⑧ are velocity responses, the signals of ③, ⑥ and ⑨ are output forces with constant acceleration-deceleration profiles of 10 mm/s, 5.0 mm/s, and 2.5 mm/s, respectively. The forces in the actual experiment are calculated values and not the values actually measured. It seems that the tracking error of velocity are almost zero. From this figure, it is seen that the output forces are different during constant velocity and the force of 2.5 mm/s is the largest in all cases. The moving force generally needs a big one where velocity is large. The reason is influence of viscous friction. When the velocities are decreasing, output forces have not decreased and when the velocities are increasing, output forces have not increased.

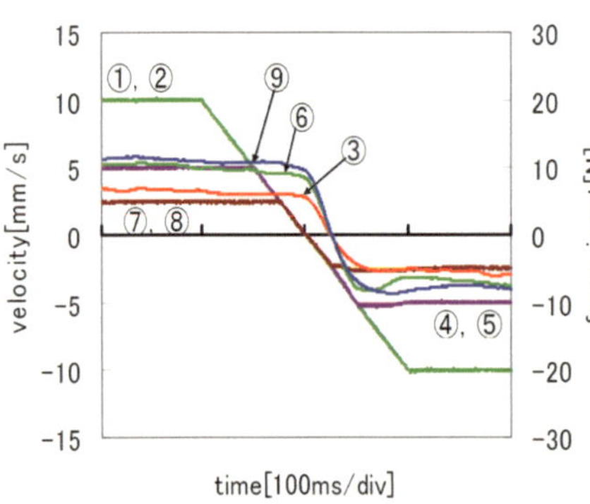

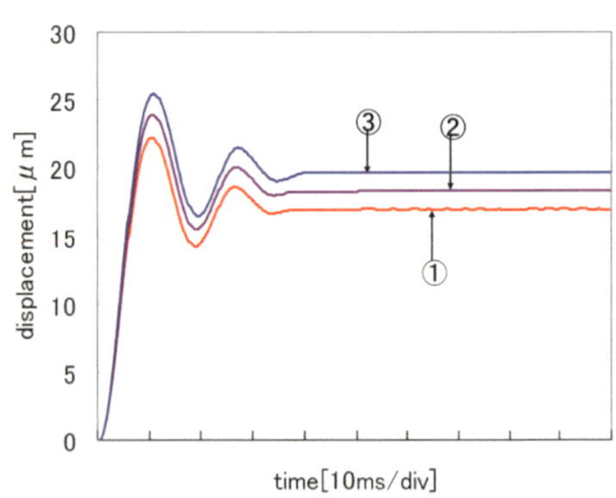

Fig. 16. The nonlinear behavior.
(left side: table motion at the end of a stroke, right side: spring-like behavior)

Further, when the output forces are set to zero, the spring- like behavior occurs at the end of a stroke, as shown in Fig.16 (right side). In this figure, the signals of ①, ② and ③ are the displacement, the command velocities which are 10 mm/s, 5.0 mm/s, and 2.5 mm/s, respectively. At values of low command velocities, the spring-like behaviors produce large displacement. The displacement, which exceeds 15μm can negatively influence precision point to point control. The frequency of vibration was observed to be 40 Hz. The spring-like characteristic behavior is thought to be due to the elastic deformation between balls and rails in the ball guide-way. Thus, friction is a natural phenomenon that is quite hard to model description by on-line identification, and is not yet completely understood. Particularly, it is known to have a bad influence in a tracking response at the low-velocity movement. Next, in this table drive system with such a nonlinear characteristic, we evaluate the effectiveness of the proposed compensation method.

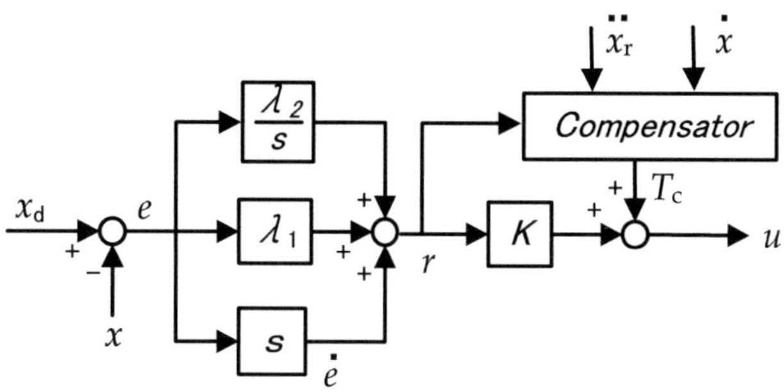

Fig. 17. The block diagram of the proposed method.

Fig. 17 shows a block diagram of the proposed control method, which consists of a PID controller (λ_1, λ_2, k), the proposed nonlinear compensator, T_c is disturbance compensation force. The control input u is given as follows

$$e = x_d - x \tag{19}$$

$$r = \dot{e} + \lambda_1 e + \lambda_2 \frac{e}{s} \tag{20}$$

$$\dot{x}_r = \dot{x}_d + \lambda_1 e + \lambda_2 \frac{e}{s} \tag{21}$$

$$u = kr + \frac{r}{\delta + |r|}(M_{\max}|\ddot{x}_r| + D_{\max}|\dot{x}| + F_{\max}) \tag{22}$$

The PID controller is tuned using the normal procedure, where a signal x_d is input reference, a signal x is displacement, a signal e is tracking error and s means Laplace transfer operator.

3.2.2 Experimental results

To show the effectiveness of the proposed method, experiments were carried out. Digital implementation was assumed in experimental setup. The sampling time of experiments was 0.25 ms. Parameters of PID controller was chosen as λ_1=125[1/s], λ_2=5208[1/s], k=62.5[1/s]. These parameters are adjusted from the ideal values which is determined by the triple multiple roots condition. Here, the force conversion fixed constant is included in K. The parameters of proposed method was chosen as same value of PID controller and was chosen as δ=0.5. The value of M_{max}, D_{max} and F_{max} were set as five times of M, D, F of the slide table which measured beforehand, respectively. To evaluate the tracking errors at the end of stroke, we used three kind of moving velocities. Figs. 18, 19, 20 show the comparison results of tracking errors in the case of state velocity are 10 mm/s, 5 mm/s, 2.5 mm/s, respectively. In these figures, ① is the velocity reference, ② is the velocity response without compensation, ③ is the same one with compensation, ④ is the tracking error without compensation, ⑤ is the same one with compensation, ⑥ is the force output without compensation, ⑦ is the same one with compensation, respectively.

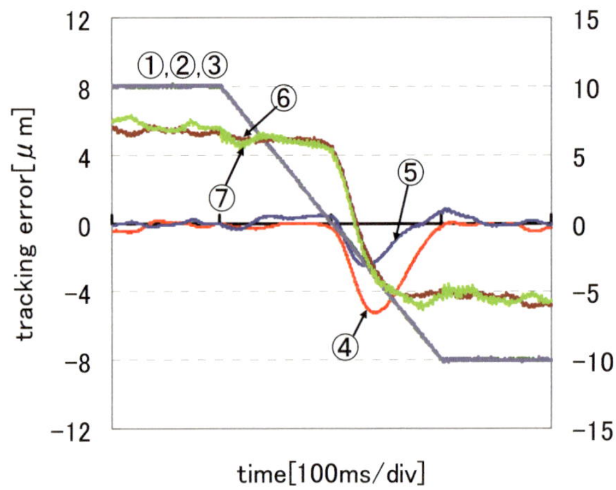

Fig. 18. The comparison results of tracking errors in the case of state velocity are 10mm/s.

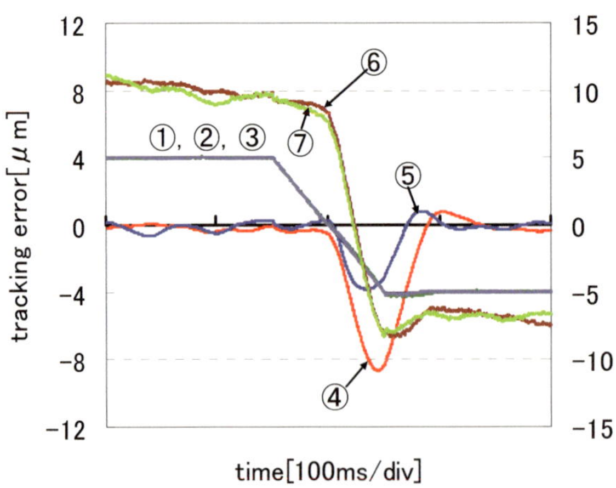

Fig. 19. The comparison results of tracking errors in the case of state velocity are 5mm/s.

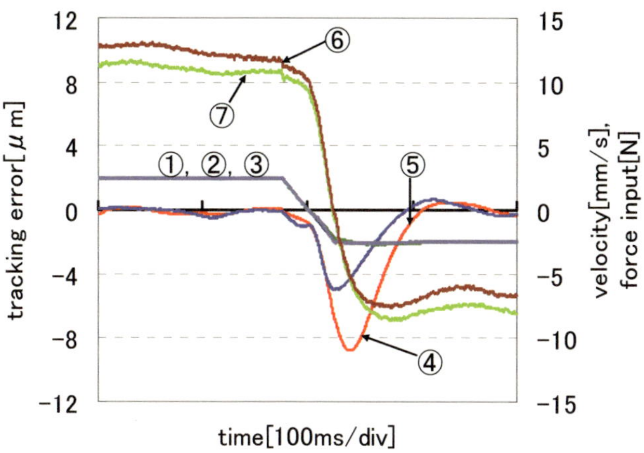

Fig. 20. The comparison results of tracking errors in the case of state velocity are 2.5mm/s.

Moving velocity	10mm/s	5mm/s	2.5mm/s
Without compensation	−5.2 μm	−8.6 μm	−8.8 μm
With compensation	−2.8 μm	−3.8 μm	−4.9 μm

Table 2. The comparison results of tracking errors.

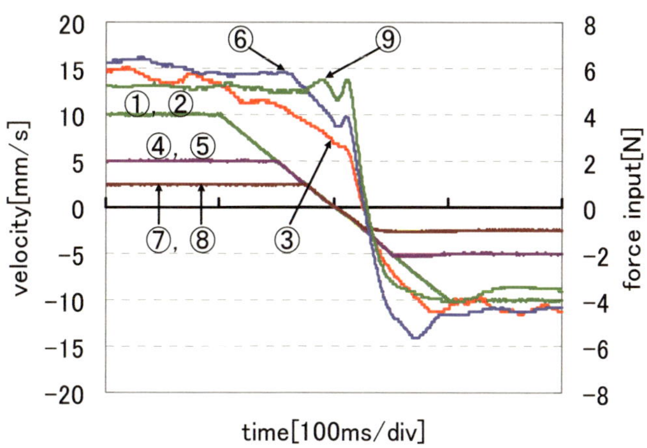

Fig. 21. The compensate force inputs T_c among three cases of constant velocity.

It is obvious that the tracking errors of the case with compensation are reduced by more than 2/3 compared to the case of without compensation at the end of a stroke. Table 2 shows the tracking errors at the end of stroke. The errors are greatly reduced by our proposed compensation method. Thus, the proposed method is judged to have better performance accuracy. Fig. 21 shows the compensate force inputs T_c among three cases of constant velocity. In Fig. 21, the signals of ①, ④ and ⑦ are velocity references, the signals of ②, ⑤ and ⑧ are velocity responses, the signals of ③, ⑥ and ⑨ are compensate forces of 10 mm/s, 5.0 mm/s, and 2.5 mm/s, respectively. It is clear that the compensation forces are similar to the nonlinear behaviors of Stribeck effect at the end of a stroke.

3.3 A table drive system using synchronous piezoelectric device driver

For the future applications of an electron beam (EB) apparatus for the semiconductor industry, a non-resonant ultrasonic motor is the most attractive device for a stage system instead of an electromagnetic motor, because the power source of the stage system is required for non-magnetic and vacuum applications. Next, we evaluated a stepping motion and tracking motion using a synchronous piezoelectric device driver.

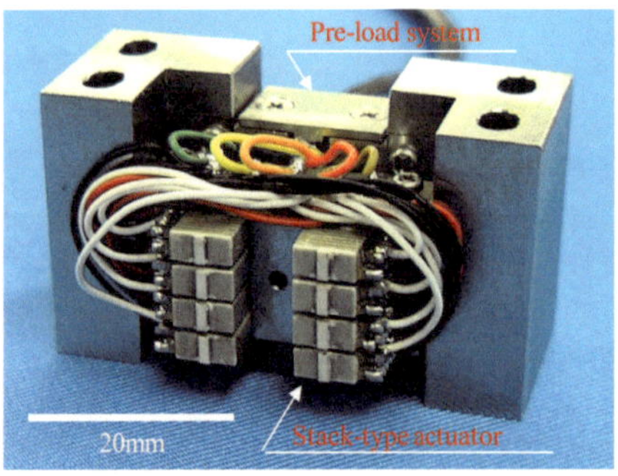

material	$Pb(Zr,Ti)O_3$
density	7.8×10^3kg/m^3
dimention	6.0mm × 3.0mm × 0.6mm
expand	660×10^{-12}m/V
shear	1010×10^{-12}m/V
layer	4(shear) × 4(expand)

Fig. 22. The photograph and specifications of SPIDER.

3.3.1 Experimental system

Fig. 22 shows a photograph of SPIDER (Synchronous Piezoelectric Device Driver) and its specifications. Fig. 23 shows the experimental setup which consists of the following parts. The control system was implemented using a Pentium IV PC with a DIO board and a counter board. The control input was calculated by the controller, and its value was translated into an appropriate input for the SPIDER through the DIO board , parallel-serial transfer unit, and drive unit. The position of the positioning table was measured by a position sensor with a resolution of 100 nm. The sensor's signal was provided as a feedback signal. The sampling period was 0.5 ms. The table was mounted on a driving rail. The weight of the moving part of the positioning table was approximately 1.2 kg. The friction tip was in contact with the side of the table. The longitudinal feed of the table was 100 mm. The positioning precision of this system depends on the resolution of the position sensor, and the best precision is less than 1 nm. The parallel-serial transfer unit translated the parallel data into serial data. The drive unit was a voltage generator for the piezoelectric actuator of SPIDER. Fig. 24 shows the motion of the SPIDER. The SPIDER has eight stacks and each stack consists of an extensible and shared piezoelectric element. The behavior of each stack is similar to that of a leg in ambulatory animals or human beings. Despite the limitation in the strokes of stack, the table can move endlessly. The motion sequence of the stacks is as follows (The sequence starts from the top of the left side. In this case, the table's direction of motion was to the right)

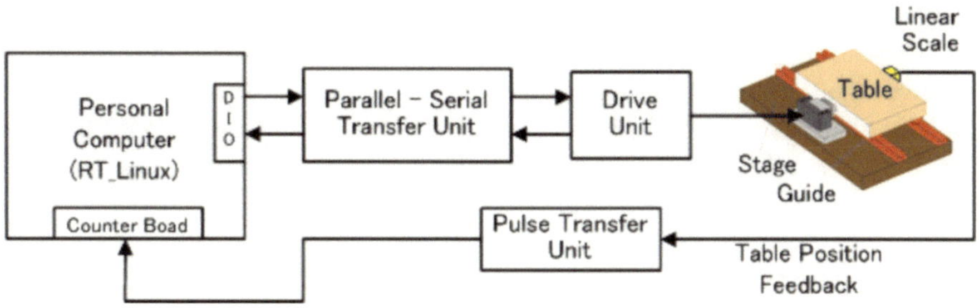

Fig. 23. The experimental system for SPIDER.

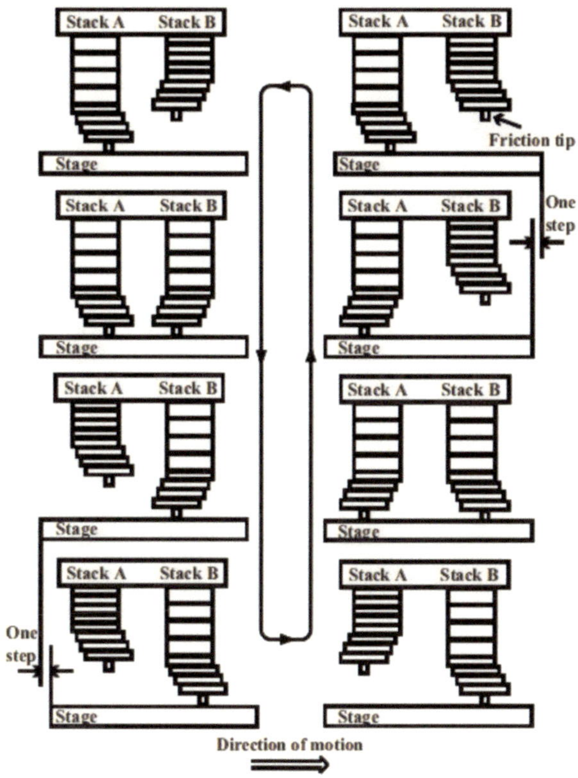

Fig. 24. Operating sequence of SPIDER.

1. Deform stack B to the counter direction of the motion of the table;
2. Expand stack B to contact the stage;
3. Retract stack A;
4. Deform stack B to the forward direction of the motion of the table; then the stage move to the forward direction at one step;
5. Deform stack A to the counter direction of the motion of the table;
6. Expand stack A to contact the stage;
7. Retract stack B;
8. Deform stack A to the forward direction of the motion of the table; then the stage move in the forward direction in one step.

Continuous displacement of the positioning table can be given by repeating this sequence periodically. In addition, it is possible to be fast at speed of the motion of positioning table by increasing the amplitude of frequency and/or the voltage of this period. As we can see the positioning table is driven by the scratching and friction force via the SPIDER system. However, it is known that friction causes stick slip behavior; therefore, the frictional force is a major problem in precision positioning systems.

Next, we observe the nonlinear characteristic of the SPIDER system. To investigate the nonlinear characteristic of the SPIDER system, the open-loop control responses are measured. Fig. 25 shows the control input u; Fig. 26 shows the experimental open-loop responses of the positioning table displacement. These responses are measured five times. As we can see in Fig. 26, despite increasing the control input, the positioning table did not move during 0.2 seconds. Then we can regard that the SPIDER system exhibits time-delay phenomena. Fig. 27 shows the control input versus the displacement of the positioning table. As we can see in Fig.27, despite the control input being monotonically increasing / decreasing between a negative and positive value with equal magnitude, the displacement of positioning table shows strong hysteresis characteristics. Therefore, it seems that this SPIDER system can be regarded as a nonlinear system and it is very difficult to control the displacement of positioning table using only a linear control strategy. It is known that the deteriorating influence of friction is a major problem in many precision positioning systems. To remove as much of influence of friction as possible is very important.

Next, we applied our proposed control method with the nonlinear compensator for this system and inspected the effect of the method. First, the results of stepping motion which is one pulse motion are described. Secondly, the results of tracking motion of which amplitude of position references are 1 mm and 10 mm are described.

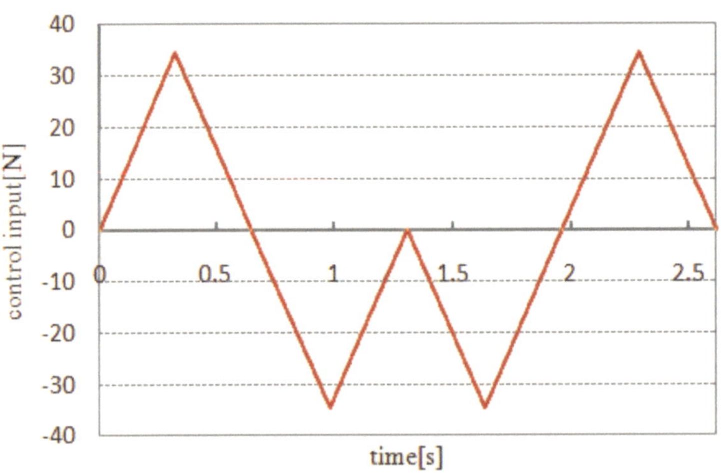

Fig. 25. The control input (0N→32N→0N→-32N→0N→-32N→0N→32N→0N).

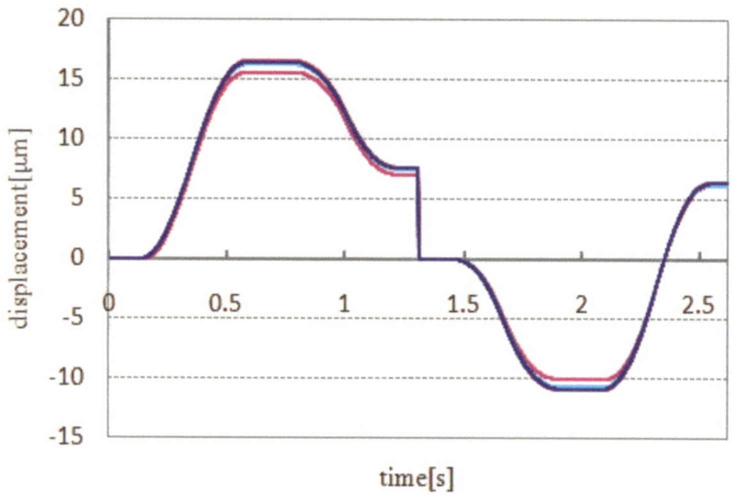

Fig. 26. The open-loop responses of SPIDER.

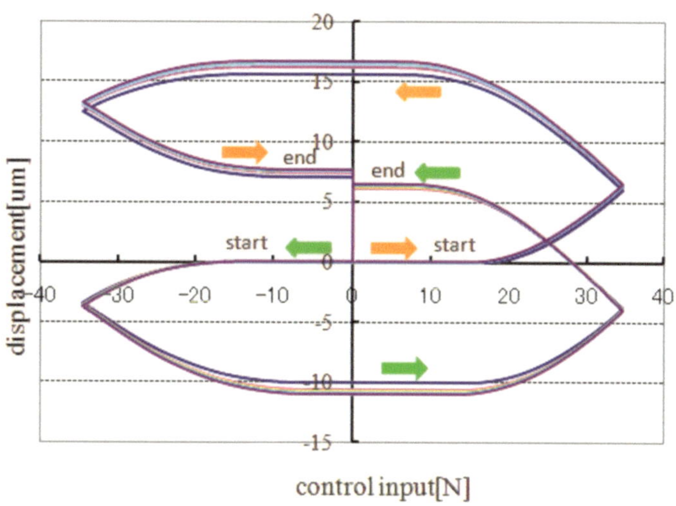

Fig. 27. The hysteresis characteristics of SPIDER.

3.3.2 Experimental results

To show the effectiveness of the proposed method, experiments were carried out. Fig. 28 shows a block diagram of the P,PI/I-P+FF control method with proposed compensator. Digital implementation was assumed in experimental setup. The sampling time of experiments was 0.5 ms. Parameters of P,PI/I-P+FF controller was chosen as in Table 3. The value of M_{max}, D_{max} and F_{max} were set as five times of M, D, F of the slide table which measured beforehand, respectively. The control input u is calculated as follows

$$e_1 = x_d - x \tag{23}$$

$$e_2 = K_p e_1 - \dot{x} + \alpha \dot{x}_d \tag{24}$$

$$r = K_v \left\{ \left(1 + \frac{K_i}{s} \right) e_2 - \beta \left(K_p e_1 + \alpha \dot{x}_d \right) \right\} \tag{25}$$

$$\ddot{x}_r = \alpha \ddot{x}_d (1 - \beta) + K_p \dot{e}_1 (1 - \beta) + K_i e_2 \tag{26}$$

$$T_c = \frac{r}{\delta + |r|} (M_{\max} |\ddot{x}_r| + D_{\max} |\dot{x}| + F_{\max}) \tag{27}$$

$$u = K_f r + T_c \tag{28}$$

To evaluate the proposed control method, the two position references were applied. First, the experimental results of stepping motion using conventional P,PI/I-P+FF control and proposed control are shown in Fig. 29. In the figure, ① is the position reference, ② is the table displacement and ③ is the positioning error. It is obvious that the positioning errors are greatly reduced using proposed control method compared to the conventional control

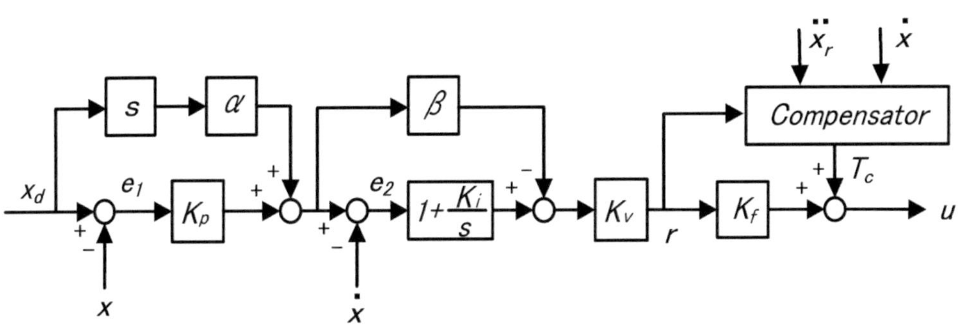

Fig. 28. Block diagram of the proposed control method.

Position loop gain K_p	30[1/s]
Velocity loop gain K_v	188[1/s]
Velocity integral gain K_i	200[1/s]
Velocity feed-forward gain α	1.0
PI/I-P change constant β	0.0
Force constant K_f	946.2[1/N]
Maximum Mass constant M_{max}	6.0[kg]
Maximum viscous coefficient D_{max}	50[Ns/m]
Maximum disturbance constant F_{max}	48[N]
Chattering reject constant δ	1000

Table 3. Setting parameters of SPIDER.

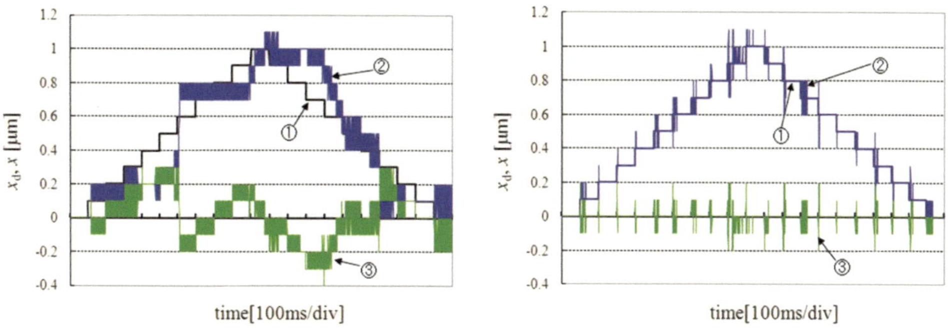

Fig. 29. The experimental results of stepping motion.
(left side: conventional control, right side: proposed control)

method. Note, because the resolution of the position sensor is 100 nm by this experiment, the positioning resolution becomes 100 nm, but can raise positioning resolution to 0.6 nm if we improve the resolution of the position sensor. Secondly, Figs. 30, 31 showed the results of tracking motion of which amplitude of position references was 1 mm or 10 mm. In Figs. 30, 31, ① is the position reference, ② is the stage displacement and ③ is the tracking error, respectively. In these figures, the tracking errors of stroke end of the stage were expanded. Compared to each response, the results of proposed control were slightly improved to the results of conventional control. In order to investigate these results, the force input of the conventional control and proposed control were shown in Fig. 32. Both force inputs had a lot of vibrations to compensate undesired friction. In conventional control, the tracking error of stroke end was changed slowly in Fig. 32 (left side). On the other hand, in proposed control, the tracking error of stroke end was tracked to zero roughly in Fig. 32 (right side). Compared to these force inputs, the force input of proposed control was quickly and smoothly changed at the marked point. Therefore, we considered the tracking error of the proposed control method was improved.

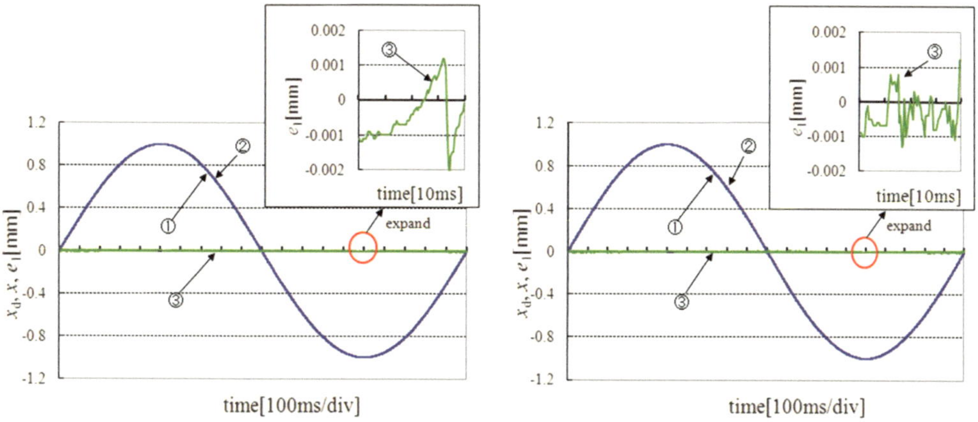

Fig. 30. The experimental results of tracking motion with 1mm moving.
(left side: conventional control, right side: proposed control)

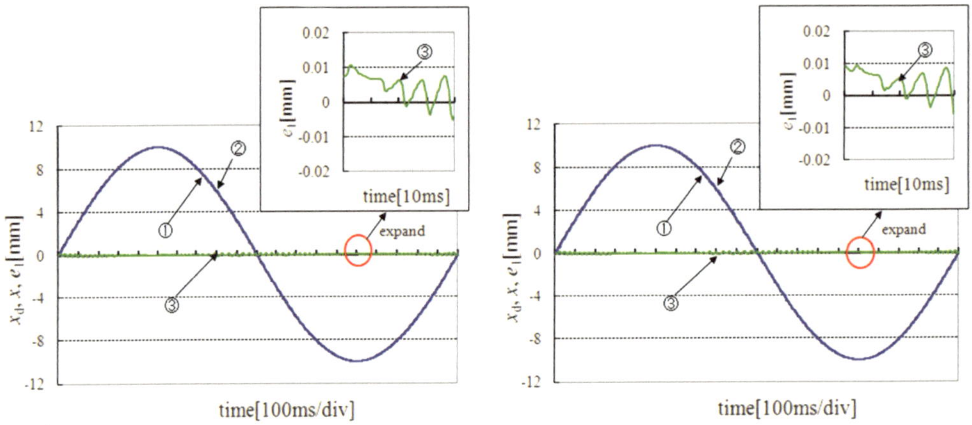

Fig. 31. The experimental results of tracking motion with 10mm moving.
(left side: conventional control, right side: proposed control)

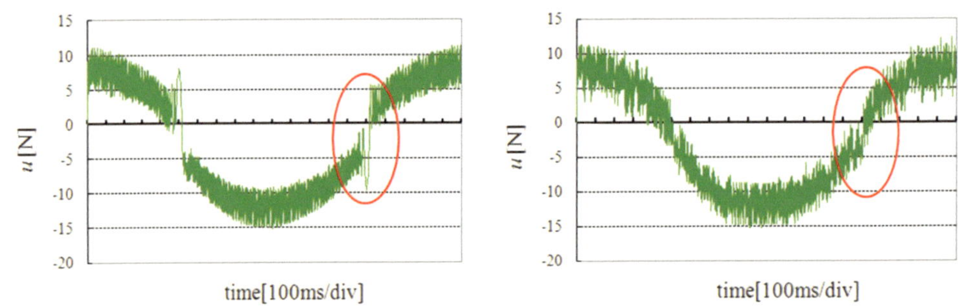

Fig. 32. The force input of tracking motion with 1mm moving.
(left side: conventional control, right side: proposed control)

4. Conclusion

In this chapter, we propose a new PID control method that includes a nonlinear compensator that it is easy to understand for a PID control designer. The algorithm of the nonlinear compensator is based on sliding mode control with chattering compensation. The effect of the proposed control method is evaluated with three kinds of single-axis slide system experimentally. The first experiment system is two slider tables comprised of an AC servo motor, a coupling and a ball-screw, and the second one is a slide table using an AC linear motor and the third one is a slide table using synchronous piezoelectric device driver. By the first experiments, it is evaluated using single-axis slide system comprised of full closed feedback via point-to-point control response and tracking control response when load characteristics of the control target change. The experimental results indicate that the proposed control method has robustness in a high-speed, high-precision positioning response and a low speed tracking response when acceleration/deceleration of position reference change or the load characteristics of the control target change. By the second experiments, it is evaluated using a linear motor driven slider system via tracking control at low-velocity, and the resolution of this system is 10nm. The tracking error of our proposed control method is reduced by more than 2/3 compared to the case of the conventional PID control method at the end of a stroke. By the third experiments, it is evaluated a stepping motion and trajectory tracking motion using a synchronous piezoelectric device driver. The positioning error of our proposed control is reduced by more than 2/3 compared to the case of the conventional PID control at a stepping motion.

Future studies will address the robustness and the control parameters tuning of the proposed compensation method. Furthermore, we want to evaluate the proposed method at an industrial robot used in a car assembly line and a twin linear drive table used in a semiconductor production device.

5. Acknowledgement

We greatly appreciate to Dr. Kosaka's help on the experiments of a synchronous piezoelectric device driver. This research was partially supported by Japan Grant-in-Aid for Scientific Research (C) (19560244).

6. References

Canudas-de-Wit, C. Olsson, H. Astrom-K, J. & Lischinsky, P. (1995). A New Model for Control of System with Friction, *IEEE transaction on automatic control*, Vol.40, No.3, pp.419-425

Egashira, Y. Kosaka, K. Iwabuchi, T. Kosaka, T. Baba, T. Endo, T. Hashiguchi, H. Harada, T. Nagamoto, K. Watanabe, M. Yamakawa, T. Miyata, N. Moriyama, S. Morizono, Y. Nakada, A. Kubota, H. & Ohmi, T. (2002). Sub-Nanometer Resolution Ultrasonic Motor for 300mm Wafer Lithography Precision Stage, *Japanese Journal of Applied Physics*, vol.41, no.9, pp.5858-5863

Fujimoto, T. Tsuruta, K. & Uchida, Y. (2006). Motion Control of a Ball-Screw Slide System Via Multi-segment Sliding Mode Control Method, *Proceedings of the 8th MOVIC*, pp.842-847

Futami, S. Furutani, A. & Yoshida, S. (1990). Nanometer positioning and its microdynamics, *Nanotechnology*, 1, pp.31-37

Harashima, F. & Hashimoto,H. (1986). Sliding Mode Control, *Computorol*, No.13, pp.72-78.

Iwasaki, M. Shibata, T. & Matsui, N. (2000). GMDH-Based Autonomous Modeling and Compensation for Nonlinear Frication in Table Drive System, *IEEJ transaction*, vol.120-C, no.1, 2000, pp.20-26

Kojima, T. (2004). Robot Control, *published by Corona co.jp*

Kosaka, K. Iwabuchi, T. Baba, T. Endo, T. Hashiguchi, Furukawa, H. Egashira, Y. Hashimoto, S. Touge, M, Uozumi, K. Nakada, A. Kubota, H. & Ohmi, T. (2006)."Wear Reduction Method for Frictionally Fast Feeding Piezoactuator, *Japanese Journal of Applied Physics*, vol.45, no.2A, pp.1005-1011

Kuwon, T. Sul, S. Nakamura, H. & Tsuruta, K. (2006). Identification of Mechanical Parameters for Servo Drive, *Proceedings of the IEEE/IAS 41st Annual meeting*, IAS22p4, Florida, USA

Lischinsky, P. Canudas-de-Wit, C. Morel, & G. (1999). Friction compensation for an Industrial Hydraulic Robot, *IEEE Control Systems*, Februaly, pp.25-32

Nonami, K. & Den, H. (1994). Sliding Mode Control, *published by Corona co.jp*

Otsuka, J. & Masuda, T. (1998). The influence of nonlinear spring behavior of rolling elements on ultra-precision positioning control systems, *Nanotechnology*, 9, pp.85-92

Sato, K. Tsuruta, K. & Ishibe, T. (2007). A Design of Adaptive Control for Systems with Input Nonlinearities and Its Experimental Evaluations, *T.SICE*, Vol.43, No.3, pp.221-226.

Sato, K. Tsuruta, K. & Shoji, A. (2005). Adaptive Friction Compensation Control of Linear Slider with Considering a Periodic Reference Signal, *IEEJ Transactions on Industry Applications*, Vol.125, No.11, pp.1022-1029

Sato, K. Tsuruta, K. & Mukai, H. (2007). A robust adaptive control for robotic manipulator with input torque uncertainty, *Proceedings of the SICE Annual Conference 2007*, Sept. 17-20, pp1293-1298

Sato, K. Ishibe, T. Tsuruta, K. (2007). A design of adaptive H^{∞} control for positioning mechanism system with input nonlinearities, *Proceedings of the 16th IEEE International Conference on Control Applications Part of IEEE Multi-conference on Systems and Control*, pp.152-157

Sato, K. Nishijima, D. Tsuruta, K. (2006). Adaptive $H\infty$ control method with frictions compensation and disturbance rejection for robotic manipulators, *Proceedings of the* IEEE International Conference on Control Applications, pp.1031-1036

Sato, K. Mukai, H. & Tsuruta, K. (2008). An adaptive H^{∞} control for robotic manipulator with compensation of input torque uncertainty, *Preprints of the 17th IFAC World Congress*, pp. 8919-8924

Suzuki, T. (2001). Adaptive Control, *published by Corona.co.jp*

Tsuruta, K. Murakami, T. Futami, S. & Sumimoto, T. (2000). Genetic Algorithm (GA) Based Modeling of Non-linear Behavior of Friction of a Ball Guide Way, *Proceedings of the 6th International Workshop on Advanced Motion Control*, pp.181-186

Tsuruta, K. Murakami, T. & Futami, S. (2000). Nonlinear Frictional Behavior at Low Speed in a Ball Guide Way, *Proceedings of the International Tribology Conference Nagasaki 2000*, pp1847-1852

Tsuruta, K. Murakami, T. & Futami, S. (2003). Nonlinear Friction Behavior of Discontinuity at Stroke End in a Ball Guide Way, *Journal of the Japan Society for Precision Engineering*,Vol.69,No.12, pp.1759-1763

Tsuruta, K. Sato, K. Ushimi, N. & Fujimoto, T. (2007). High Precision Positioning Control for Table Drive System using PID Controller with Nonlinear Friction Compensator, *Proceedings of the 4th International Conference on Leading Edge Manufacturing in 21st Century Fukuoka*, pp.173-176

Tsuruta, K. Sato, Sawada, S. Kosaka, K. & Schilling, K. (2007). A Comparison of Nonlinear Friction Compensations for a High Precision Stage using Synchronous Piezoelectric Device Driver, *Asian Symposium for Precision Engineering and Nanotechnology 2009*, 1E-10-2191

Utkin, V, I. (1977). Variable Structure Systems with Sliding Mode, *IEEE Transaction on Automatic Control*, Vol.AC-22, pp.212-222

9

PID Tuning: Robust and Intelligent Multi-Objective Approaches

Hassan Bevrani[1] and Hossein Bevrani[2]
[1]*University of Kurdistan,*
[2]*University of Tabriz,*
Iran

1. Introduction

The proportional-integral-derivative (PID) control structures have been widely used in industrial applications due to their design/structure simplicity and inexpensive cost. The success of the PID controllers depends on the appropriate choice of their parameters. In practice, tuning the PID parameters/gains is usually realized by classical, trial-and-error approaches, and experienced human experts, which they may not capable to achieve a desirable performance for complex real-world systems with high-order, time-delays, nonlinearities, uncertainties, and without precise mathematical models.

On the other hand, the most of real-world control problems refer to multi-objective control designs that several objectives such as stability, disturbance attenuation and reference tracking with considering practical constraints must be simultaneously followed by a controller. In such cases, using a single norm based performance criteria to evaluate the robustness of resulted PID-based control systems is difficult and multi-objective tuning solutions are needed.

This chapter introduces three effective robust and intelligent multi-objective methodologies for tuning of PID controllers to improve the performance of the closed-loop systems in comparison of conventional PID tuning approaches. The introduced tuning strategies are based on mixed H_2/H_∞, multi-objective genetic algorithm (GA), fuzzy logic, and particle swarm optimization (PSO) techniques. Indeed, these robust and intelligent techniques are employed as optimization engines to produce the PID parameters in the control loops with performance indices near to the optimal ones.

Numerical examples on automatic generation control (AGC) design in multi-area power systems are given to illustrate the mentioned methodologies. It has been found that the controlled systems with proposed PID controllers have better capabilities of handling the large scale and complex dynamical systems.

2. Mixed H_2/H_∞-based PID tuning

Mixed H_2/H_∞ provides a powerful control design to meet different specified control objectives. However, it is usually complicated and not easily implemented for the real industrial applications. Recently, some efforts are reported to make a connection between the theoretical mixed H_2/H_∞ optimal control and simple classical PID control (Takahashi et

al., 1997; Chen et al., 1998; Bevrani & Hiyama, 2007). (Takahashi et al., 1997) has used a combination of different optimization criteria through a multiobjective technique to tune the PI parameters. A genetic algorithm (GA) approach to mixed H_2/H_∞ optimal PID control is given in (Chen et al., 1998). (Bevrani & Hiyama, 2007) has addressed a new method to bridge the gap between the power of optimal mixed H_2/H_∞ multiobjective control and PI/PID industrial controls. In this work, the PI/PID control problem is reduced to a static output feedback control synthesis through the mixed H_2/H_∞ control technique, and then the control parameters are easily carried out using an iterative linear matrix inequalities (ILMI) algorithm.
In this section, based on the idea given in (Bevrani & Hiyama, 2007), the interesting combination of different objectives including H_2 and H_∞ tracking performances for a PID controller has been addressed by a systematical, simple and fast algorithm. The multiobjective PID control problem is formulated as a mixed H_2/H_∞ static output feedback (SOF) control problem to obtain a desired PID controller. The developed ILMI algorithm in (Bevrani & Hiyama, 2007) is used to tune the PID control parameters to achieve mixed H_2/H_∞ optimal performance.

2.1 PID as a SOF control

Consider a general system (G(s)) with u and y_o variables as input and output signals. Assume that it is desirable to stabilize the system using a PID controller. Here, it will be shown that the PID control synthesis can be easily transferable to a SOF control problem. The main merit of this transformation is in possibility of using the well-known SOF control techniques to calculate the fixed gains, and once the SOF gain vector is obtained, the PID gains are ready in hand and no additional computation is needed.
In a given PID-based control system, the measured output signal (y_o) performs the input signal for the controller which can be written as follows

$$u = k_P y_o + k_I \int y_o d\tau + k_D \frac{dy_o}{dt} \tag{1}$$

where k_P, k_I and k_D are constant real numbers. Therefore, by generalizing the system description to include the y_o, its integral and derivative as a new measured output vector (y), the PID control problem becomes one of finding a SOF that satisfied the prescribed performance requirements. In order to change (1) to a simple SOF control as

$$u = Ky \tag{2}$$

Equation (1) can be written as follows

$$u = [k_P \quad k_I \quad k_D]\begin{bmatrix} y_o & \int y_o d\tau & \frac{dy_o}{dt} \end{bmatrix}^T \tag{3}$$

Therefore, y in (2) can be generalized to the following form (Fig. 1).

$$y = \begin{bmatrix} y_o & \int y_o d\tau & \frac{dy_o}{dt} \end{bmatrix}^T \tag{4}$$

Since, the ideal differentiator used in (1), (3) and (4) is unrealizable, a real differentiator should be applied in practice. Although most of PID controllers in use have the derivative part switched off, proper use of the derivative action can improve the stability and help maximize the integral gain for a better performance. For real implementation, ideal differintiator ($k_D s$) can be approximated as ($k_D s/(\lambda k_D s+1)$), where λ is a small number. The effect of real and approximated differentiator on the closed-loop dynamics are discussed in PID control literature.

2.2 ILMI-based H_2/H_∞ SOF design

A general control scheme using mixed H_2/H_∞ control technique is shown in Fig. 2. G(s) is a linear time invariant system with the given state-space realization in (5). The matrix coeificients are constants and it is assumed the system to be stabilizable via a SOF system. Here, x is the state variable vector, w is disturbance and other external input vector, y is the augmented measured output vector and K is the controller. The output channel z_2 is associated with the LQG aspects (H_2 performance) while the output channel z_∞ is associated with the H_∞ performance.

$$\begin{aligned}
\dot{x} &= Ax + B_1 w + B_2 u \\
z_\infty &= C_\infty x + D_{\infty 1} w_i + D_{\infty 2} u \\
z_2 &= C_2 x + D_{21} w + D_{22} u \\
y &= C_y x + D_{y1} w
\end{aligned} \tag{5}$$

Assume $T_{z_\infty w}$ and $T_{z_2 w}$ are the transfer functions from w to z_∞ and w to z_2, respectively; and consider the following state-space realization for the closed-loop system. After defining the appropriate H_∞ and H_2 control outputs (z_∞ and z_2) for the system, it will be easy to determine matrix coefficients (C_∞, $D_{\infty 1}$, $D_{\infty 2}$) and (C_2, D_{21}, D_{22}).

$$\begin{aligned}
\dot{x} &= A_c x_i + B_{1_c} w \\
z_\infty &= C_{\infty c} x + D_{\infty_c} w \\
z_2 &= C_{2c} x + D_{2_c} w \\
y &= C_{y_c} x + D_{y_c} w
\end{aligned} \tag{6}$$

A mixed H_2/H_∞ SOF control design can be expressed as following optimization problem:
Optimization problem: Determine an admissible SOF law K, belong to a family of internally stabilizing SOF gains K_{sof},

$$u = Ky, \; K \in K_{sof} \tag{7}$$

such that

$$\inf_{K \in K_{sof}} \left\| T_{z_2 w_2} \right\|_2 \text{ subject to } \left\| T_{z_\infty w_1} \right\|_\infty < 1 \tag{8}$$

The following lemma gives the necessary and sufficient condition for the existence of the H_2 based SOF controller to meet the following performance criteria.

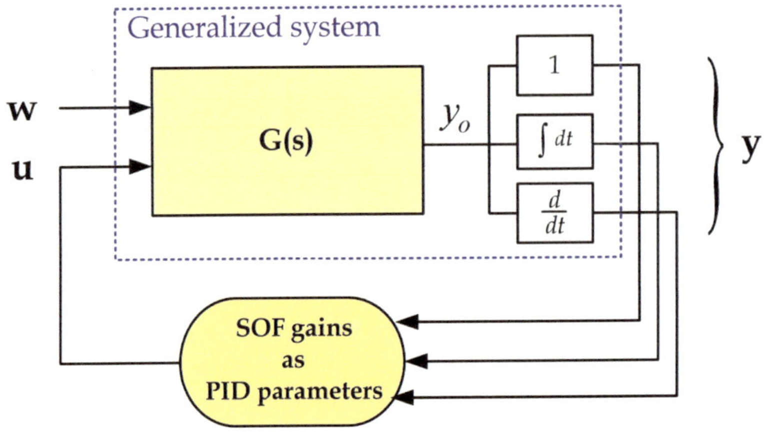

Fig. 1. PID as SOF control.

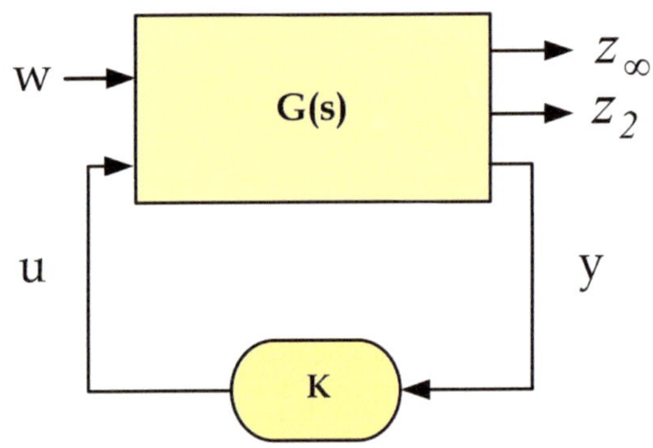

Fig. 2. Closed-loop system via mixed H_2/H_∞ control.

$$\left\|T_{z_2 w_2}\right\|_2 < \gamma_2 \tag{9}$$

where, γ_2 is the H_2 optimal performance index, which demonstrates the minimum upper bound of H_2 norm and specifies the disturbance attention level.
The H_2 and H_∞ norms of a transfer function matrix $T(s)$ with m lines and n columns, for a MIMO system are defined as:

$$\|T(s)\|_2 = \sqrt{\sum_{j=1}^{n}\sum_{i=1}^{m}\|T_{ij}(s)\|_2^{\ 2}} \tag{10}$$

$$\|T(s)\|_\infty = \underset{w}{Sup}\max\sigma[T(jw)] \tag{11}$$

where, σ is represents the singular values of $T(jw)$.
Lemma 1, (Zheng et al., 2002):
For fixed (A, B_1, B_2, C_y, K), there exists a positive definite matrix X which solves inequality

$$(A+B_2KC_y)X+X(A+B_2KC_y)^T+B_1B_1^T<0$$
$$X>L_C \tag{12}$$

to satisfy (9), if and only if the following inequality has a positive definite matrix solution,

$$AX+XA^T-XC_y^TC_yX$$
$$+(B_2K+XC_y^T)(B_2K+XC_y^T)^T+B_1B_1^T<0 \tag{13}$$

where L_C in (12) denotes the controllability gramian matrix of the pair (A_c, B_{1c}) and can be related to the H_2 norm presented in (9) as follows.

$$\left\|T_{z_2 w_2}\right\|_2^2=trace(C_{2c}L_CC_{2c}^T) \tag{14}$$

It is notable that the condition that $A+B_2KC_y$ is Hurwitz is implied by inequality (12). Thus if

$$trace(C_{2c}XC_{2c}^T)<\gamma_2^2 \tag{15}$$

the requirement (9) is satisfied.

Lemma 2, (Cao et al., 1998)

The system (A, B, C) is stabilizable via static output feedback if and only if there exists P>0, X>0 and K satisfying the following quadratic matrix inequality

$$\begin{bmatrix} A^TX+XA-PBB^TX-XBB^TP+PBB^TP & (B^TX+KC)^T \\ B^TX+KC & -I \end{bmatrix}<0 \tag{16}$$

In the proposed control strategy, to design the PI/PID multiobjective controller, the obtained SOF control problem to be considered as a mixed H_2/H_∞ SOF control problem. Then to solve the yielding nonconvex optimization problem, which cannot be directly achieved by using LMI techniques, an ILMI algorithm is developed.

The optimization problem given in (8) defines a robust performance synthesis problem where the H_2 norm is chosen as a performance measure. Recently, several LMI-based methods are proposed to obtain the suboptimal solution for the H_2, H_∞ and/or H_2/H_∞ SOF control problems. It is noteworthy that using lemma 1, it is difficult to achieve a solution for (13) by the general LMI, directly. Here, to get a simultaneous solution to meet (9) and H_∞ constraint, and to get a desired solution for the above optimization problem, an ILMI algorithm is introduced which is well-discussed in (Bevrani & Hiyama, 2007). The developed algorithm formulates the H_2/H_∞ SOF control through a general SOF stabilization. In the proposed strategy, based on the generalized static output stabilization feedback lemma (lemma 2), first the stability domain of gain vector (PID parameters) space, which guarantees the stability of the closed-loop system, is specified. In the second step, the subset of the stability domain in the PID parameter space in step one is specified so that minimizes the H_2 performance indix. Finally and in the third step, the design problem is reduced to find a point in the previous subset domain, with the closest H2 performance index to the optimal one which meets the H_∞ constraint. In summary, the proposed algorithm searches a desired mixed H_2/H_∞ SOF controller $K \in K_{sof}$ within a family of H_2 stabilizing controllers K_{sof}, such that

$$\left|\gamma_2^* - \gamma_2\right| < \varepsilon \,,\; \gamma_\infty = \left\|T_{z_\infty w_I}\right\|_\infty < 1 \tag{17}$$

where ε is a small real positive number, γ_2^* is H_2 performance corresponded to the H_2/H_∞ SOF controller K_i and γ_2 is the reference optimal H_2 performance index provided by application of standard H_2/H_∞ dynamic output feedback control. The key point is to formulate the H_2/H_∞ problem via the generalized static output stabilization feedback lemma such that all eigenvalues of $(A+BKC)$ shift towards the left half-plane through the reduction of a, a real negative number, to close to feasibility of (8). Infact, the a shows the pole region for the closed-loop system. The developed ILMI algorithm is summarized in Fig. 3 (Bevrani & Hiyama, 2007; Bevrani, 2009). The application of above methodology in automatic generation control for a multi-area power system is given in section 4.

3. Multi-objective GA-based PID tuning

3.1 Intelligent methodologies

The intelligent technology offers many benefits in the area of complex and nonlinear control problems, particularly when the system is operating over an uncertain operating range. Generally for the sake of control synthesis, nonlinear systems are approximated by reduced order dynamic models, possibly linear, that represent the simplified dominant systems' characteristics. However, these models are only valid within specific operating ranges, and a different model may be required in the case of changing operating conditions. On the other hand, classical and nonflexible PID designs may not represent desirable performance over a wide range of operating conditions. Therefore, more flexible and intelligent PID synthesis approaches are needed.

In recent years, following the advent of modern intelligent methods, such as artificial neural networks (ANNs), fuzzy logic, multi-agent systems, GAs, expert systems, simulated annealing, Tabu search, particle swarm optimization, Ant colony optimization, and hybrid intelligent techniques, some new potentials and powerful solutions for PID tuing have arisen.

In control configuration point of view, the most proposed intelligent based PID tuning mechanisms are used for tuning the parameters of existing fixed structure PID controller as conceptually shown in Fig. 4. In Fig. 4, it is assumed that the system is controllable and can be stabilized via a PID controller. Here, the applied intelligent technique performs an automatic tuner. The initial values for the parameters of the fixed-structure controller (k_P , k_I and k_D gains in PID) must first be defined. The trial-error and the widely used Ziegler-Nichols tuning rules are usually employed to set initial gain values according to the open-loop step response of the plant. The intelligent technique collects information about the system response and recommends adjustments to be made to the PID gains. This is an iterative procedure until the fastest possible critical damping for the controlled system is achieved. The main components of the intelligent tuner include a response recognition unit to monitor the controlled response and extract knowledge about the performance of the current PID gain setting, and an embedded unit to suggest suitable changes to be made to the PID gains.

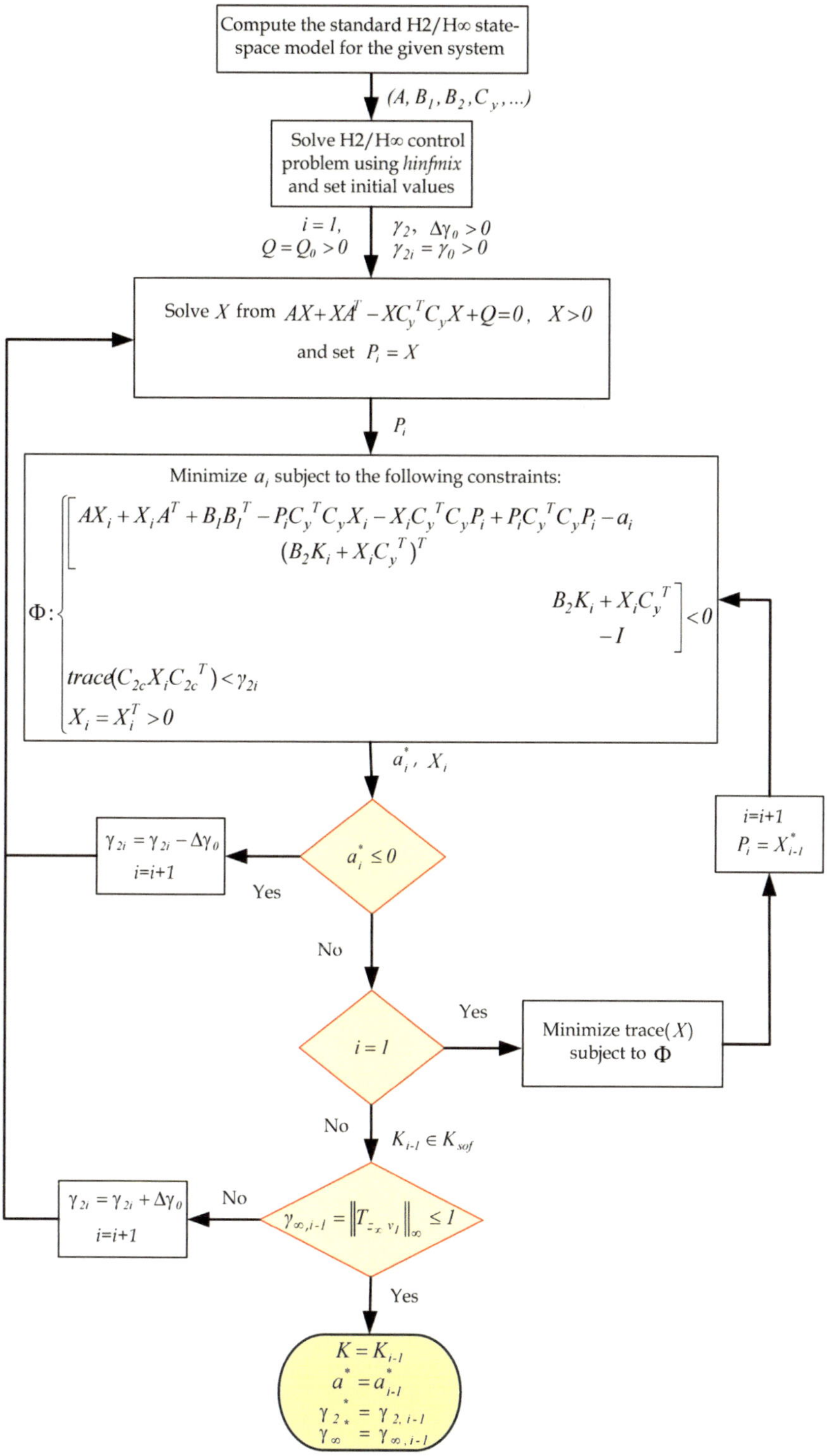

Fig. 3. Developed ILMI algorithm.

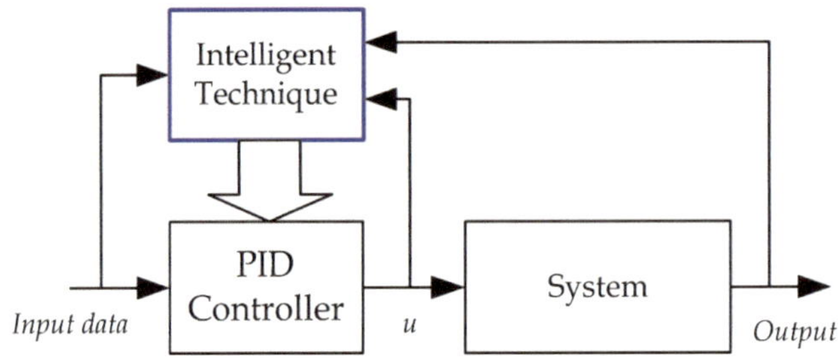

Fig. 4. Common configurations for intelligent-based PID designs.

3.2 Genetic algorithm

Genetic algorithm (GA) is a searching algorithm which uses the mechanism of natural selection and natural genetics; operates without knowledge of the task domain, and utilizes only the fitness of evaluated individuals. The GA as a general purpose optimization method has been widely used to solve many complex engineering optimization problems, over the years. In Fact, GA as a random search approach which imitates natural process of evolution is appropriate for finding global optimal solution inside a multidimensional searching space. From random initial population, GA starts a loop of evolution processes in order to improve the average fitness function of the whole population. GAs have been used to adjust parameters for different control schemes, e.g. integral, PI, PID, sliding mode control, or variable structure control (Bevrani & Hiyama, 2007). The overall control framework for PID controllers is shown in Fig. 5.

Genetic algorithm (GA) is capable of being applied to a wide range of optimization problems that guarantees the survival of the fittest. Time consumption methods such as trial and error for finding the optimum solution cause to the interest on the meta-heuristic method such as GA. The GA becomes a very useful tool for tuning of parameters in PI/PID based control systems.

GA mechanism is inspired by the mechanism of natural selection where stronger individuals would likely be the winners in a competing environment. Normally in a GA, the parameters to be optimized are represented in a binary string. A simplified flowchart for GA is shown in Fig. 6. The *cost function* which determines the optimization problem represents the main link between the problem at hand (system) and GA, and also provides the fundamental source to provide the mechanism for evaluating of algorithm steps. To start the optimization, GA uses randomly produced initial solutions created by random number generator. This method is preferred when a priori information about the problem is not available. There are basically three genetic operators used to produce a new generation. These operators are *selection, crossover,* and *mutation*. The GA employs these operators to converge at the global optimum. After randomly generating the initial population (as random solutions), the GA uses the genetic operators to achieve a new set of solutions at each iteration. In the selection operation, each solution of the current population is evaluated by its fitness normally represented by the value of some objective function, and individuals with higher fitness value are selected (Bevrani & Hiyama, 2011).

Different selection methods such as stochastic selection or ranking-based selection can be used. In selection procedure the individual chromosome are selected from the population for the later recombination/crossover. The fitness values are normalized by dividing each one by the sum of all fitness values named selection probability. The chromosomes with higher selection probability have a higher chance to be selected for later breeding.

The crossover operator works on pairs of selected solutions with certain crossover rate. The crossover rate is defined as the probability of applying crossover to a pair of selected solutions (chromosomes). There are many ways to define the crossover operator. The most common way is called the *one-point crossover*. In this method, a point (e.g, for given two binary coded solutions of certain bit length) is determined randomly in two strings and corresponding bits are swapped to generate two new solutions.

Mutation is a random alteration with small probability of the binary value of a string position, and will prevent GA from being trapped in a local minimum. The coefficients assigned to the crossover and mutation specify number of the children. Information generated by fitness evaluation unit about the quality of different solutions is used by the selection operation in the GA. The algorithm is repeated until a predefined number of generations has been produced. Unlike the gradient-based optimization methods, GAs operate simultaneously on an entire population of potential solutions (chromosomes or individuals) instead of producing successive iterates of a single element, and the computation of the gradient of the cost functional is not necessary (Bevrani & Hiyama, 2011).

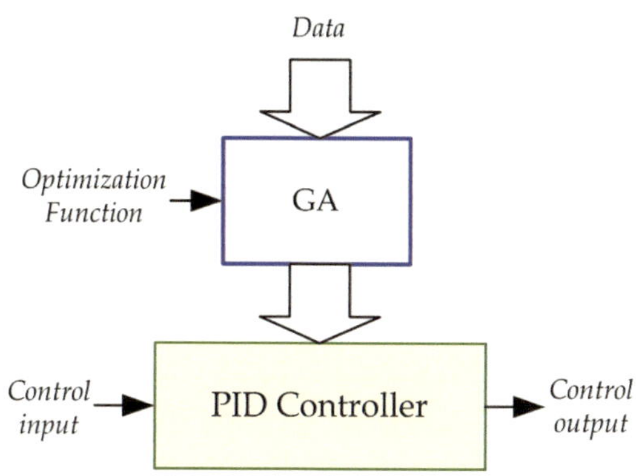

Fig. 5. GA-based PID tuning scheme.

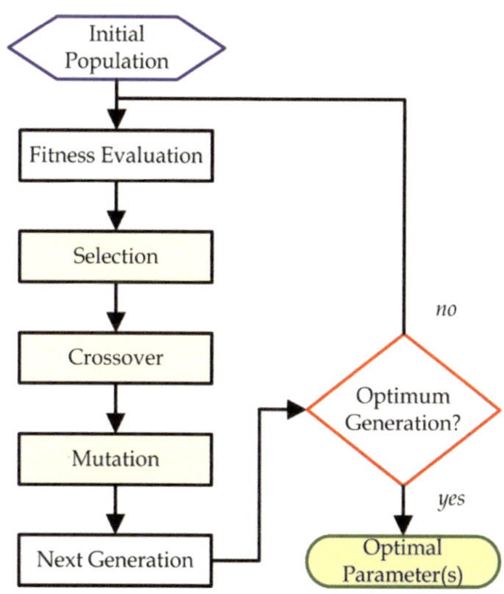

Fig. 6. A simplified GA flowchart.

Several approaches are given for the analysis and proof of the convergence behavior of GAs. The proof of convergence is an important step towards a better theoretical understanding of GAs. Some proposed methodologies are based on building blocks idea and schema theorem (Thierens & Goldberg, 1994; Holland, 1998; Sazuki, 1995).

3.3 Multi-objective GA-based tuning mechanism

The majority of PID control design problems are inherently multi-objective problems, in that there are several conflicting design objectives which need to be simultaneously achieved in the presence of determined constraints. If these synthesis objectives are analytically represented as a set of design objective functions subject to the existing constraints, the synthesis problem could be formulated as a multi-objective optimization problem.

In a multi-objective problem unlike a single optimization problem, the notation of optimality is not so straightforward and obvious. Practically in most cases, the objective functions are in conflict and show different behavior, so the reduction of one objective function leads to the increase in another. Therefore, in a multi-objective optimization problem, there may not exist one solution that is best with respect to all objectives. Usually, the goal is reduced to set compromising all objectives and determine a trade-off surface representing a set of nondominated solution points, known as *Pareto-optimal* solutions. A Pareto-optimal solution has the property that it is not possible to reduce any of the objective functions without increasing at least one of the other objective functions (Bevrani & Hiyama, 2011).

Mathematically, a multi-objective optimization (in form of minimization) problem can be expressed as,

$$\begin{aligned} &Minimize\ y = f(x) = \{f_1(x), f_2(x), \ldots, f_M(x)\} \\ &Subject\ to\ g(x) = \{g_1(x), g_2(x), \ldots, g_l(x)\} \le 0 \end{aligned} \tag{18}$$

where $x = \{x_1, x_2, \ldots, x_N\} \in X$ is the vector of decision variables in the decision space X, $y = \{y_1, y_2, \ldots, y_N\} \in Y$ is the objective vector in the objective space. Practically, since there could be a number of Pareto-optimal solutions and the suitability of one solution may depends on system dynamics, environment, the designer's choice, etc., finding the center point of Pareto-optimal solutions set may be desired.

GA is well suited for solving of multi-optimization problems. In the most common method, the solution is simply achieved by developing a population of Pareto-optimal or near Pareto-optimal solutions which are nondominated. The x^i is said to be nondominated if there does not exist any x^j in the population that dominates x^i. Nondominated individuals are given the greatest fitness, and individuals that are dominated by many other individuals are given a small fitness. Using this mechanism, the population evolves towards a set of nondominated, near Pareto-optimal individuals (Fonseca & Fleming, 1995). The multi-objective GA methodology is conducted to optimize the PID parameters. Here, the control objective is summarized to minimize the error signal in the control system. To achieve this goal and satisfy an optimal performance, the parameters of the PID controller can be selected through minimization of following objective function:

$$ObjFnc = \int_0^L |e(\tau)| d\tau \quad ; \quad e(t) = y(t) - y_r(t) \tag{19}$$

where, *ObjFnc* is the objective function of control system, *L* is equal to the simulation time duration (sec), $y_r(t)$ is the reference signal, and $|e(t)|$ is the absolute value of error signal at

time t. Following using multi-objective GA optimization technique to tune the PID controller and find the optimum value of objective function (18), the fitness function (*FitFunc*) can be also defined as objective control function. Each GA individual is a double vector presenting PID parameters. Since, a PID controller has three gain parameters, the number of GA variables could be $N_{\text{var}} = 3$. The population should be considered in a matrix with size of $m \times N_{\text{var}}$; where the m represents individuals.
The basic line of the algorithm is derived from a GA, where only one replacement occurs per generation. The selection phase should be done, first. Initial solutions are randomly generated using a uniform random number of PID control parameters. The crossover and mutation operators are then applied. The crossover is applied on both selected individuals, generating two childes. The mutation is applied uniformly on the best individual. The best resulting individual is integrated into the population, replacing the worst ranked individual in the population. This process is conceptually shown in Fig. 7.

4. Application to AGC design

To illustrate the effectiveness of the introduced PID tuning strategies decribed in sections 2 and 3, the autumatic genertion control (AGC) synthesis for an interconnected three control areas power system, is considered as an example. AGC in a power system automaticaly minimizes the system frequency deviation and tie-line power fluctuation due to imballance between total generation and load, following a disturbance. AGC has a fundamental role in modern power system control/operation, and is well-disscussed in (Bevrani 2009, Bevrani & Hiyama 2011). The power system configuration, data and parameters are given in (Rerkpreedapong et al., 2003). Each control area is approximated to a 9th order linear system which includes three generating units.

4.1 Mixed H_2/H_∞ approach

According to (5), the state-space model for each control area can be calculated as follows:

$$\begin{aligned} \dot{x}_i &= A_i x_i + B_{1_i} w_i + B_{2_i} u_i \\ z_{\infty i} &= C_{\infty i} x_i + D_{\infty 1_i} w_i + D_{\infty 2_i} u_i \\ z_{2i} &= C_{2i} x_i + D_{21_i} w_i + D_{22_i} u_i \\ y_i &= C_{yi} x_i + D_{y1_i} w_i \end{aligned} \qquad i = 1, 2, 3 \tag{20}$$

y_i is the measured output (performed by area control error-ACE and its derivative and integral), u_i is the control input and w_i includes the perturbed and disturbance signals in the given control area.
The H_2 controlled output signals in each control area includes Δf_i, ACE_i and ΔP_{ci} which are frequency deviation, ACE (measured output) and governor load setpoint, respectively.
The H_2 performance is used to minimize the effects of disturbances on area frequency, ACE and penalize fast changes and large overshoot in the governor load set-point. The H_∞ performance is used to meet the robustness against specified uncertainties and reduction of its impact on the closed-loop system performance (Bevrani, 2009). First, a mixed H_2/H_∞ dynamic controller is designed for each control area, using *hinfmix* function in the LMI control toolbox of MATLAB software. In this case, the resulted controller is dynamic type, whose order is the same as size of generalized plant model. Then, according to the tuning

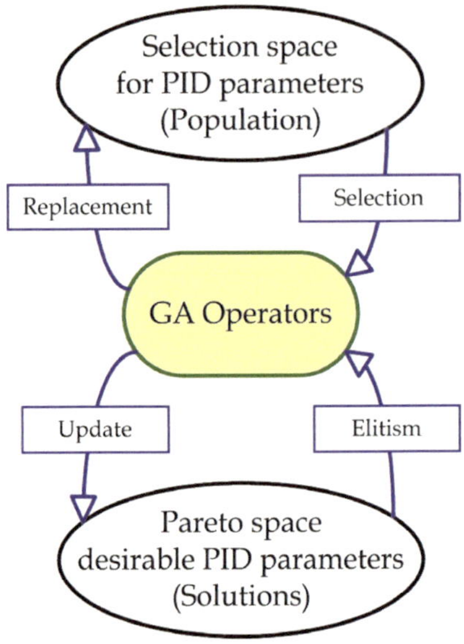

Fig. 7. Multi-objective GA for tuning of PID parameters.

methodology described in section 2, a set of three decentralized robust PID controllers are designed. Using developed ILMI algorithm, the controllers are obtained following several iterations. The proposed control parameters, the guaranteed optimal H_2 and H_∞ indices (γ_{2i} and $\gamma_{\infty i}$) for dynamic/PID controllers, and simulation results are shown in section 4.3.
It is noteworthy that here the design of dynamic controller is not a gole. However, the performance indeces of robust dynamic controller are used as valid (desirable) refrences to apply in the developed ILMI algorithm. It is shown that although the proposed ILMI approach gives a set of much simpler controllers (PID) than the dynamic H_2/H_∞ design, however they holds robustness as well as dynamic H_2/H_∞ controllers.

4.2 GA approach

The multi-objective GA-based tuning goal is summarized to minimize the area control error (ACE) signals in the interconnected control areas. Usally, the ACE signal is a linear combination of frequency deviation and tie-line power change (Bevrani, 2009). To achieve this goal, the objective function in a control is considered as

$$ObjFnc_i = \sum_{t=0}^{L} \left| ACE_{i,t} \right| \tag{21}$$

where, $\left| ACE_{i,t} \right|$ is the absolute value of ACE signal for area i at time t, and the fitness function is defined as follows,

$$ObjFnc(.) = \left[ObjFnc_1, ObjFnc_2, ..., ObjFnc_n \right] \tag{22}$$

Here, the number of GA variables is $N_{\text{var}} = 3n$, where n is the number of control areas.

4.3 Simulation results

The above described tuning approaches are applied to the 3-control area power system example. Fig. 8 shows the closed-loop response (ACE signals) for three areas, in the presence of simultaneous 0.1 pu step load disturbances, and 20% decrease in inertia constant and damping coefficient as uncertainties in all areas. Simulation results demonstrate that the GA-based tuning method is able to track the load fluctuations and meet robustness for a serious load disturbances as well as robust mixed H_2/H_∞ tuning methodology. Interested readers can find more time domain simulations for various load disturbance scenarios in (Bevrani & Hiyama, 2011).

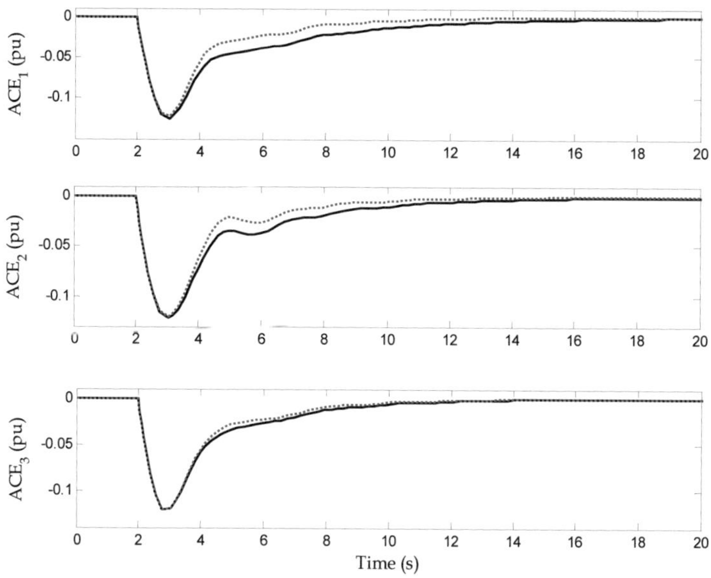

Fig. 8. Closed-loop system response; solid (GA), dotted (ILMI).

A new combination of these two tuning approaches is also introduced in (Bevrani & Hiyama, 2011), which uses the GA to achieve the same robust performance indices ($\gamma_2^*, \gamma_\infty^*$) as obtained via mixed H_2/H_∞ control technique. In the proposed approach, the GA is employed as an optimization engine to produce the PID controllers with performance indices near to optimal ones.

5. Fuzzy logic and PSO-based PID tuning

5.1 Overall framework

Nowadays, fuzzy logic because of simplicity, robustness and reliability is used in almost all fields of science and technology, including solving a wide range of control and tuning problems. Unlike the traditional tuning methodologies, which are essentially based on the linearized mathematical models of the controlled systems, the fuzzy-based tuning technique tries to tune the controller parameters directly based on the measurements, long-term experiences and the knowledge of domain experts/operators.

This section addresses a new intelligent methodology using a combination of fuzzy logic and particle swarm optimization (PSO) techniques to tune the parameters of PID controllers. The control parameters, K_P, K_I and K_D, are automatically tuned using fuzzy rules, according to the on-line information. The PSO technique is used to find optimal values for membership functions parameters of the fuzzy logic scheme. The overall control framework is shown in Fig. 9.

5.2 Tuning scheme

As already mentioned, to improve the performance of PID controllers against changing of operating condition and system parameters, a fuzzy-based tuning mechanism can be able to adapt the PID parameters during the system operation and according to the on-line information. Such controllers are generally known as *Two-level Controllers*, or *Gain Scheduling PID Controllers*. In a two-level PID controller, usually the lower level controller (PID controller) performs fast direct control and higher level controller (fuzzy logic system as a supervisor) performs low speed supervision.

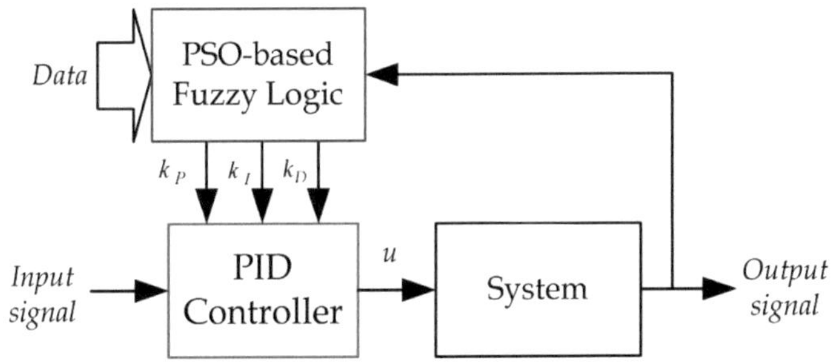

Fig. 9. Fuzzy logic for tuning of PID controller.

In the two-level Fuzzy-PID controller, direct control of the system (lower level) composed of a simple PID controller that generates the control signal $u(t)$ to apply to the plant as follows:

$$u(t) = k_P e(t) + k_I \int e(t)dt + k_D \frac{d}{dt} e(t) \tag{23}$$

where error signal $e(t)$ is used as input signal of the PID controller. Also, the fuzzy logic system acts as supervisor of PID controller performance and real-time tuning of its parameters according to system operating conditions.

Fig. 10, shows the structure of supervisory fuzzy system which is composed of four blocks. The fuzzification block represents the process of making crisp quantity into fuzzy. In fact, the fuzzifier converts the crisp input to a linguistic variable using the membership functions stored in the fuzzy knowledge base. Fuzzines in a fuzzy set is characterized by the *membership functions*. Using suitable membership functions, the ranges of input and output variables are assigned with linguistic variables. These variables transform the numerical values of the fuzzy unit input to the fuzzy quantities. These linguistic variables specify the quality of the control.

The concepts associated with a database are used to characterize fuzzy rules and a fuzzy data manipulation in fuzzy logic system. A lookup table is made based on discrete universes defines the output for all possible combinations of the input signals. A fuzzy system is characterized by a set of linguistic statements in the form of 'IF-THEN' rules. Fuzzy conditional statements make the rules or the rule set of the fuzzy system. Finally, the *Inference engine* uses the IF-THEN rules to convert the fuzzy input to the fuzzy output. On the other hand, defuzzifier converts the fuzzy output of the inference engine to crisp using membership functions analogous to the ones used by the fuzzifier. For *defuzzification* process, commonly center of sums, mean-max, weighted average and centroid methods are employed to defuzzify the fuzzy incremental control law (Bevrani & Hiyama, 2011).

Generally, fuzzy logic design for a dynamical system involves the following four main steps:

Step 1: Understanding of the system dynamic behavior and characteristics. Define the states and input/output variables and their variation ranges,

Step 2: Identify appropriate fuzzy sets and membership functions. Create the degree of fuzzy membership function for each input/output variable and complete fuzzification,

Step 3: Define a suitable inference engine. Construct the fuzzy rule base, using the control rules that the system will operate under. Decide how the action will be executed by assigning strengths to the rules, and

Step 4: Determine defuzzification method. Combine the rules and defuzzify the output.

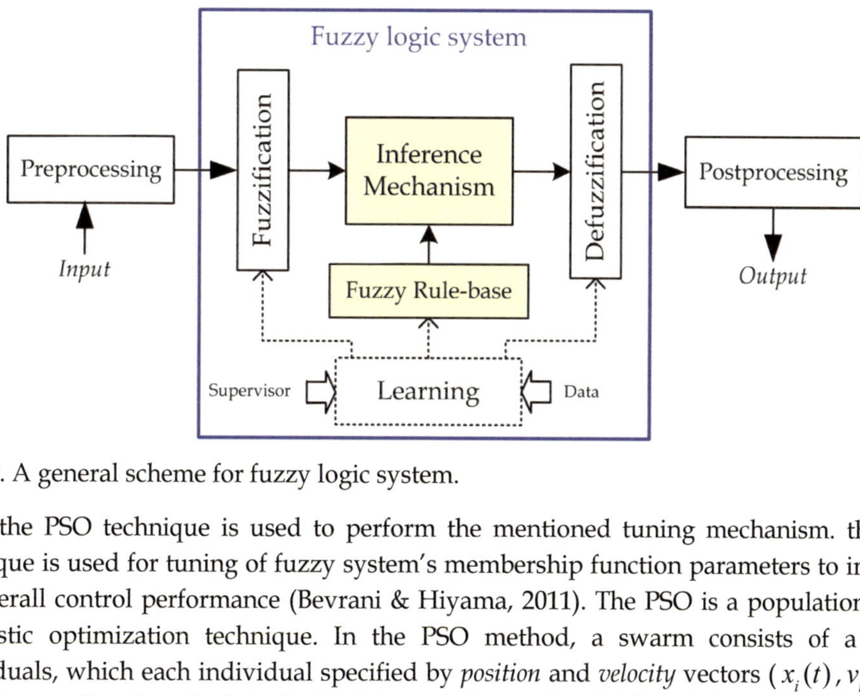

Fig. 10. A general scheme for fuzzy logic system.

Here, the PSO technique is used to perform the mentioned tuning mechanism. the PSO technique is used for tuning of fuzzy system's membership function parameters to improve the overall control performance (Bevrani & Hiyama, 2011). The PSO is a population based stochastic optimization technique. In the PSO method, a swarm consists of a set of individuals, which each individual specified by *position* and *velocity* vectors ($x_i(t), v_i(t)$) at each time or iteration. Each individual is named as a *"particle"* and the position of every particle represents a potential solution to the under study optimization problem. In an n-dimensional solution space, each particle treated as a *n-dimensional* space vector and the position of the *i*-th particle is presented by $v_i = (x_{i1}, x_{i2}, \cdots, x_{in})$; then it flies to a new position by velocity represented by $v_i = (v_{i1}, v_{i2}, \cdots, v_{in})$. The best position for *i*-th particle represented by $p_{best,i} = (p_{best,i1}, p_{best,i2}, \cdots, p_{best,in})$ is determined according to the best value obtained for the specified objective function.

Furthermore, the best position found by all particles in the population (global best position), can be represented as $g_{best} = (g_{best,1}, g_{best,2}, \cdots, g_{best,n})$. In each step, the best particle position, global position, and the corresponding objective function values should be saved. For the next iteration, the position x_{ik} and velocity v_{ik} corresponding to the k-th dimension of i-th particle can be updated using following equations:

$$v_{ik}(t+1) = w.v_{ik} + c_1.rand_{1,ik}(p_{best,ik}(t) - x_{ik}(t)) + c_2.rand_{2,ik}(g_{best,k}(t) - x_{ik}(t)) \tag{24}$$

$$x_{ik}(t+1) = x_{ik}(t) + v_{ik}(t+1) \tag{25}$$

where, $i=1, 2, \ldots, n$ is the index of particles, w is the inertia weight, $rand_{1,ik}$ and $rand_{2,ik}$ are random numbers in interval [0 1], c_1 and c_2 are learning factors, and t represents the iterations.

Usually, a standard PSO algorithm contains the following steps (Daneshmand, 2010):

Step 1: All particles are initialized via a random solution. In this step, each particle position and associated velocity are set by randomly generated vectors. Dimension of position should be generated within a specified interval, and the dimension of velocity vector should be also generated from a bounded domain using uniform distributions.

Step 2: Compute the objective function for the particles.

Step 3: Compare the value of the objective function for the present position of each particle with the value of objective function corresponding to the pre-specified best position, and replace the pre-specified best position by the present position, if it provides a better result.

Step 4: Compare the value of the objective function for the present best position with the value of the objective function corresponding to the global best position, and replace the present best position by the global best position, if it provides a better result.

Step 5: Update the position and velocity of each particle according to equations (24) and (25).

Step 6: Stop algorithm if the stop criterion is satisfied. Otherwise, go to step 2.

5.3 Application example

In order to investigate the efficiency of the proposed PID tuning strategy, a computer simulation has been conducted to design of PID-based AGC system for the standard 39-bus 10-generator test system, including three wind farms (Daneshmand, 2010). The obtained results are compared with the conventional fuzzy logic-based AGC system.

Here, ACE is considered as input signal, and the provided control signal, $u(t)$ is used to change the set points of AGC participant generating units. To track a desirable AGC performance in the presence of high penetration wind power in a multi-area power system, a decentralized fuzzy logic based PID control design is proposed. Decreasing the frequency deviations due to fast changes in output power of wind turbines, and limiting tie-lines power interchanges in an acceptable range, following disturbances, are the main goals of this effort.

The *Mamdani* type inference system is applied, and symmetric 7-segments triangular membership functions are used for input and output variables. The membership functions are defined as zero (ZO), large negative (LN), medium negative (MN), small negative (SN), small positive (SP), medium positive (MP), and large positive (LP).

In order to reach fast response from the fuzzy system, all membership functions considered as triangular with the following mathematical definition:

$$\mu_X(x_i) = \max\left(0, 1 - \left|\frac{x - x_i}{c}\right|\right) \tag{26}$$

where, x and c are the mean and spread of the fuzzy set X, respectively; and x_i is a crisp variable. Fuzzy rule base is the basis of fuzzy logic operation to map input space to the output space. Here, a rule base including 49 fuzzy rules is considered (Table 1). The rule base works on vectors composed of *ACE* and its gradient *dACE*.
Since fuzzy rules are stated in terms of linguistic variables, crisp inputs should be also mapped to linguistic values using Fuzzification. The antecedent part of the rules composed of two parts, combined with fuzzy "*AND*" operators. The combination is done based on interpreting the "*AND*" operator by "*Minimum*" operation. Considering (26), the antecedent part of above statement may be defined as follows:

$$\mu_{(ACE\ AND\ dACE)}(x, y) = \min(\mu_{ACE}(x), \mu_{dACE}(y)) \tag{27}$$

where $\mu_{(ACE\ AND\ dACE)}(x)$ is the membership value of antecedent part, and μ_{ACE} and μ_{dACE} are the membership values of *ACE* and *dACE*, respectively.
Similarly, for computing the consequent of each rule, the membership function of "*Mamdani Minimum*" implication method can be represented by

$$\mu_{MP} = \min(\mu_{(ACE\ AND dACE)}, \mu_{\Delta P_c}) \tag{28}$$

where μ_{MP} denotes the membership function resulted by "*Mamdani Minimum*" implication, and $\mu_{(ACE\ AND\ dACE)}$ is the membership value of the related antecedent part.
In order to combine rules and make a decision based on the all given rules, the *sum* method is used. Finally, for converting output fuzzy set of the fuzzy system to a crisp value the *centroid* method is used for defuzzification (Daneshmand, 2010).
Each set of input membership functions can be specified by one parameter, $ACE_{\max}$ for ACE and $dACE_{\max}$ for dACE. Also, for control output variables, lower and upper limits should be specified for PID parameters of each controller. Therefore, totally eight parameters should be optimized for membership functions using PSO algorithm.
For the sake of PSO algorithm in the present AGC design, the number of particles, particles size, $v_{\min}$, $v_{\max}$, c_1, and c_2 are chosen as 10, 6, -0.5, 0.5, 2.8, and 1.3, respectively. Following use of PSO algorithm, the optimal values for membership function parameters can be easily obtained.
To investigate the performance of the proposed control strategy, a network with the same topology as the well-known IEEE 10 generators 39-bus system is considered as a test system. The system consists of 10 generators, 19 loads, 34 transmission lines, and 12 transformers. The power system is divided to three control areas. Single-line diagram, simulation parameters for the generators, loads, lines, and transformers of the test system are given in (Daneshmand, 2010). The desined PID controllers are responsible for producing appropriate control action signals according to the measured *ACE* signals and their time derivatives (*dACE*).
For the present case study, the installed capacity includes 582.57 MW of conventional generation and 68.4 MW of average wind power generation (10% penetration). To demonstrate the effectiveness of the proposed control design, some nonlinear simulations

are performed in the SimPower environment of MATLAB software. In the simulations, the performance of the closed loop system using the designed fuzzy logic based controllers are compared with well-tuned conventional PID controllers.

As a serious test condition, three load disturbances (step increase in demand) are applied to control areas as simultaneous 6.66 pu step load increase in each area at 5 sec. All unitized values in this paper are given based on the value of the largest generator nominal power, i.e. 150 MW. The simulation results are shown in Fig. 11. The *ACE* signals of the closed loop system for all areas are presented, following the applied load disturbances. These figures show the superior performance of the proposed fuzzy logic based AGC schemes to the conventional PID-based AGC designs in deriving area control error and frequency deviation close to zero. The PID parameters for conventional PID controllers in three control areas are listed in Table 2.

		dACE						
		LN	MN	SN	ZO	SP	MP	LP
	LN.	LP	LP	LP	MP	MP	SP	ZO
ACE	NM	LP	MP	MP	MP	SP	ZO	SN
	SN	LP	MP	SP	SP	ZO	SN	MN
	ZO	MP	MP	SP	ZO	SN	MN	MN
	S P	MP	SP	ZO	SN	SN	MN	LN
	MP	SP	ZO	SN	MN	MN	MN	LN
	LP	ZO	SN	MN	MN	LN	LN	LN

Table 1. Fuzzy rule base.

Area	k_P	k_I	k_D
I	-0.852	-1.724	-0.001
II	-0.579	-0.950	-0.013
III	-0.971	-1.900	-0.007

Table 2. Conventional PID parameters.

6. Conclusion

Most of real-world control problems refer to multi-objective control designs that several objectives such as stability, disturbance attenuation and reference tracking with considering practical constraints must be simultaneously followed usually by a simple PID controller. In such cases, multi-objective based tuning approaches are needed. In this direction, the present chapter addresses three powerfull robust/intelligent multi-objective methodologies to improve the performance of PID-based control systems.

The proposed approaches use mixed H_2/H_∞, multi-objective genetic algorithm (GA), fuzzy logic, and particle swarm optimization (PSO) techniques as optimization tools for optimal tuning of PID parameters. Numerical examples on AGC design in multi-area power systems are given to illustrate the effectiveness of tuning methods.

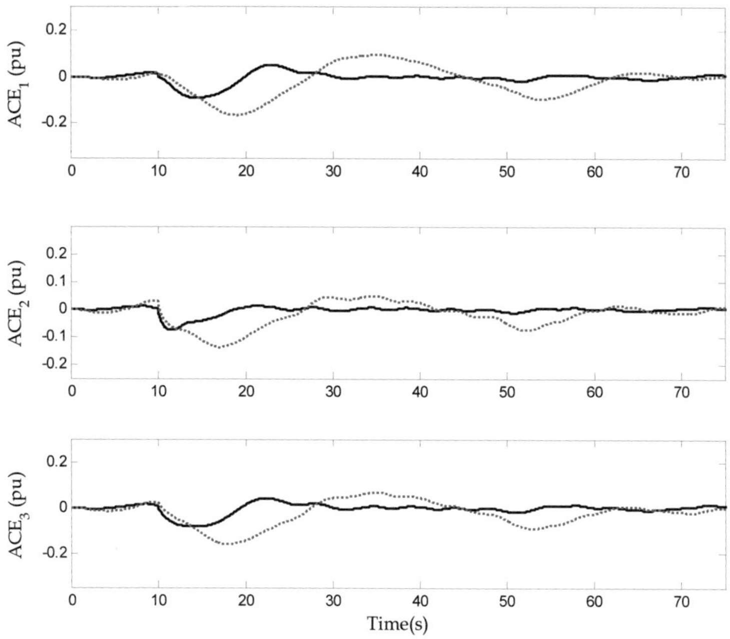

Fig. 11. ACE signals; proposed PID scheme (solid), conventional PID design (dotted).

7. References

Bevrani, H. & Hiyama, T. (2007) Multiobjective PI/PID Control Design Using an Iterative Linear Matrix Inequalities Algorithm. *International Journal of Control, Automation, and Systems*, Vol.5, No.4, pp. 117-127.

Bevrani, H. (2009) *Robust Power System Frequency Control*. Springer, ISBN : 9780387848778, New York, USA.

Bevrani, H. & Hiyama, T. (2011) *Intelligent Automatic Generation Control*. CRC Press (Taylor & Francis Group), ISBN : 9781439849538, New York, USA.

Cao, Y.Y. ; Lam, J. ; Sun, Y. X. & Mao, W. J. (1998) Static output feedback stabilization: an ILMI approach. *Automatica*, Vol. 34, No. 12, pp. 1641-1645.

Chen, B. S. ; Cheng, Y. M. & Lee, C. H. (1998) A genetic approach to mixed H_2/H_∞ optimal PID control. *IEEE Control Systems*, Vol. 15, No. 5, pp. 51-60.

Daneshmand, P. R. (2010) *Power system frequency control in the presence of wind turbines*, MSc. dissertation, Department of Electrical and Computer engineering, University of Kurdistan, Sanandaj, Iran, 2010.

Fonseca, C.M. & Fleming, P.J. (1995) Multiobjective optimization and multiple constraint handling with evolutionary algorithms -part I: a unified formulation. *IEEE Trans. Syst. Man & Cybernetics, A*, Vol.28, No.1, pp. 26-37.

Holland, J. H. (1998) Application in natural and artificial systems. The MIT Press, Combridge, Massachusetts, USA.

Rerkpreedapong, D ; Hasanovic, A. & Feliachi, A. (2003) Robust load frequency control using genetic algorithms and linear matrix inequalities. *IEEE Trans. On Power Systems*, Vol. 18, No. 2, pp. 855-861.

Rudolph, G (1997) Convergence properties of evolutionary algorithms. Kovac, Hamburg

Sazuki, J (1995) A markov chain analysis on simple genetic algorithms. *IEEE Transactions on Systems, Man and Cybernetics*, Vol.25, No.4, pp. 655-659.

Takahashi, R. H. C. ; Peres, P. L. D. & Ferreira, P. A. V. (1997) Multiobjective H_2/H_∞ guaranteed cost PID design. *IEEE Control Systems*, Vol. 17, No. 5, pp. 37-47.

Thierens, D. & Goldberg, D. E. (1994) Convergence models of genetic algorithm selection schemes. *Parallel problem solving from nature-PPSN III*, pp.119-129.

Zheng, F. ; Wang, Q. G. & Lee, H. T. (2002) On the design of multivariable PID controllers via LMI approach. *Automatica*, Vol. 38, pp. 517-526.

Part 2

Implementation and PID Control Applications

10

Pole-Zero-Cancellation Technique for DC-DC Converter

Seiya Abe, Toshiyuki Zaitsu, Satoshi Obata,
Masahito Shoyama and Tamotsu Ninomiya
International Centre for the Study of East Asian Development,
Texas Instruments Japan Ltd., Kyushu University, Nagasaki University,
Japan

1. Introduction

Many types of electric equipments are digitized in recent years. However, the configuration of switch mode power supply is still only analog circuit because the analog circuit is held down to low cost. The digitized system is operated on the basis of a processor. When the switch mode power supply is treated as a part of the system, it is difficult that switch mode power supply inhabit alone in the system as the analog-circuit. Therefore, the digitization of the switch mode power supply is necessary to harmonize with other electronic circuits in the system. So far, various examinations have been discussed about digitally controlled switch mode power supplies[1-5]. However, important parameters such as the switching frequency were impractical because the performance of processor was not so good. Recently, due to the development of the semiconductor manufacture technology, the performance of processor such as DSP and FPGA is developed remarkably. Hence, the expectation of the practical realization in the digitally controlled switch mode power supply becomes higher.

So far, in many case on digitally controlled switch mode power supply, the control system is constructed by very complicated, difficult modern control theory (nonlinear control theory) such as adaptive control or predictive control.

Moreover, also in the most popular and easiest control method such as PID control, the design method is not so clear, and the optimal design is difficult[6, 7].

On the other hand, there are two methods of controller design. One is the digital direct design. The other is the digital redesign. The digital redesign method converts the analog compensator which is designed on s-region into digital compensator. The digital redesign method has some advantages. For example, the control system is designed from classical control theory (linear control theory).

Therefore, many experiences and design techniques of the conventional analog compensator can be utilized. Moreover, from the practical stance, the digital redesign method is more realistic than digital direct design.

This paper investigates the digitally controlled switch mode power supply by means of classical control theory. Especially, the interesting control technique which is cancelled the transfer function of the converter by using pole-zero-cancellation technique is introduced. This technique is very simple and stability design of converter system is very easy.

Furthermore, the arbitrary frequency characteristics can be created by introducing a new frequency characteristic. Here, the design method and system stability of the proposed control technique is examined by using buck converter as a simple example.

2. Converter analysis

For the design of the control system, it is necessary to grasp correctly the characteristics of the converter in detail. The buck converter as a controlled object is shown in Fig. 1. The dynamic characteristics of buck converter can be derived by applying the state space averaging method[8,9]. The transfer function of duty to output voltage of buck converter is derived following equation;

$$G_{dv}(s) = \frac{\Delta V_o(s)}{\Delta D(s)} = \frac{G_{dvo}(s)}{P(s)} \tag{1}$$

where;

$$P(s) = \frac{s^2}{\omega_o^{\ 2}} + s\frac{2\delta}{\omega_o} + 1 \tag{2}$$

$$G_{dvo}(s) = \left(\frac{s}{\omega_{esr}} + 1\right)\frac{R}{R + r_L}V_i \tag{3}$$

$$\omega_o = \sqrt{\frac{R + r_L}{LC(R + r_c)}} \tag{4}$$

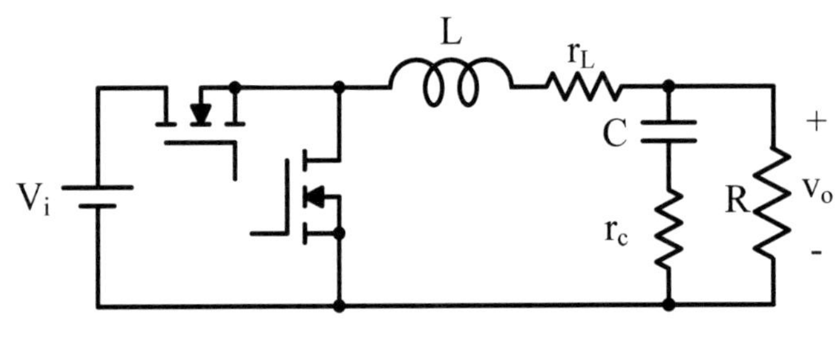

Fig. 1. Synchronous buck converter.

$$\delta = \frac{L + C\{Rr_c + r_L(R + r_c)\}}{2\sqrt{LC(R + r_c)(R + r_L)}} \tag{5}$$

$$\omega_{esr} = \frac{1}{Cr_c} \tag{6}$$

Figure 2 shows the block diagram of analog system. From, Fig. 2, the loop gain of analog controlled converter can be derived following equation;

$$T(s) = \frac{\Delta V_o(s)}{\Delta V_o^*(s)} = \frac{G_{dvo}(s)}{P(s)} \cdot G_c(s) \cdot K \cdot K_s \cdot PWM \tag{7}$$

where;
Gc(s) : Transfer function of phase compensator
K : DC gain of error amp.
Ks : Sense gain of output voltage
PWM : transfer gain of voltage to duty

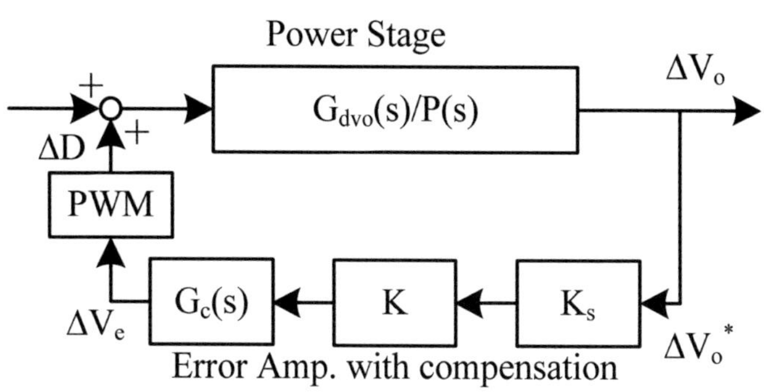

Fig. 2. Block diagram of analog system.

In order to evaluate the validity of the analytical result, the experimental circuit is implemented by means of the specifications and parameters shown in Table 1.

Symbol	Description	Value
V_i	Input Voltage	12V
V_o/I_o	Load Condition	2.5V/5A
L	Filter Inductor	22μH
C	Filter Capacitor	470μF
r_L	DC Resistance of L	100mΩ
r_c	ESR of C	10mΩ
R	Load Resistance	1Ω
K_s	Sense Gain	0.32
K	Feedback DC Gain	5
PWM	PWM Gain	0.5
fs	Switching Frequency	100kHz

Table 1. Circuit parameters and specifications.

Figure 3 shows the loop gain of the buck converter with p-control in analog control. As shown in Fig. 3, the analytical and experimental results are agreed well. However, as shown in Fig. 4, the big difference is shown in phase characteristics at high frequency side between analog control and digital control.

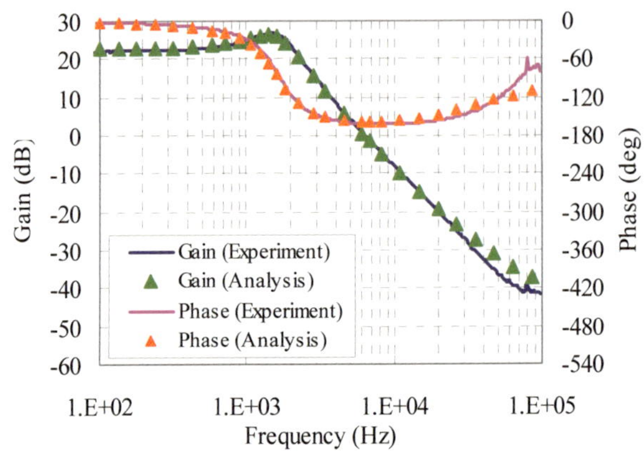

Fig. 3. Frequency response of loop gain (analog control).

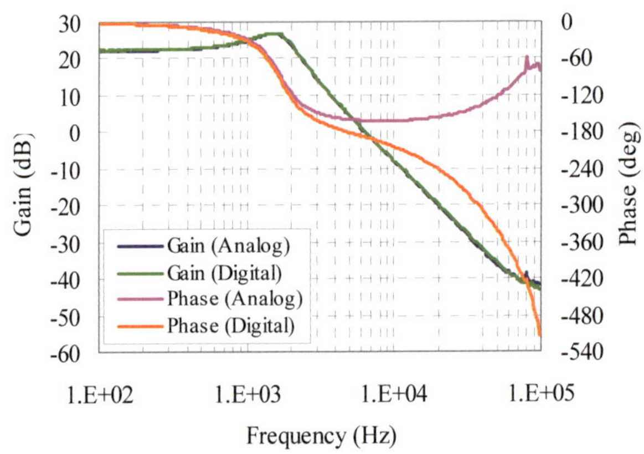

Fig. 4. Frequency response comparison of analog control and digital control (Experiment).

In digital control system, the output voltage as a detected signal is converted to digital signal by AD converter, after that the converted signal is calculated by DSP. Next, the calculated signal decides the duty ratio of next switching period. Hence, the information of the output voltage as the detected signal at certain switching period is reflected into the duty ratio of the next switching period.

Therefore, the dead time element He(s) is included into the control loop as shown in Fig. 5. From Fig. 5, the loop gain of digital controlled system can be derived following equation;

$$T(s) = \frac{\Delta V_o(s)}{\Delta V_o^*(s)} = \frac{G_{dvo}(s)}{P(s)} \cdot G_c(s) \cdot H_e(s) \cdot K \cdot K_s \cdot PWM \tag{8}$$

where;

$$H_e(s) = e^{-sT_{sample}} \tag{9}$$

Gc(s) : Transfer function of phase compensator
K : DC gain of error amp.
Ks : Sense gain of output voltage
PWM : transfer gain of voltage to duty
He(s) : Dead time component of digital controller
Tsample : Sampling period
Figure 6 shows the frequency response of dead time element He(s). As shown in Fig. 6, the gain characteristic does not depend on frequency and it is constant.

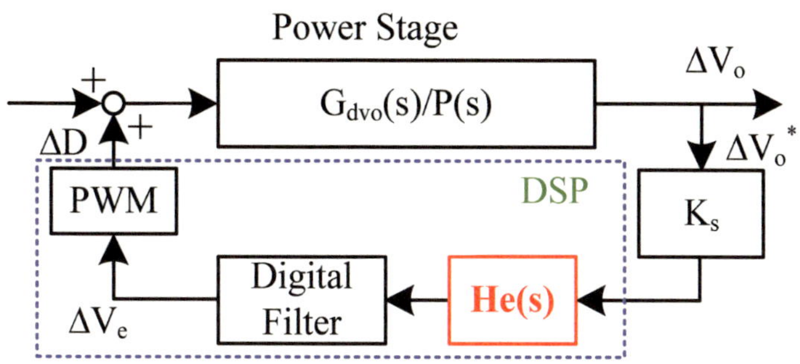

Fig. 5. Block diagram of digital system.

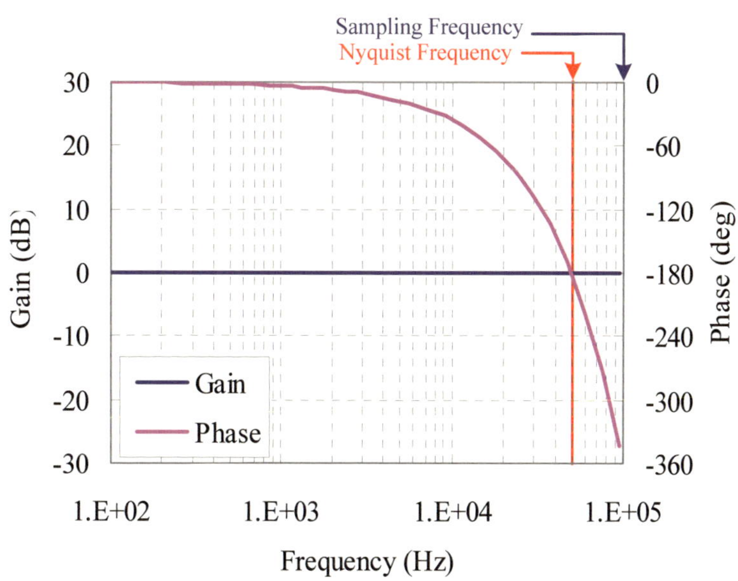

Fig. 6. Frequency response of dead time element He(s).

On the other hand, phase characteristic depends on frequency. The phase is rotated around 180 degrees at Nyquist frequency (=f/2), and it is rotated around 360 degrees at switching

frequency (sampling frequency). From these results, the phase is drastically rotated at high frequency side by the influence of dead time element He(s). In order to evaluate these discussions, the experimental circuit is implemented by means of the specifications and parameters shown in Table 1. Moreover, the experimental result is compared with analytical result. Figure 7 shows the loop gain of the buck converter with p-control in digital control. As shown in Fig. 7, the analytical and experimental results are agreed well. In analog control system, the phase characteristic of frequency response is improved at higher frequency side by the influence of ESR-Zero as shown in Fig. 4, and the system has stable operation.
On the other hand, in digital control system, the phase characteristic of frequency response is drastically rotated by the influence of the dead time element He(s) as shown in Fig. 7. As a result, the phase margin disappears, and the system becomes unstable.
In digital control system, the phase rotation is larger than analog control system by the influence of the dead time element He(s), so the phase compensation is necessary to keep the system stability.

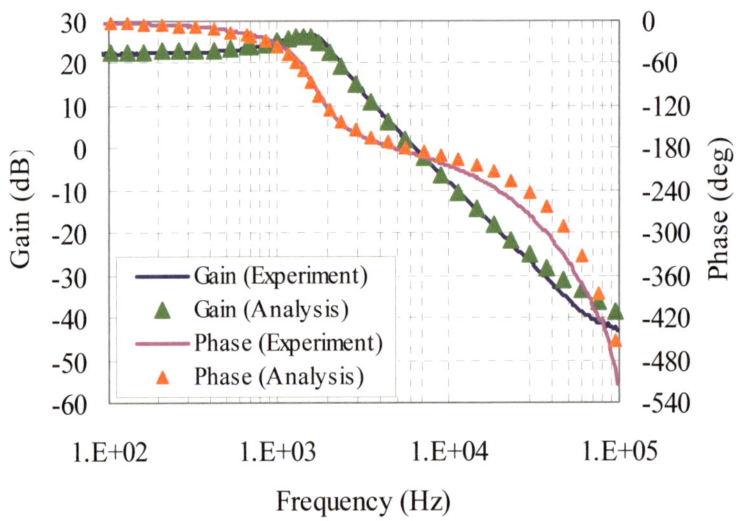

Fig. 7. Frequency response of loop gain (digital control).

3. Conventional phase compensation (Phase lead-lag compensation)

The phase compensation is usually used to improve the system stability. There is various phase compensation. Here, the phase lead-lag compensation is used as the most popular compensation. The digital filter is designed by digital redesign method. The transfer function of phase lead-lag compensation is given by following equation;

$$G_c(s) = \frac{\Delta v_e}{\Delta v_o^*} = \frac{K_c\left(\frac{s}{\omega_{z1}}+1\right)\left(\frac{s}{\omega_{z2}}+1\right)}{\left(\frac{s}{\omega_{p1}}+1\right)\left(\frac{s}{\omega_{p2}}+1\right)} \tag{10}$$

The digital filter can be realized by means of the bilinear transformation.

$$s = \frac{2}{T_{sample}} \cdot \frac{1 - z^{-1}}{1 + z^{-1}} \tag{11}$$

$$G_c(z) = \frac{\Delta v_e}{\Delta v_o^*} = k \frac{z^{-2}B_2 + z^{-1}B_1 + B_0}{z^{-2}A_2 + z^{-1}A_1 + A_0} \tag{12}$$

where;

$$k = K_c \frac{\omega_{p1}\omega_{p2}}{\omega_{z1}\omega_{z2}} \tag{13}$$

$$A_0 = \frac{4}{T_{sample}{}^2} + \frac{2(\omega_{p1} + \omega_{p2})}{T_{sample}} + \omega_{p1}\omega_{p2} \tag{14}$$

$$A_1 = -\frac{8}{T_{sample}{}^2} + 2\omega_{p1}\omega_{p2} \tag{15}$$

$$A_2 = \frac{4}{T_{sample}{}^2} - \frac{2(\omega_{p1} + \omega_{p2})}{T_{sample}} + \omega_{p1}\omega_{p2} \tag{16}$$

$$B_0 = \frac{4}{T_{sample}{}^2} + \frac{2(\omega_{z1} + \omega_{z2})}{T_{sample}} + \omega_{z1}\omega_{z2} \tag{17}$$

$$B_1 = -\frac{8}{T_{sample}{}^2} + 2\omega_{z1}\omega_{z2} \tag{18}$$

$$B_2 = \frac{4}{T_{sample}{}^2} - \frac{2(\omega_{z1} + \omega_{z2})}{T_{sample}} + \omega_{z1}\omega_{z2} \tag{19}$$

The determination of the compensator parameter is various. Here, these parameter decide from phase margin. Figure 8 shows the analytical result of loop gain frequency response with phase lead-lag compensation. Where, Kc=10000, fp1=0.03Hz, fz1=1.3kHz, fp2=20kHz, fz2=1.5kHz. As shown in Fig. 8, this system has the stable operation, and then the bandwidth is around 5.5kHz, the phase margin is around 45 degrees. Figure 9 shows the experimental result of loop gain frequency response with phase lead-lag compensation. In this case, the bandwidth is around 5kHz, and the phase margin is around 45 degrees. Moreover, the analytical and experimental results are agreed well. Thus, the observation of control object frequency response is needed in classical control theory (linear control theory).

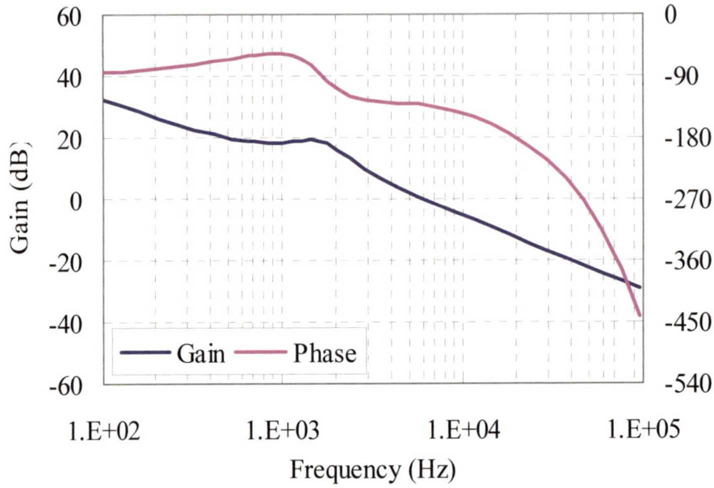

Fig. 8. Frequency response of loop gain with phase lead-lag compensation (analytical result).

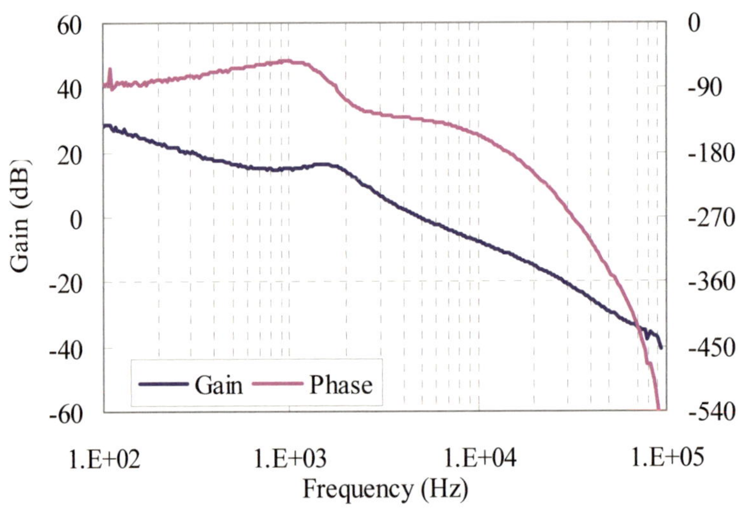

Fig. 9. Frequency response of loop gain with phase lead-lag compensation (experimental result).

Moreover, much experience and knowledge are needed for controller design, because many parameters in compensator should be decided. Therefore, the design method is not so clear and depends on knowledge and experience, and the optimal design is difficult.

The controller design becomes very simple if the controller design is enabled without considering the frequency response of the converter as the control object.

4. Principle of PZC technique

Reduction of the phase rotation is very important for system stability. Especially in the second order system, the phase is drastically rotated around 180 degrees at resonance peak. The stability of the system is improved remarkably if the phase rotation can be reduced.

This paper proposes the control technique which is cancelled the transfer function of the converter power stage by means of pole-zero-cancellation method. The phase rotation and gain change can be suppressed by cancelling the converter power stage characteristics. Furthermore, new characteristic can be designed in the system as the arbitrary transfer function. Figure 10 shows the block diagram of converter system including the pole-zero-cancellation technique.

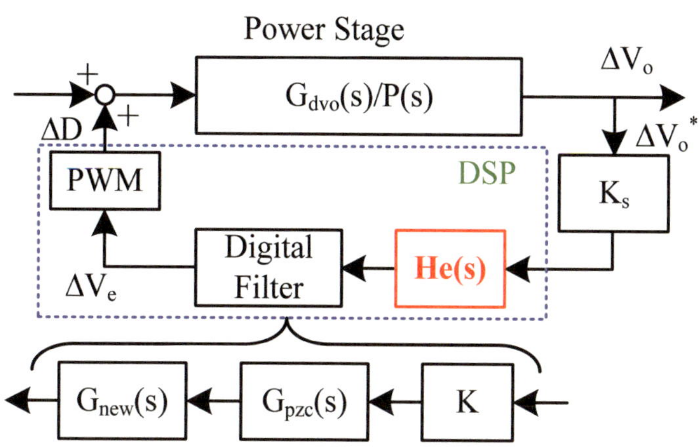

Fig. 10. Block diagram of digital system with PZC control.

From Fig. 10, the transfer function of compensator part is given following equation;

$$G_c(s) = G_{new}(s) \cdot G_{pzc}(s) \tag{20}$$

The Gnew(s) is the arbitrary transfer function. This transfer function decides the frequency response of converter system. Here, the Gnew(s) is defined as first-order low pass filter.

$$G_{new}(s) = \frac{K_c}{\dfrac{s}{\omega_c} + 1} \tag{21}$$

In buck converter case, the resonance peak and ESR-Zero are cancelled. The phase rotation of 180 degree is reduced by cancelling resonance peak. The transfer function of the pole-zero-cancellation Gpzc(s) is given following equation;

$$G_{pzc}(s) = \frac{\dfrac{s^2}{\omega_o^{\ 2}} + s\dfrac{2\delta}{\omega_o} + 1}{\dfrac{s}{\omega_{esr}} + 1} \tag{22}$$

Moreover, the transfer function of the compensator is given following equation;

$$G_c(s) = K_c \frac{\dfrac{s^2}{\omega_o^{\ 2}} + s\dfrac{2\delta}{\omega_o} + 1}{\left(\dfrac{s}{\omega_{esr}} + 1\right)\left(\dfrac{s}{\omega_c} + 1\right)} \tag{23}$$

The digital filter can be realized by means of the bilinear transformation (Eq. 11) as following equation;

$$G_c(z) = \frac{\Delta v_e}{\Delta v_o^{\ *}} = k\frac{z^{-2}B_2 + z^{-1}B_1 + B_0}{z^{-2}A_2 + z^{-1}A_1 + A_0} \tag{24}$$

where;

$$k = K_c \tag{25}$$

$$A_0 = \frac{4/\omega_{esr}\omega_c}{T_{sample}^{\ 2}} + \frac{2(1/\omega_{esr} + 1/\omega_c)}{T_{sample}} + 1 \tag{26}$$

$$A_1 = -\frac{8/\omega_{esr}\omega_c}{T_{sample}^{\ 2}} + 2 \tag{27}$$

$$A_2 = \frac{4/\omega_{esr}\omega_c}{T_{sample}^{\ 2}} - \frac{2(1/\omega_{esr} + 1/\omega_c)}{T_{sample}} + 1 \tag{28}$$

$$B_0 = \frac{4/\omega_o^{\ 2}}{T_{sample}^{\ 2}} + \frac{4\delta/\omega_o}{T_{sample}} + 1 \tag{29}$$

$$B_1 = -\frac{8/\omega_o^{\ 2}}{T_{sample}^{\ 2}} + 2 \tag{30}$$

$$B_2 = \frac{4/\omega_o^{\ 2}}{T_{sample}^{\ 2}} - \frac{4\delta/\omega_o}{T_{sample}} + 1 \tag{31}$$

Figure 11 shows the frequency response of PZC part Gpzc(s). As shown in Fig. 11, the ant resonance peak is appeared at the same frequency of power stage frequency response. Figure 12 shows the analytical result of the loop gain frequency response with PZC technique. Where, Kc=5000, fc=0.01Hz. As shown in Fig. 12, this system has the stable operation, and then the bandwidth is around 400Hz, the phase margin is around 88 degrees.

Moreover, the resonance peak and ESR-Zero are completely cancelled, and this system becomes 1st order response. From these results, the converter frequency response is completely cancelled by the influence of PZC part, and the new characteristic is created (1st order characteristic).

Figure 13 shows the experimental result of loop gain frequency response with PZC technique. In this case, the bandwidth is around 400Hz, and the phase margin is around 89 degrees. Moreover, the analytical and experimental results are agreed well.

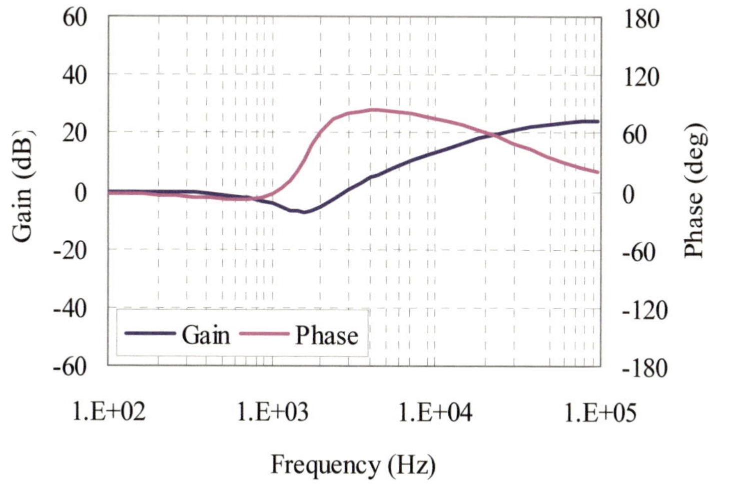

Fig. 11. Frequency response of PZC part (analytical result).

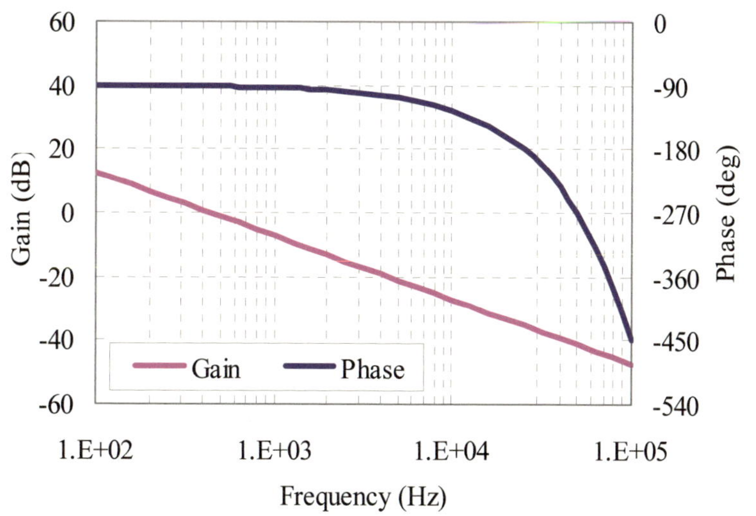

Fig. 12. Frequency response of loop gain with PZC technique (analytical result).

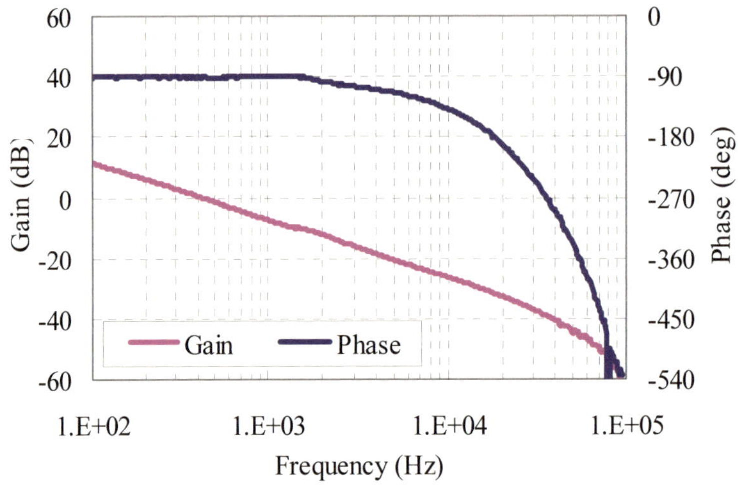

Fig. 13. Frequency response of loop gain with PZC technique (experimental result).

5. Optimal design of the new transfer function

The first order low pass filter as Gnew(s) is designed for system stability at previous section. Here, the optimization of the Gnew(s) is considered. At first, the stability margin is investigated. In this case, the integrator is included, so the phase starts -90deg. In addition, the phase is shifted by the influence of dead time element He(s) as shown in Fig. 14. Therefore, when the crossover frequency sets to f_{BW}, the phase margin can be derived as follows;

$$P_m = 90 - \frac{360}{f_s} f_{BW} \tag{32}$$

When f=fs/4, the phase margin becomes zero.
Next, the gain margin is investigated. In this case, this system has 1st order response, so the slope of gain curve becomes -20dB/dec. Therefore, the gain margin can be derived following equation by using the crossover frequency f_{BW} and fs/4.

$$G_m = 20\log_{10}\left(\frac{f_s}{4f_{BW}}\right) \tag{32}$$

From eq. (31), (32), it is clarified that the phase margin and gain margin is automatically decided by the determination of crossover frequency f_{BW}. The Gnew(s) is optimized by means of crossover frequency f_{BW}. The Gnew(s) has two coefficients, ωc and Kc. The coefficient of ωc is decided from Kc and f_{BW}.
The steady state error depends on the output impedance, especially the low frequency component of the closed loop output impedance Zoc. The open loop output impedance can be derived by applying the state space averaging method as following equation;

$$Z_o(s) = \frac{s^2 LCr_c + s\left(L + Cr_L r_c\right) + r_L}{s^2 LC + sC\left(r_L + r_c\right) + 1} \tag{33}$$

Moreover, the closed loop output impedance given from eq. (7) and (33).

$$Z_{oc}(s) = \frac{Z_o(s)}{1+T(s)} \tag{34}$$

Therefore, the low frequency component of the closed loop output impedance Zoc can be derived approximately as following equation.

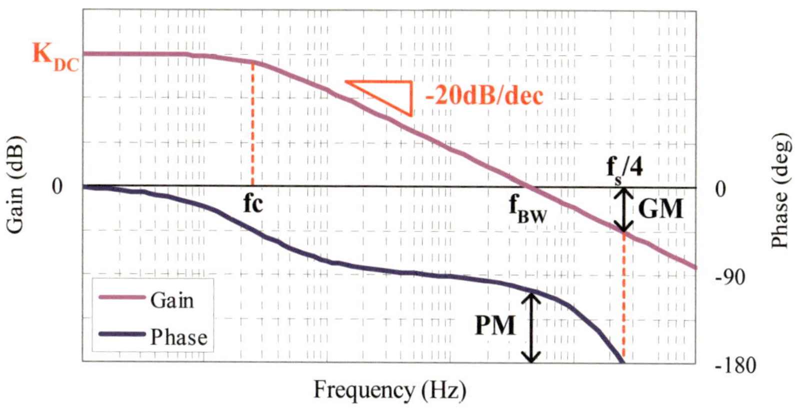

Fig. 14. Frequency response of loop gain with PZC technique for optimal filter design.

$$Z_{oc} = \frac{r_L}{K \cdot K_s \cdot K_c \cdot PWM \cdot V_{in}} \tag{35}$$

The steady state error of the output voltage ΔV is given by Zoc product output current variation ΔIo. Therefore, the coefficient Kc can be derived by determining the tolerance of the output voltage variation. From eq. (35), the coefficient of Kc can be derived approximately as following equation.

$$K_c = \frac{r_L}{Z_{oc} \cdot K \cdot K_s \cdot PWM \cdot V_{in}} \tag{36}$$

Moreover, the total DC gain K_{DC} of loop gain T(s) becomes following equation.

$$K_{DC} = 20\log_{10}\left(K \cdot K_s \cdot K_c \cdot PWM \cdot V_{in}\right) = 20\log_{10}\left(\frac{r_L}{Z_{oc}}\right) \tag{37}$$

The bandwidth f_{BW} and the coefficient of Kc are decided, and the slope of loop gain is -20dB/dec. From these parameters, the total DC gain K_{DC} can be expressed by using f_{BW} and fc as following equation.

$$K_{DC} = 20\log_{10}\left(\frac{f_{BW}}{f_c}\right) \tag{38}$$

From eq. (37), (38), the coefficient of fc is given as following equation.

$$f_c = \frac{Z_{oc}}{r_L} f_{BW} \tag{39}$$

From mentioned above discussion, the coefficients fc and Kc is optimized. Here, the crossover frequency f_{BW} is set to 10kHz. In this case, the phase margin is around 54 degrees and the gain margin is round 8dB. Moreover, the each coefficient is Kc=42, fc=25Hz. Where, the output impedance is set to around 0.25mΩ.

Figure 15 shows the analytical results of the loop gain frequency response with optimal filter design. As shown in Fig. 15, the bandwidth is around 10kHz, the phase margin is around 50 degrees. Figure 16 shows the experimental results of the loop gain frequency response with optimal filter design. In this case, the bandwidth is around 10kHz, the phase margin is around 50 degrees. Moreover, the analytical and experimental results are agreed well.

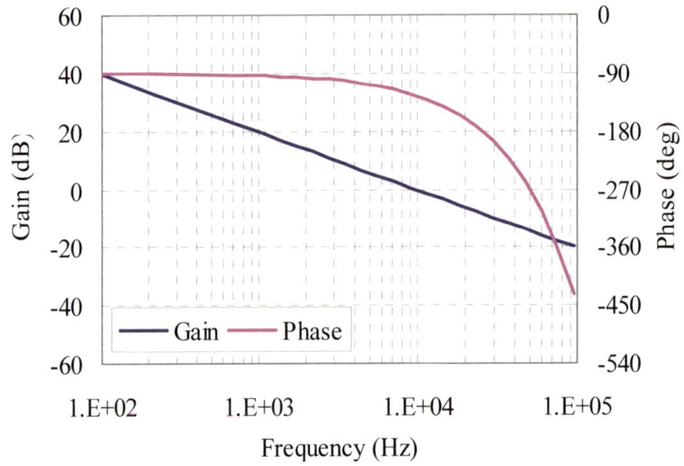

Fig. 15. Optimal design of loop gain (analytical result).

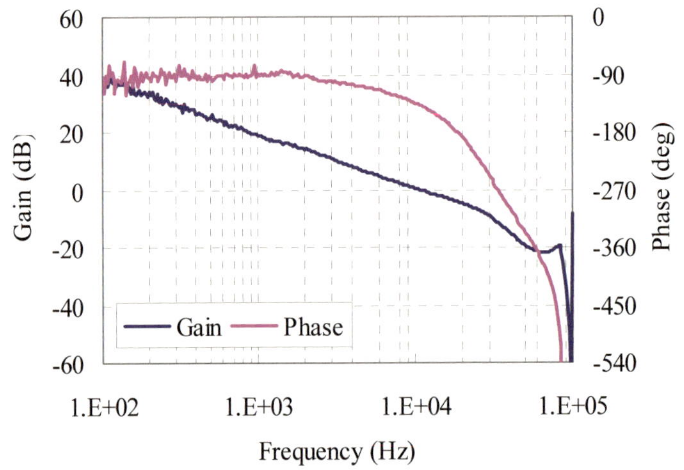

Fig. 16. Optimal design of loop gain (experimental result).

Next, the transient response of the conventional phase lead-lag compensation and the PZC technique are measured using experimental circuit of 2.5V/5A during the step load transition from 1A to 4A (10A/μs). Figure 17, 18 shows the transient response of the conventional phase lead-lag compensation and PZC technique, respectively. In phase lead-lag comensation case, the output voltage drop is around 320mV and the transient time to the steady state is around 400μs. On the other hand, in the case with PZC technique, the output voltage drop is around 160mV and the transient time to the steady state is around 200μs as shown in Fig. 15, and the transient response is improved.

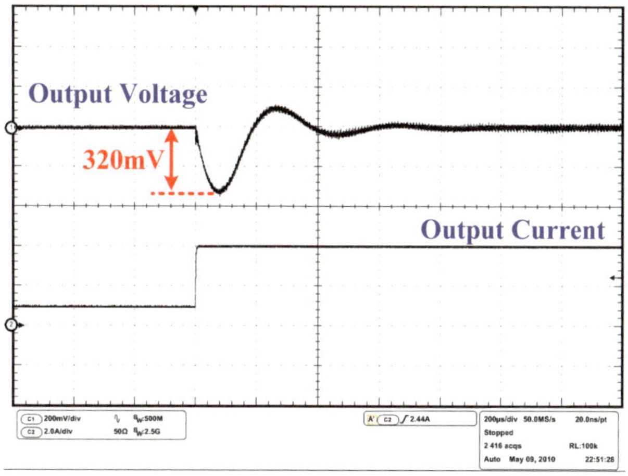

Fig. 17. Transient response (Phase lead-lag compensation).

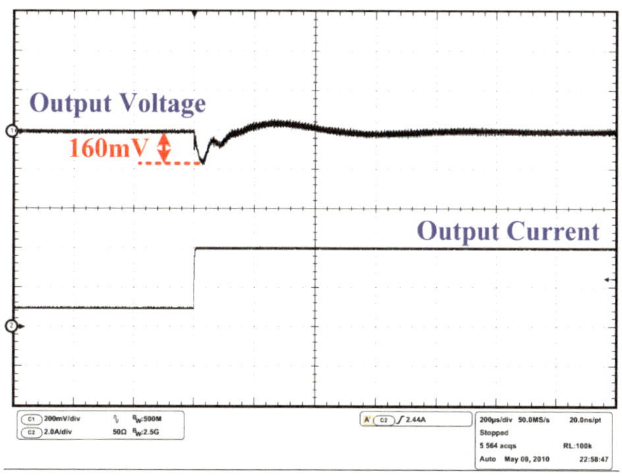

Fig. 18. Transient response (PZC technique).

6. Parameter tolerance

Here, the actual system implementation is discussed. So far, Conductive Polymer Aluminum Solid Capacitor (CPASC) is usually used as the output capacitor of low output voltage converter. However, the Ceramic chip capacitor is recently used by the demand of diminution and thinness. The issue of Ceramic chip capacitor is that the capacitance is changed by the applied voltage. Conventionally, the controller is designed by means of power stage frequency response, and it is designed to have some stability margin. However, when the capacitance is changed by the output voltage, the power stage frequency response is changed. Then, the whole system frequency response is changed. Hence, the stability margin is changed, and then the system may become unstable. Moreover, the transient response becomes worse. As a result, prospective performance is not provided.

In order to keep the system stability, it is necessary to understand correctly the characteristics of capacitance variation in detail. Figure 19 shows the experimental measurements of capacitance vs. applied voltage.

The capacitors are used as follows;

Sample 1: CPASC

Nominal value : 470μF

Rated voltage : 10V

Sample 2: Ceramic chip capacitor

Nominal value : 100μF (5 parallel, Total : 500μF)

Rated voltage : 6.3V

As shown in Fig. 19, the capacitance is almost flat in CPASC. On the other hand, the capacitance is drastically changed in Ceramic chip capacitor. In this case, the capacitance variation is around 60%. When the applied voltage is 0V, the capacitance is 410μF, and when the applied voltage is 3.5V, the capacitance is 220μF. As mentioned above, when the capacitance is changed, the system stability is also changed.

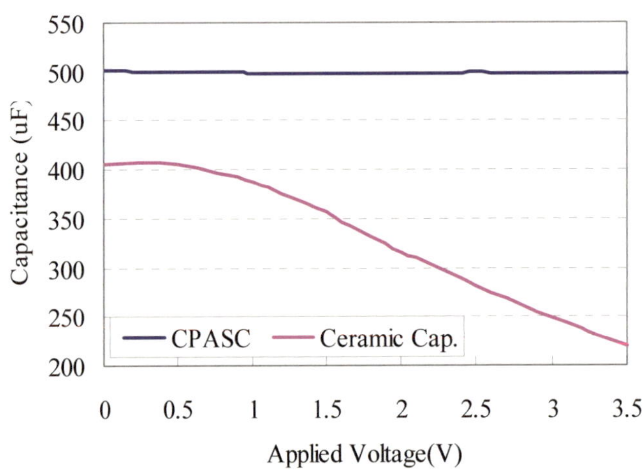

Fig. 19. Applied voltage vs. capacitance.

Figure 20 shows the analytical result of stability margin vs. applied voltage. Initially, the stability margin is set "9dB GM and 50deg PM" at CPASC. As shown in Fig. 20, the stability margin is flat for all voltage range at CPASC. On the other hand, the stability margin is reduced when the applied voltage becomes higher. At applied voltage 2.5V, the stability margin is changed form "9dB GM and 50deg PM" to "3dB GM and 25deg PM". Finally, when the applied voltage is 3.5V, the stability margin becomes limited.

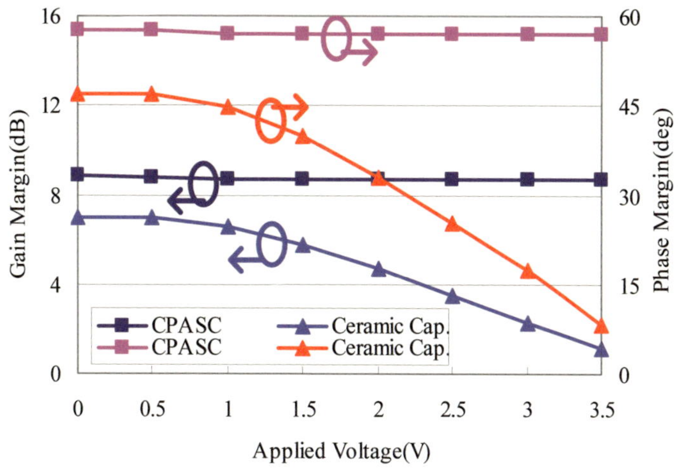

Fig. 20. Applied voltage vs. stability margin.

Figure 21 shows the analytical result of loop gain when the output voltage is 3.5V. Figure 19 has big difference compared with Fig. 15 as an initial condition. As shown in Fig. 19, the anti-resonance peak is appeared at around 1.8kHz. This anti-resonance peak is the influence of Gpzc(s).

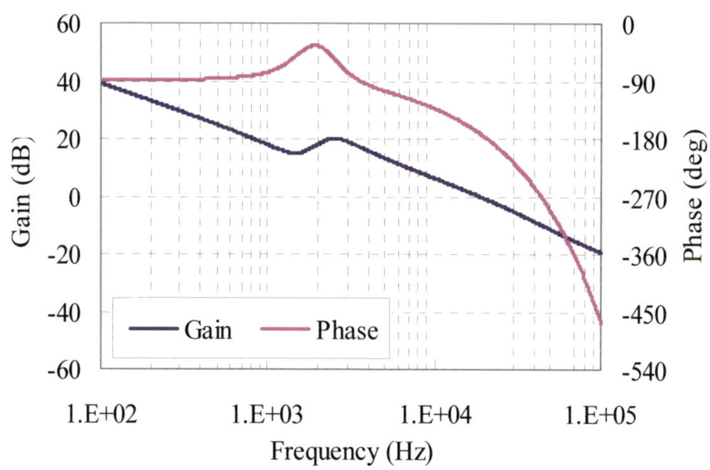

Fig. 21. Lop gain with PZC control when capacitance changes (analytical result).

This anti-resonance peak is cancelled by resonance peak of the power stage, essentially. However, the anti-resonance peak is appeared on frequency response because of the power stage resonance peak is shifted by the influence of parameter variation. Moreover, the resonance peak is appeared at around 2.5kHz. This resonance peak is power stage resonance peak.

In this case, the bandwidth is changed from 10kHz to 20kHz, and the stability margin becomes very few. The performance of the system is greatly affected by the parameter variation in this way. Therefore, the parameter tracking is needed to keep the system performance.

There are two methods of parameter tracking. One is perfect tracking method. Another is simplified tracking. The influence of parameter variation is completely cancelled by the perfect tracking method.

However, the accurate detection of the several mV high frequency voltage is very difficult. So, the perfect tracking is not available solution. Here, the simplified tracking method is examined. The data table is used in the simplified tracking method. Figure 22 shows the experimental measurements of capacitance vs. stability margin.

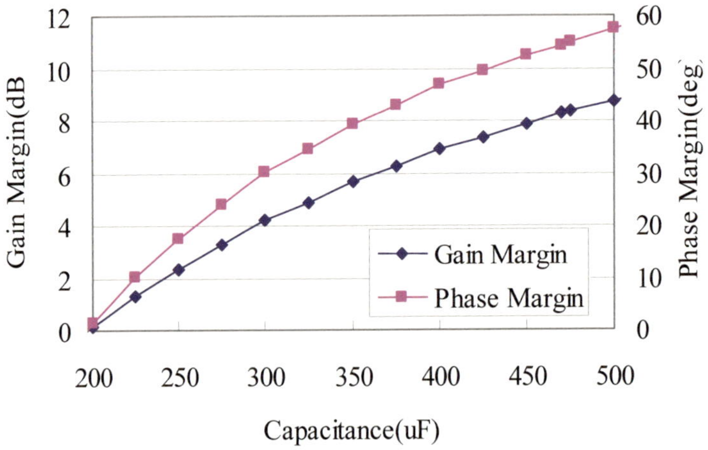

Fig. 22. Capacitance vs. stability margin.

From Fig. 19 and Fig. 22, The designed parameters are listed in Table 2. The auto parameter tracking can be realized by implementation of data table to DSP.

No.	Voltage Range (V)	Capacitance (μF)
1	0 - 0.5	500
2	0.5 - 1.0	400
3	1.0 - 1.5	350
4	1.5 - 2.0	310
5	2.0 - 2.5	270
6	2.5 - 3.0	240
7	3.0 - 3.5	200

Table 2. Parameter list.

Figure 23 shows the experimental result of loop gain when the output voltage is 3.5V. As shown in Fig. 23, the anti-resonance peak at around 1.8kHz is reduced. Moreover, the resonance peak at around 2.5kHz is also reduced. In this case, the bandwidth is around 10kHz, and the stability margin is improved. From these results, for parameter tracking, the system characteristics are kept initial conditions.

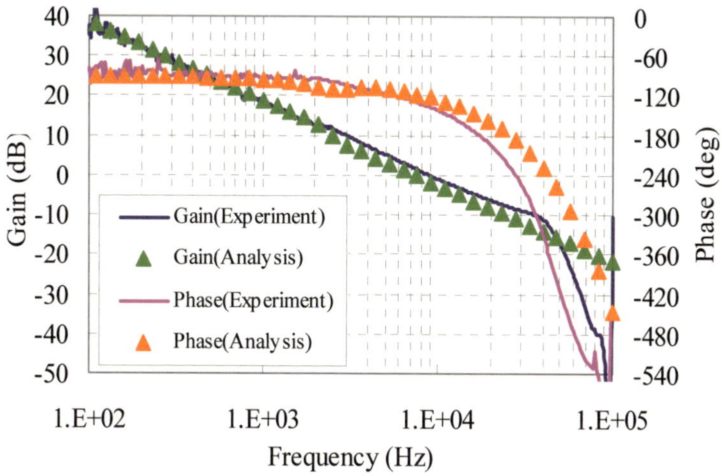

Fig. 23. Lop gain with parameter tracking.

7. Conclusions

This paper proposes the interesting control technique which is cancelled the transfer function of the converter by means of pole-zero-cancellation technique. This technique is very simple, and easy to stability design of converter system. Furthermore, the arbitrary frequency characteristics can be created by introducing a new frequency characteristic. Especially, optimal design of first-order low pass filter is considered and, the design method and system stability of the proposed control technique is examined analytically and experimentally by using buck converter. Furthermore, the parameter tracking is also examined.

As a result, the effectiveness of proposed control technique is confirmed. Moreover, it is confirmed that the characteristic cancellation of the converter can be realized very easy and can be set the arbitrary characteristic. Furthermore, the effective of parameter tracking is also confirmed.

8. References

[1] Philip T. Krein, "Digital Control Generations -- Digital Controls for Power Electronics through the Third Generation," IEEE PEDS'07, pp P-1-P5, 2007

[2] A. Kelly and K. Rinne, "Control of DC-DC Converters by Direct Pole Placement and Adaptive Feedforward Gain Adjustment," IEEE APEC'05, pp - , 2005.

[3] A. Kelly, K. Rinne,"A Self-Compensating Adaptive Digital Regulator for Switching Converters Based on Linear Prediction," IEEE APEC'06, pp 712-718, 2006.

[4] Y. Wen, S. Xiao, Y. Jin, I. Batarseh, "Adaptive Nonlinear Compensation for Asymmetrical Half Bridge DC-DC Converters," IEEE APEC'06, pp 731-736, 2006.

[5] L. Guo, J. Y. Hung, and R. M. Nelms, "Digital controller design for buck and boost converters using root locus," IEEE IECON'03, pp. 1864-1869, 2003.

[6] H. Guo, Y. Shiroishi, and O. Ichinokura, "Digital PI controller for high frequency switching DC/DC converter based on FPGA," IEEE INTELEC'03, pp-536-541, 2003.

[7] M. He, J. Xu, "Nonlinear PID in Digital Controlled Buck Converters," IEEE APEC'07, pp 1461-1465, 2007.

[8] R.D. Middlebrook, S. Cuk, "A General Unified Approach to Modeling Switching-Converter Power Stages," IEEE Power Electronics Specialists Conference (PESC) 1976, pp. 18-34.

[9] T. Ninomiya, M. Nakahara, T. Higashi, K. Harada, "A Unified Analysis of Resonant Converters," IEEE Transactions on Power Electronics Vol. 6. No. 2. April 1991, pp. 260-270.

11

Air-Conditioning PID Control System with Adjustable Reset to Offset Thermal Loads Upsets

Takanori Yamazaki[1], Yuji Yamakawa[2], Kazuyuki Kamimura[3] and Shigeru Kurosu[4]
[1]Oyama National College of Technology,
[2]Univ. of Tokyo,
[3]National Institute for Environmental Studies,
[4]Research Inst. "Crotech",
Japan

1. Introduction

The heating, ventilating, and air-conditioning (HVAC) systems have huge different characteristics in control engineering from chemical and steel processes. One of the characteristics is that the equilibrium point (or the operating point) usually varies with disturbances such as outdoor temperature (or weather conditions) and thermal loads. The variations of the operating point intend to vary parameters of a plant model. Thus, the HVAC control systems are extremely difficult to obtain an exact mathematical model (Kasahara 2000). Proportional-plus-integral (PI) controllers have been by far the most common control strategy as the complexity of the control problem increased (Åström 1995).
Today, a variable air volume (VAV) system has been universally accepted as means of achieving energy efficient and comfortable building environment. While the VAV control strategies provide a high quality environment for building occupants, the VAV system analysis rarely receives the attention it deserves. As a result, basic control strategies for the VAV system have remained unchanged up to now (Hartman 2003).
In addition, applying the model predictive control method to the HVAC systems, the control performance has been highly improved by pursuing the deviation from the operating point (Taira 2004). According to this report, recognizing the deviation from the operating point and calculating the optimal control inputs about the newly obtained operating point on next sampling time, the control system gives better responses than the traditional feedback control system.
Motivated by these considerations in these reports, we consider the room temperature and humidity controls using the adjustable resets which compensate for thermal loads upsets. One of the primary objectives of the HVAC systems is to maintain the room air temperature and humidity at the setpoint values to a high quality environment for building occupants. The room temperature and humidity control systems may be represented in the same block-diagram form as single-variable, single-loop feedback control systems because this interaction is weak relative to the desired control performance.

In some applications, disturbances can be estimated in advance before they entered the plant. Particularly, in the HVAC systems, it is possible that the outdoor thermometer detects sudden weather changes and the occupant roughly anticipates thermal loads upsets. Using this information, disturbances can be offset by the compensation of the reset, which is the exactly same function as an integral (I) control action. In the previous paper, the compensation method of the reset for PID controllers was proposed and the control system for room air temperature was often effective in reducing thermal loads upsets (Yamakawa 2010).

In this paper, of special interest to us is how to tune PID parameters more effective for the room temperature and humidity control. And the control performances for compensation of the adjustable reset are compared with the traditional method of the fixed reset. Namely, obtaining the approximate operating point using outdoor temperature and thermal loads profiles and adjusting the reset, the stabilization of the control system will be improved. The validation simulations will be demonstrated in terms of three performance indices such as the integral values of the squared errors, total control input, and PID control input.

2. Plant and control system

In this paper, we consider only the cooling mode of operation in summer and therefore refer to this system as a room air cooling system. The definition of variables in Equations is described in NOMENCLATURE.

2.1 Dynamics of air-conditioning system

To explore the application of PID controllers to the room temperature and humidity control system, we consider a single-zone cooling system, as shown in Figure 1. It is due to the fact that cooling and heating modes are found to perform nearly the same under most circumstances. The controlled room (the controlled plant) measures 10 m by 10 m by 2.7 m and is furnished with an air-handling unit (AHU) consisting of the cooling coil and the humidifier to control room air temperature and humidity. In general, since the responses of the AHU are faster than those of the controlled room, the dynamics of the AHU may be neglected for all practical purposes. Thus, as will be seen later, this rough assumption may be fairly validated. The model, however, possesses the important elements (the controlled room and the AHU) to analyze the air-conditioning system.

With this system, the room air temperature (θ) and relative humidity (φ) are measured with a thermometer and a hygrometer (sensors). The output signals from the sensors are amplified and then fed back to the PID controllers. Using the errors defined as the differences between the setpoint value (θ_r and φ_r) and the measured values of the controlled variables (θ and φ), the PID controllers generate the control inputs for the actuators (the supply air damper and the humidifier) so that the errors are reduced. The AHU responds to the control inputs (f_s and x_s (is adjusted by humidifier h)) by providing the appropriate thermal power and humidity to the supply airflow. Air enters the AHU at a warm temperature, which decreases as air passes the cooling coil, and then the humidifier supplies steam to cooled air if necessary. This occurs in a momentary period because there are a lot of times when the humidifier is not running. In this AHU, a dehumidifier is not installed, so an excessive demand for humidity is difficult to achieve.

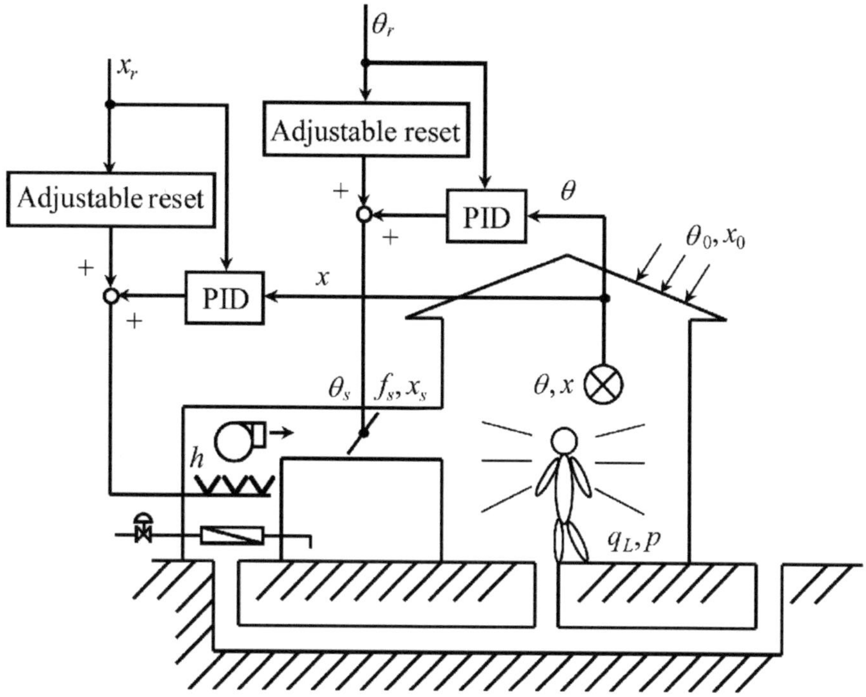

Fig. 1. Overall structure of a single-zone cooling system.

2.1.1 Room temperature model

Simplifying this thermal system to be a single-zone space enclosed by an envelope exposed to certain outdoor conditions is of significant interest to treat the fundamental issues in control system design (Zhang 1992, Matsuba 1998, Yamakawa 2009). This simplified thermal system (the room temperature model) can be obtained by applying the principle of energy balance,

$$C\frac{d\theta}{dt} = w_s(\theta_s - \theta) + \alpha(\theta_0 - \theta) + q_L \quad (1)$$

where

C = overall heat capacity of air-conditioned space [kJ/K],
α = overall transmittance-area factor [kJ/min K],
q_L = thermal load from internal heat generation [kJ/min],
$w_s = \rho_a c_p f_s$ [kJ/min K], which is heat of supply air flowrate,
ρ_a = density of air [kg/m^3],
c_p = specific heat of air [kJ/kg K],
f_s = supply air flowrate [m^3/min].

The physical interpretation of Equation 1 is that the rate of change of energy in the room is equal to the difference between the energy supplied to and removed from the room. The first term on the right-hand side is the heat loss which is controlled by the supply air flowrate. The second term is the heat gain through the room envelope, including the warm air infiltration due to the indoor-outdoor temperature differential. The third term is the

thermal loads from the internal heat generation and the infiltration. In this simplified model, any other uncontrolled inputs (e.g., ambient weather conditions, solar radiation and inter-zonal airflow, etc) are not considered.
It should be noted that all variables such as θ, θ_s, θ_0, q_L and w_s in Equation 1 are obviously the function of a time t. For the sake of simplicity the time t is not presented. When realizing a digital controller, a deadtime exists between the sampling operation and the outputting time of control input, thus w_s, namely f_s, includes a deadtime L_P.
These plant parameters have been obtained by experimental results (National Institute for Environment Studies in Tsukuba, Yamakawa 2009). The room dynamics can be approximated by a first-order lag plus deadtime system from the experimental data (Åström 1995, Ozawa 2003). Thus, the plant dynamics including the AHU and the sensor can be represented by,

$$P(s) = \frac{K_P}{T_P s + 1} e^{-L_P s} = \frac{0.64}{18s + 1} e^{-2.4s} . \tag{2}$$

Comparing to Equation 1, the plant gain (K_P) and the time constant (T_P) can be given by,

$$K_P = \frac{\theta_s}{w_s + \alpha},\ T_P = \frac{C}{w_s + \alpha},\ w_s = \rho_a c_p f_s . \tag{3}$$

Therefore, K_P and T_P change with the control input (the supply air flowrate f_s). Similarly, it is assumed that L_P changes with the control input. Namely,

$$L_P = \frac{L_{P0}}{w_s + \alpha}, \tag{4}$$

where L_{P0} is determined so that L_P is equal to 2.4 [min] when f_s is equal to 50 [%]. From L_P = 2.4 [min], $w_s = \rho_a c_p f_s$ = 10.89 [kJ/min K] and α = 9.69 [kJ/min K], L_{P0} can be obtained to be equal to 49.4 [kJ/K]. It is easily be found that these parameters are strongly affected by the operating points. Carrying out an open-loop experiment in the HVAC field to measure K_P, T_P and L_P is one way to get the information needed to tune a control loop.
To get some insight into the relations between Equation 1 and Equation 2, we will describe a bilinear system in detail (Yamakawa 2009). Introducing small variations about the operating points and normalizing the variables, Equation 1 has been transformed to a bilinear system with time delayed feedback. A parametric analysis of the stability region has been presented.
The important conclusion is that the stability analysis demonstrated the validity of PID controllers and there was no significant advantage in analyzing a bilinear system for VAV systems. It was fortunate that the linear system like a first-order lag plus a deadtime system derived in Equation 2 often satisfactorily approximated to the bilinear system derived in Equation 1. The linear system is an imaginary system, but it does represent it closely enough for some particular purpose involved in our analysis.
Certainly the linear model derived in Equation 2 can be used to tune the PID controller and the physical model derived in Equation 1 can be used for numerical simulations. Over the range upon which this control analysis is focused, the relations between Equation 1 and Equation 2 are determined to be sufficiently close.

2.1.2 Room humidity model

The room humidity model can be derived by applying the principle of mass balance,

$$V\frac{dx}{dt} = f_s(x_s - x) + \frac{n}{\rho_a}p \tag{5}$$

where

V = room volume (10×10×2.7[m³])
x = absolute humidity of the room [kg/kg (DA)]
x_s = absolute humidity of the supply air [kg/kg (DA)]
p = evaporation rate of a occupant (0.00133 [kg/min])
n = number of occupants in the room [-].

Equation 5 states that the rate of change of moisture in the room is equal to the difference between the moisture removed from and added to the room. The first term expresses a dehumidifying effect by the supply air flowrate. The second term is the moisture due to the occupants in the room. The absolute humidity x can be converted to the relative humidity φ as described in the next section.

In the same way as the room temperature model, the humidity model can be approximated by a first-order lag plus deadtime system as shown in Equation 2. Thus, the plant dynamics concerned with the room humidity model can be represented by,

$$P'(s) = \frac{K_{Ph}}{T_{Ph}s+1}e^{-L_{Ph}s} = \frac{1.0}{13.5s+1}e^{-2.4s}. \tag{6}$$

The gain constant K_{Ph} and the time constant T_{Ph} are given by,

$$K_{Ph} = \frac{f_s}{f_s} = 1,\ T_{Ph} = \frac{V}{f_s}. \tag{7}$$

Thus, K_{Ph} and T_{Ph} change with the supply air flowrate as same as those represented in the room temperature model. Similarly, the deadtime L_{Ph} is assumed to be changed with the supply air flowrate. Thus,

$$L_{Ph} = \frac{L_{Ph0}}{f_s}, \tag{8}$$

where L_{Ph0} is the constant. The deadtime L_{Ph} of the humidity model is assumed in the same way as one of the temperature model. Thus, the deadtime L_{Ph0} can be calculated by $L_{Ph}\times f_s$ = 2.4×8.33 = 19.99.

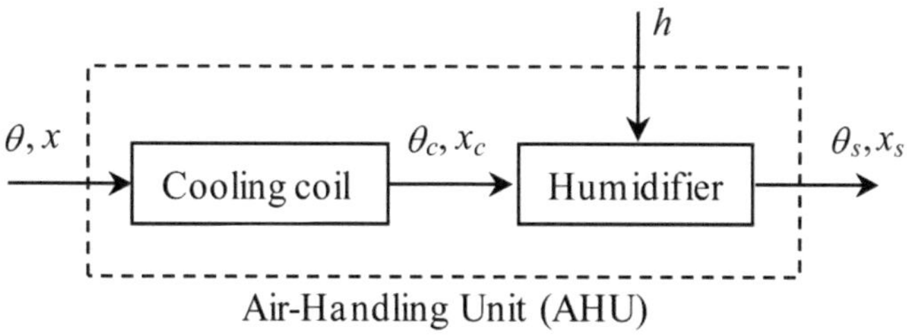

Fig. 2. Block diagram for AHU.

The room humidity can be determined by regulating the moisture of the supply air to the room. This implies that the room humidity can be indirectly controlled. Similarly the first-order lag plus a deadtime model by Equation 6 can be used to tune the PID controller and the physical model by Equation 5 can be used in numerical simulations. It does not mean that Equation 5 and 6 are mathematically equivalent.

2.1.3 Air-handling unit (AHU) model

Figure 2 shows the simple block diagram for the AHU that conditions supply air for the room. Air brought back to the AHU from the room is called return air. The portion of the return air discharged to the outdoor air is exhaust air, and a large part of the return air reused is recirculated air. Air brought in intentionally from the outdoor air is outdoor air. The outdoor air and the recirculated air are mixed to form mixed air, which is then conditioned and delivered to the room as supply air.

The AHU consists of a cooling coil, a humidifier, and a fan to control supply air temperature (θ_s) and humidity (x_s). The mixed air enters the cooling coil at a given temperature θ, which decreases as the air passes through the cooling coil. The temperature of the air leaving the cooling coil is θ_c. Since the responses of the cooling coil and the humidifier are significantly faster than those of the room (a principal controlled plant), it can be generally assumed that the cooling coil and the humidifier are static systems. Namely, it is common for the cooling coil to be controlled to maintain the supply air temperature at a setpoint value (θ_{sr}). Thus, the temperature (θ_c) and the absolute humidity (x_c) of the cooling coil can be given by;

$$\begin{aligned} \theta_c &= \theta_{sr} \\ x_c &= \begin{cases} x_{si} & (p_w \le p_{ws}) \\ \dfrac{0.622 p_{ws}}{P - p_{ws}} & (p_w > p_{ws}) \end{cases} \end{aligned} \tag{9}$$

where θ_{sr} is the setpoint of the supply air temperature, p_w is the partial pressure of water vapor, p_{ws} is the partial pressure of saturated vapor at temperature, P (=101.3 [kPa]) is the total pressure of mixed air, and x_{si} is the absolute humidity of the air entering the cooling coil. The humidity is divided into two calculations depending on the difference between p_{ws} and p_w. This constraint means that the relative humidity does not exceed 100 %.

The humidifier is the most important actuator to control the room relative humidity (φ) for heating mode in winter. Nevertheless, we are interested here in examining control characteristics in the operation mode of cooling. Note that the control input $h(t)$ does not have strong effect on the room relative humidity (φ) in cooling mode. From the energy and mass balances, the dynamics of the humidifier can be described by,

$$\begin{aligned} C_{ad}\frac{d\theta_s}{dt} &= w_s(\theta_c - \theta_s) + \alpha_d(\theta_0 - \theta_s) + q_B + q_d \\ V_d\frac{dx_s}{dt} &= f_s(x_c - x_s) + \frac{h}{\rho_a} \end{aligned} \tag{10}$$

where

C_{ad} = overall heat capacity of humidifier space [kJ/K],

V_d = room volume of humidifier [m³],
α_d = overall transmittance-area factor [kJ/min K],
q_B = fan load (59.43 [kJ/min]),
q_d = load by humidifier ($(190.1 - 1.805\theta_h)h$) [kJ/min]), and
h = rate of moist air produced in the humidifier.

Considering the steady-state of the dynamics of the humidifier, the supply air temperature θ_s and the supply air absolute humidity x_s can be obtained by,

$$\theta_s = \frac{c_p \rho_a f_s \theta_c + \alpha_d \theta_0 + q_B + q_d}{c_p \rho_a f_s + \alpha_d}$$
$$x_s = x_c + \frac{h}{f_s \rho_a} \tag{11}$$

As can be seen in Equation 11, the supply air temperature (θ_s) can be influenced by the humidifier (h), so that the errors in the reset (f_{s0}) can be produced. Thus, the control performance may be deteriorated.

The air flowrate from the outdoor air is considered 25% of the total supply air flowrate. This ratio will be held constant in this study. Note that the pressure losses and heat gains occurring in the duct have negligible effects on the physical properties of air for simplification. The absolute humidity of mixed air entering the cooling coil can be described by,

$$f_s x_{si} = 0.25 f_s x_0 + 0.75 f_s x\,. \tag{12}$$

where x_0 and x are the absolute humidity of outdoor air and of indoor air, respectively. All the actual values of the plant parameters used in the numerical simulations are listed in Table 1. Since we assume that the supply air temperature for the cooling coil can be controlled so as to maintain the setpoint value (θ_{sr}) of the supply air temperature, the energy-balance of mixed air is not needed to consider.

C	370.44 [kJ/K]
V	270 [m³]
c_p	1.3 [kJ/kg K]
ρ_a	1.006 [kg/m³]
α	9.69 [kJ/min K]
α_d	0.1932 [kJ/min K]
q_L	121.72 [kJ/min]
f_{smax}	16.66 [m³/min]
f_{smin}	0.00 [m³/min]
h_{max}	0.33 [m³/min]
h_{min}	0.00 [m³/min]
θ_{sr}	13.1 [°C]

Table 1. Summary of significant parameters in the development of the room and the AHU

2.1.4 Calculation of relative humidity

In this section, the conversion from the absolute humidity to the relative humidity is briefly explained. The relative humidity is derived from the air temperature and the absolute humidity of the air (ASHRAE 1989; Wexler and Hyland 1983).
First, the air temperature must be converted to the absolute temperature as,

$$\Theta_a = \theta_a + 273.15 , \tag{13}$$

where θ_a is the air temperature, and Θ_a is the absolute temperature of the air.
Second, to evaluate the supply air temperature θ_c reaches its dew-point temperature, the two partial pressures p_w and p_{ws} can be conveniently defined. The partial pressure of water vapor p_w can be obtained by,

$$p_w = \frac{Px_i}{0.622 + x_i} , \tag{14}$$

where x_i is the absolute humidity of water vapor and P is the total pressure of mixed air (101.3 [kPa]). And, the partial pressure p_{ws} of saturated vapor at temperature Θ_a can be given by,

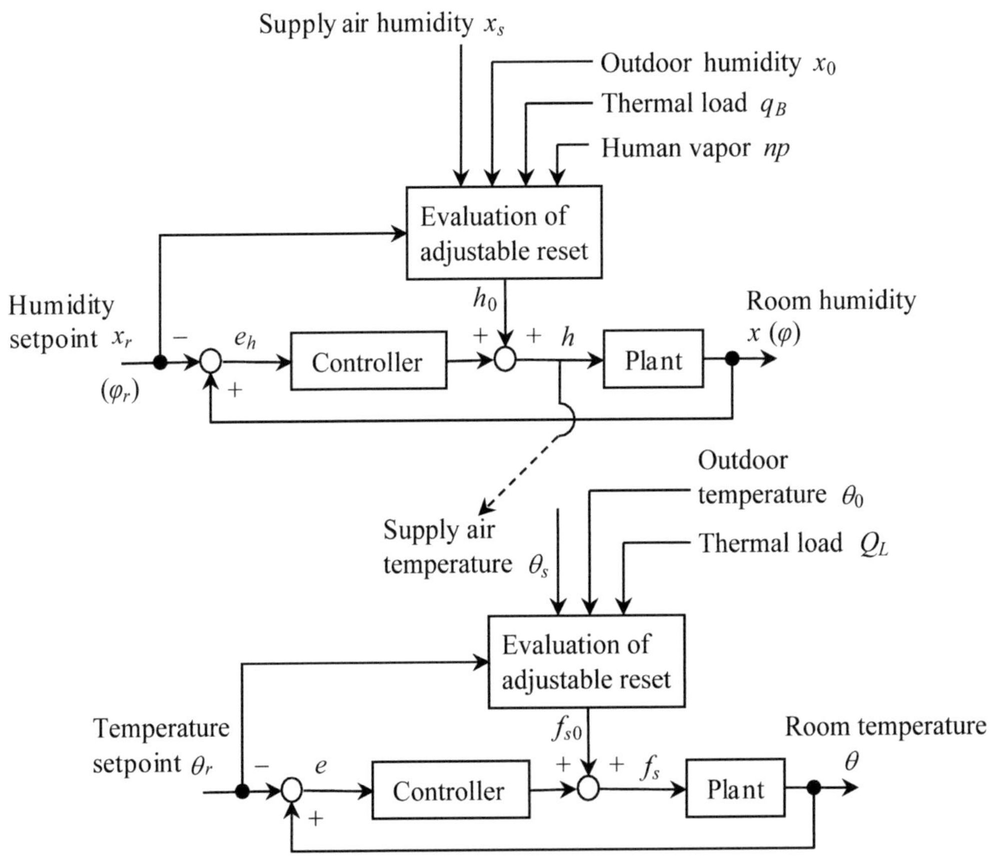

Fig. 3. Overall of the temperature-humidity control system.

$$\ln(10^3 p_{ws}) = -0.58002206\times10^4 / \Theta_a + 0.13914993\times10 \\ -0.48640239\times10^{-1}\Theta_a + 0.41764768\times10^{-4}\Theta_a^{\ 2} \\ -0.1445293\times10^{-7}\Theta_a^{\ 3} + 0.65459673\times10\times\ln\Theta_a \quad (15)$$

Finally, the relative humidity φ for the room can be given by,

$$\varphi = \frac{p_w}{p_{ws}}\times100 \, . \quad (16)$$

2.2 Control system

Figure 3 shows a block diagram of the room temperature and humidity control systems using adjustable resets which compensate for thermal loads upsets. In this figure, signals appear as lines and functional relations as blocks. The primary controlled plant is the room. The cooling coil, the humidifier and the damper are defined as the secondary controlled plants (to produce appropriate actuating signals). The following control loops are existed in our room temperature and humidity control system:

- Room air temperature control system
- Room air humidity control system

The control outputs of interests are room air temperature (θ) and relative humidity (φ). In order to maintain room air temperature and humidity in desirable ranges, traditional PID controllers have been used to reduce component costs. The control inputs that vary according to the control actions are the supply air flowrate (f_s) and the rate of moist air produced in the humidifier (h), which will be discussed in more detail.

2.2.1 Room temperature control system

Taking the PID control algorithm into account, one of control inputs, related to the room air temperature (θ) can be given by,

$$f_s(t) = k_p e(t) + k_i\int_0^t e(\tau)d\tau + k_d\frac{de(t)}{dt} + f_{s0}(t) \quad (17)$$

where $f_{s0}(t)$ is the manual reset. In electronic controllers, the manual reset is often referred to as "tracking input". The error $e(t)$ can be defined by,

$$e(t) = \theta(t - L_P) - \theta_r, \quad (18)$$

where θ_r is the setpoint value of the room air temperature, and L_P (= 2.4 [min]) is the deadtime. The PID parameters (the proportional gain k_p, the integral gain k_i, and the derivative gain k_d) can be determined by the well-known tuning method. The inherent disadvantage of the I action, which easily causes instabilities, can be reduced by varying the reset $f_{s0}(t)$ to compensate for thermal loads upsets (disturbances). In some cases of HVAC systems, the reset $f_{s0}(t)$ can be estimated by knowledge of the plant dynamics.

Equation 17 can be given in a discrete-time system when control input and error signal are respectively assumed to be $f_s(k)$ and $e(k)$ at time kT (T is the sampling period).

$$f_s(k) = k_p e(k) + k_i T\sum_{j=0}^{k}\frac{e(j-1)+e(j)}{2} + \frac{k_d}{T}\{e(k) - e(k-1)\} + f_{s0}(t) \quad (19)$$

This is called the position algorithm because $f_s(k)$ typically represents the position of an actuator (Takahashi 1969).
From Equation 1, the operating point at its steady-state can be written:

$$w_s(\theta_s - \theta) + \alpha(\theta_0 - \theta) + q_L + Q = 0 \ \ (w_s = \rho_a c_p f_s). \tag{20}$$

The reset (f_{s0}) of the supply air flowrate can be obtained by,

$$f_{s0}(t) = \frac{q_L(t) + q_{th}(t) + \alpha(\theta_0(t) - \theta_r(t))}{c_p \rho_a(\theta_r(t) - \theta_s(t))}. \tag{21}$$

In Equation 21, the supply air temperature (θ_s), the outdoor temperature (θ_0), and the setpoint (θ_r) can easily be measured. However, thermal loads cannot be specified in advance. Thus, it is recommended that occupants must roughly estimate thermal loads to improve the control performance at adequate sampling interval. For example, three of the rough estimates for compensation can be used as:

the maximum (75%), the medium (50%), and the minimum (25%),

where 100 % means the maximum supply air flowrate 16.66 [m^3/min].At any given point of operation, the reset (f_{s0}) to offset thermal loads can be easily calculated using Equation 21. Thus, it can be concluded that the controller with lower I action is superior to that with no I action, and is also called a PD controller.

2.2.2 Room humidity control system

To control the room air relative humidity, another one of control inputs that vary according to the control actions is the rate of moist air produced in the humidifier $h(t)$. The control input can be given by,

$$h(t) = k_{ph} e_h(t) + k_{ih} \int_0^t e_h(\tau) d\tau + k_{dh} \frac{de_h(t)}{dt} + h_0(t), \tag{22}$$

where $h_0(t)$ is the reset. The error $e_h(t)$ can be defined by,

$$e_h(t) = \varphi_r - \varphi(t - L_{Ph}) \tag{23}$$

where φ_r is the setpoint value of the room air relative humidity and L_{Ph} (= 2.4 [min]) is the deadtime. The hygrometer in the room can detect the room air relative humidity (φ), but not the absolute humidity (x). Therefore, the relative humidity is used in the error $e_h(t)$ for the calculation of the control input $h(t)$. However, the humidity model can be described by the relational expression of the absolute humidity. And, the derivation of the humidity model parameters from the experimental results in terms of the relative humidity may be extremely difficult. As a result, PID parameters (proportional gain k_{ph}, integral gain k_{ih}, and derivative gain k_{dh}) must be determined by trial and error under the consideration that the absolute humidity cannot be directly measurable. In this study, for the sake of simplicity, it is assumed that the basic relation of the humidity model is invariant even if the variable in the humidity model is changed the absolute humidity into the relative humidity. For this reason, the traditional tuning method (Ziegler and Nichols 1942) for the first-order lag plus deadtime system as shown in Equation 6 (the plant parameters is described by Equations 7 and 8) can be used.

Since the supply air temperature (θ_s) can be affected by the rate of moist air produced in the humidifier (h), the reset of the supply air flowrate (f_{s0}) arising from moist air variations must be accounted. This means that good control performance for heating mode can be expected.

The reset $h_0(t)$ for the humidifier can be obtained from Equations 5, 9, 11, and 12 as follows: First, taking the humidity model (Equation 5) at the steady-state and the setpoint value x_r of the absolute humidity into account, the following equation can be obtained by,

$$f_s(x_s - x_r) + \frac{n}{\rho_a} p = 0 . \tag{24}$$

Second, substituting Equation 24 into Equation 9, 11 and 12, Equation 24 can be rewritten by,

$$f_s\left(x_c + \frac{h_0}{\rho_a f_s} - x_r\right) + \frac{n}{\rho_a} p = 0$$

$$f_s\left(0.25x_0 + 0.75x_r + \frac{h_0}{\rho_a f_s} - x_r\right) + \frac{n}{\rho_a} p = 0$$

$$0.25x_0 f_s - 0.25x_r f_s + \frac{h_0}{\rho_a} + \frac{n}{\rho_a} p = 0$$

$$\frac{h_0}{\rho_a} = 0.25x_r f_s - 0.25x_0 f_s - \frac{n}{\rho_a} p$$

$$h_0(t) = 0.25 f_{s0} \rho_a (x_r - x_0) - np \tag{25}$$

However, as will be seen in Equation 25, the first term is small in comparison to the second term and $h_0(t)$ may be negative. The adjustable reset $h_0(t)$ can be found to be nearly zero under most circumstances in the present work.

Table 2 provides PID parameters tuned by the traditional ultimate sensitivity method (Ziegler and Nichols 1942) and the empirical modified PID method.

The ultimate sensitivity method is simple and intuitive. It has been still widely used, either in its original form or in some modification. Since it only gives "ball-park" values, it is necessary to make manual tuning to obtain the desired performance. Our empirical modified PID controller can help improve the time response of a control system because thermal loads and operating conditions are changing continuously in HVAC systems.

In modified PID parameters for room air temperature control, the proportional gain (k_p) is about 80 % of that of the conventional tuning method. The integral gain (k_i) is one-fourth of that of the conventional tuning method. The derivative gain (k_d) is nearly the same as that of the conventional tuning method. In modified PID parameters for room air relative humidity control, all gains (k_{ph}, k_{ih}, and k_{dh}) are nearly one-tenth of those of the conventional tuning method.

3. Simulation results in daily operation

To illustrate the control performance of the room temperature and humidity control systems, several simulation runs are made. Representative outdoor temperature and thermal loads profiles for one-day (between 08:00 in the morning and 08:00 in the next morning) are assumed as shown in Figure 4. These profiles are based on the experimental data obtained from the National Institute for Environmental Studies in Tsukuba, Japan. In the right hand side of Figure 4, the dashed line depicts the artificial estimated value of the thermal load. At the start-up (at 08:00 in the morning), the feedback control system takes over and controls the room air temperature and relative humidity. These simulation runs are carried out under the same conditions mentioned above. Figure 5 depicts the adjustable reset (f_{s0}) of the supply air flowrate for daily operation calculated using Equation 21. The computational interval of 1 hour (60 min) for adjusting the reset is used in this control. These simulation runs are made on MATLAB which is an effective tool for field engineers in control engineering.

The following control configurations are used in our room temperature and humidity control. These abbreviations are common throughout the remainder of this paper.

	k_p	k_i (T_i)	k_d (T_d)
Conventional PID	11.65	2.55 (4.57)	13.26 (1.16)
Modified PID	8.73	0.8 (10.9)	10 (1.15)

(a) Temperature control

k_{ph}	k_{ih} (T_{ih})	k_{dh} (T_{dh})
1.22	0.26 (4.65)	1.41 (1.16)

(b) Humidity control

(T_i, T_{ih}: integral times, T_d, T_{dh}: derivative times for temperature and humidity controls, respectively)

Table 2. PID parameters.

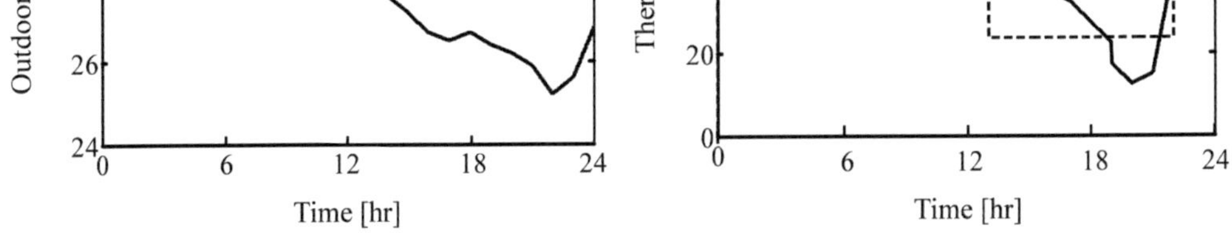

Fig. 4. Outdoor temperature and thermal loads profiles.

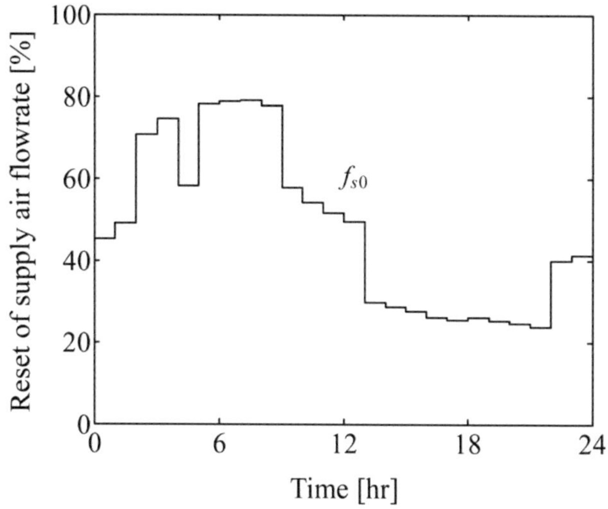

Fig. 5. Reset of supply air flowrate.

- Number of control outputs of interest
  Room temperature and humidity control
  This refers to the room air temperature and relative humidity control.
- Setpoints of control outputs

Regarding the room air temperature θ_r:

1. Fixed setpoint
   The setpoint θ_r is fixed at 24 °C for daily operation.
2. Variable setpoint
   The setpoint θ_r are varied within the range, that θ_r is set at the value (θ_0 – 4) °C where θ_0 is the outdoor temperature, and θ_r is limited to the minimum 20 °C and the maximum 28 °C.

Regarding the room air relative humidity φ_r:
The setpoint φ_r is usually fixed at 55 % for daily operation.

- Control strategies for the reset

1. Conventional PID control
   This refers to conventional PID control with the fixed reset (f_{s0} = 50 %).
2. Modified PID control
   This refers to modified PID control with the adjustable reset (Figure 5).

- Performance indices

The control performance should be evaluated by defining three performance indices.

1. ISE (the integral of squared error)
   $$\text{ISE} = \int_0^{24} e^2 dt$$
2. ICI (the integral of control input)
   $$\text{ICI} = \int_0^{24} f_s dt$$
3. IPID (the integral of control input produced in PID controller only)
   $$\text{IPID} = \int_0^{24} (f_s - f_{s0}) dt$$

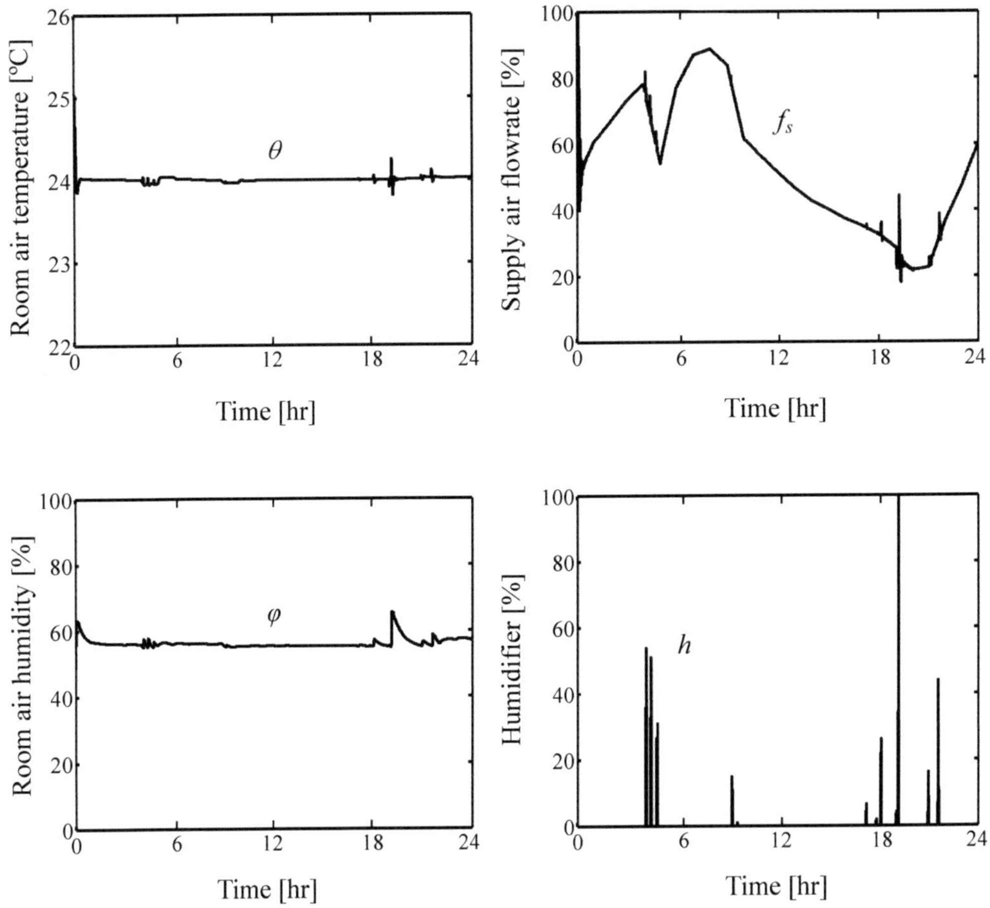

Fig. 6. Simulation results of conventional PID.

Room temperature and humidity control

Typical daily simulation results show that the conventional PID and the suitably modified PID controllers can maintain the room air temperature and relative humidity close their respective setpoints irrespective of variable thermal loads. The method of determining PID parameters for the modified controller is practical for room temperature and humidity control systems.

Fixed setpoint

Figure 6 and 7 show the responses to the fixed setpoint of the room air temperature for the cases of the conventional PID and the modified PID controls, respectively.

In Figure 6, there are sudden changes in θ and φ during the initial few hours, which then settle to setpoints. We can expect that, since the transient responses of θ and φ will also change rapidly, θ and φ are very close to their setpoints even though θ_0 and q_L are varied. The supply air flowrate illustrates instabilities locally due to humidifier working.

When looking over results of Figures 6 and 7, it should be noted that the responses (θ and φ) of the conventional PID control and the modified PID control are somewhat different.

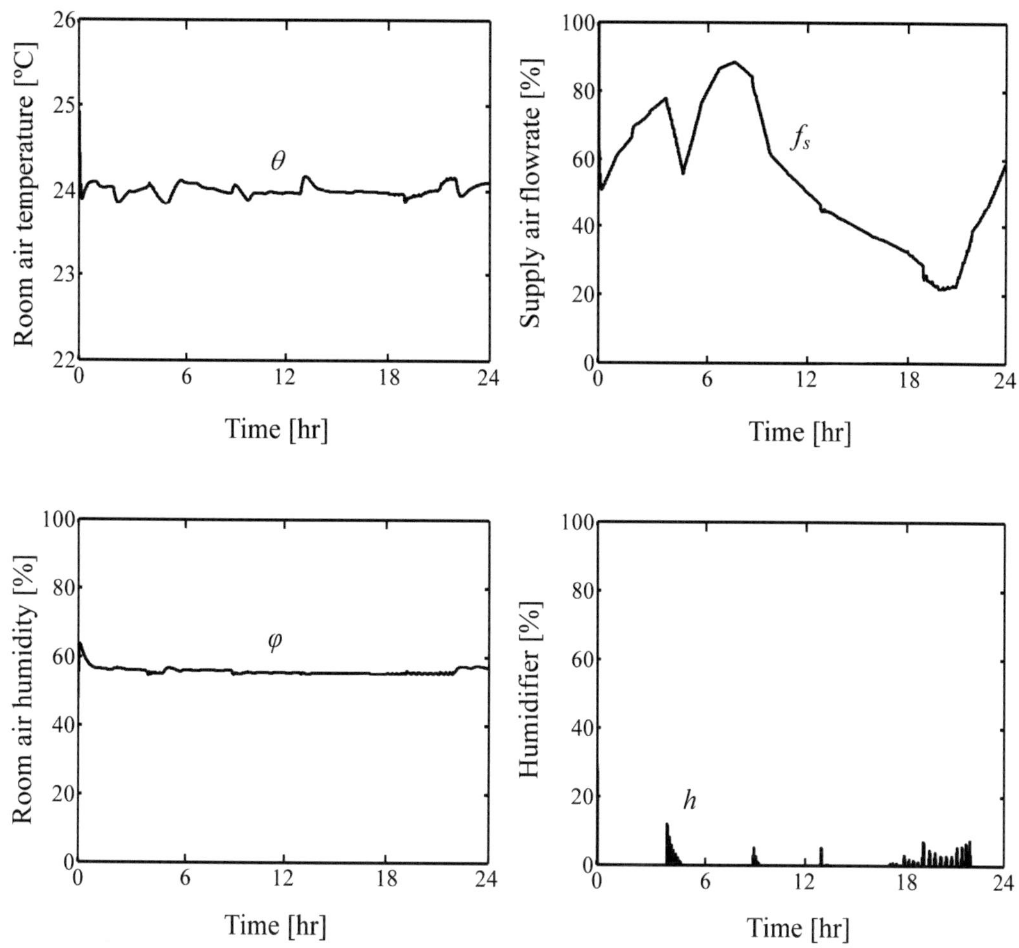

Fig. 7. Simulation results of Modified PID.

	Conventional PID	Modified PID
ISE	3.09	7.95
ICI	2.08×10^4	2.08×10^4
IPID	8742	2734

Table 3. Comparison of control performance indices to fixed setpoint.

Because the reset for the modified PID control can be adjusted very often, it becomes difficult to maintain θ and φ at the setpoints, so θ fluctuates around the setpoint. It is clear that the results for modified PD control cannot represent an improvement over those for the conventional PID control. For small values of the integral gain (k_i) for the modified PID control, θ creeps slowly towards the setpoint. However, as will be seen in the near future, this disadvantage may be clearly solved.

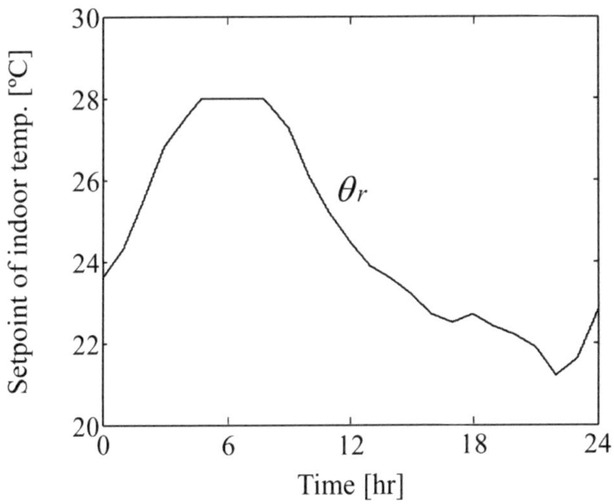

Fig. 8. Variable setpoint profile.

Table 3 shows that the results of the validation simulations in terms of three performance indices. For the ISE (tracking accuracy), it is evident that the sharply change of the reset aggravates the tracking accuracy of θ for the modified PID control, but it is enhanced by increasing the integral gain (k_i). Further investigation into the total amount of control inputs (ICI and IPID) can lead to some interesting results. It is recognized that the ICI is exactly the same for the two control strategies. The physical interpretation of this fact is that there is no difference of supply air flowrates between two control strategies. However, for the IPID, the modified PID control clearly represents an improvement over the conventional PID control. As a matter of fact, the merit of the modified PID control becomes obvious when the maximum capacity of the controller is limited.

Variable setpoint

Figure 8 depicts the setpoint profile ((θ_0 – 4) [°C]) of room air temperature depending on the outdoor temperature on a typical day. The responses to the variable setpoint for the cases of the conventional PID and the modified controls are shown in Figure 9 and 10, respectively. The room air temperature and humidity follow their respective setpoint profiles even though thermal loads are variable. It is apparent from Figure 9 that the solid areas indicate rapidly oscillating values due to hunting when the humidifier is positioned between 0 % and 100 %. Subsequently, the room air temperature can be oscillated with the occurrence of such huntings. The same trend is also apparent in the supply air flowartes.

It can be seen from Figure 10 that suitably tuned modified controller can maintain the room air temperature and humidity close to their respective setpoints suppressing such huntings. The effectiveness of the modified PID control can be confirmed. By comparing these responses with those of Figure 6 and 7, it is clear that the humidifier is turned on very often and the hunting of the room air temperature may occur simultaneously.

Fig. 9 and 10 demonstrate locally rapid oscillation of the humidifier when the indoor relative humidity φ becomes below the setpoint 55 %. This is due to the fact that the humidifier is very sensitive to control inputs. There are also many technological problems to be solved when we make positive use of the humidifier in cooling operation.

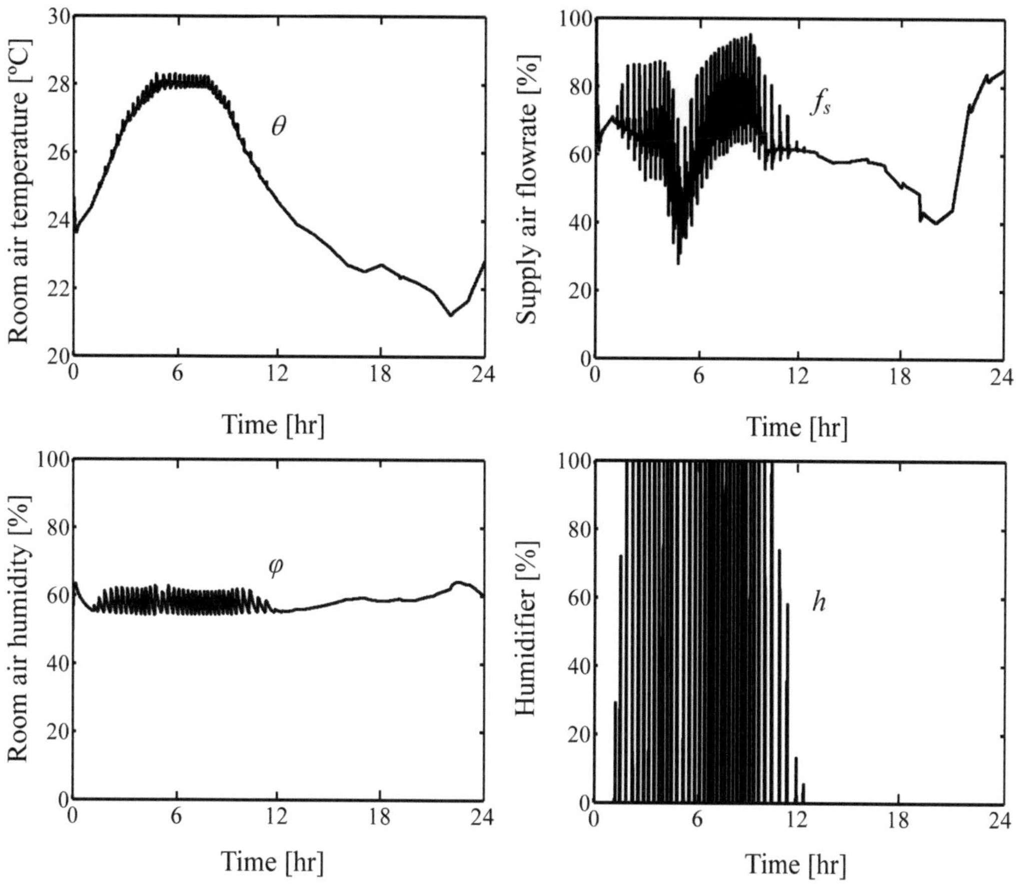

Fig. 9. Simulation results of conventional PID.

	Conventional PID	Modified PID
ISE	5.98	8.83
ICI	1.92×10^4	1.92×10^4
IPID	4276	3005

Table 4. Comparison of control performance indices to variable setpoint.

Table 4 represents the control performance indices obtained by typical daily simulation results. A comparison with Table 3 shows that there is very little difference in performance between the fixed setpoint and the variable setpoint. For the ISE, the ISE for the modified PID is larger than that for the conventional PID. This means that the I action is effective for not only elimination of offset (steady-state error) but also disturbance attenuation. Tracking accuracy and disturbance attenuation will be enhanced by selecting high integral gain.

For the ICI, it is striking that the ICI values are exactly the same for two control strategies. For the IPID, the modified PID control gives slightly better results than the conventional PID control. It is concluded that the modified PID control should be also incorporated by limiting the maximum control input available to the controller.

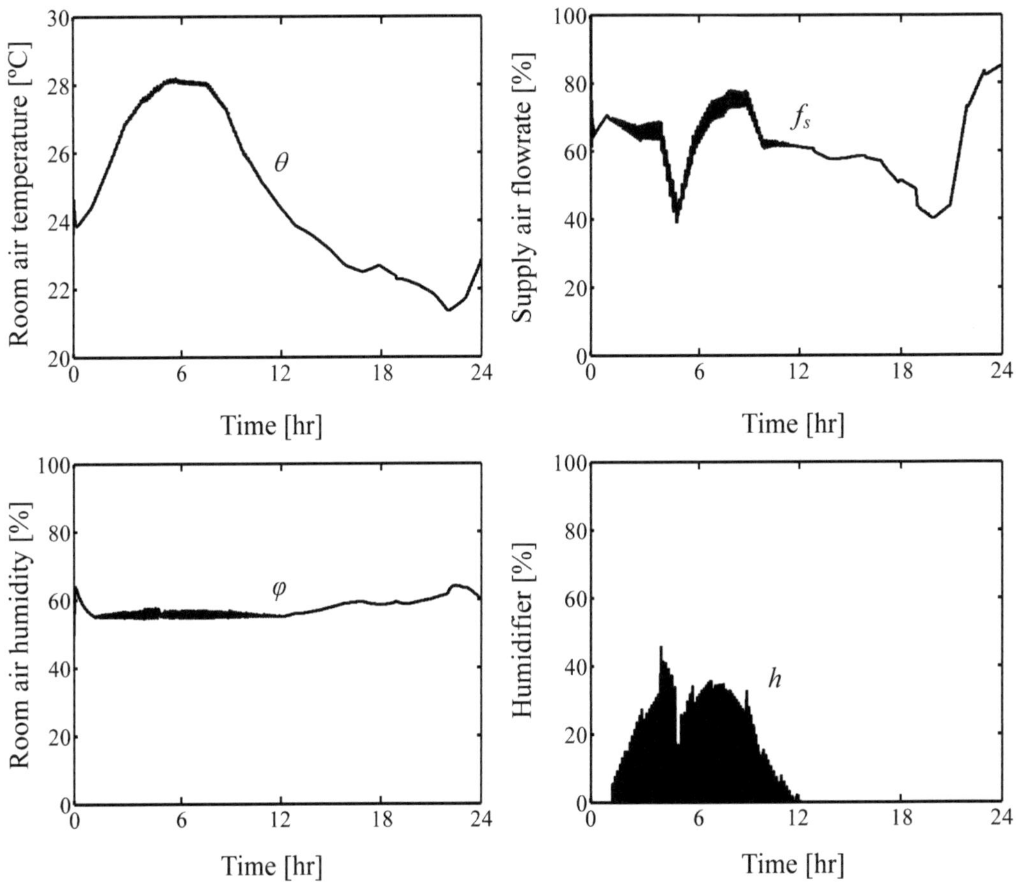

Fig. 10. Simulation results of Modified PID.

4. Conclusions

In this paper, the room temperature and humidity control systems with the conventioanl PID control using fixed reset or the modified PID control using adjustable resets which compensate for thermal loads upset are examined. The simulation results for one-day operation based on practical outdoor temperature and thermal loads profiles provide satisfactory control characteristics. The results of validation simulations are demonstrated in terms of three performance indicies (as three integrals of squared error (ISE), control input (ICI), and control input in PID controller only (IPID)).

The results obtained in this study are summarized in the following:

1. The room air temperature and humidity illustrate instabilities locally due to humidifier working.
2. By changing the setpoint of the room air temperature on the basis of the outdoor temperatures profile, the control performance can be remarkably improved.
3. In daily operation, when the reset is adjusted at every hour, the sharply change of the reset aggravate the response of the room air temperature. The response can be improved by proper selection of the computational period.

4. The proposed control strategy for the adjustable reset cannot be effective for energy-savings, but has a possibility in case that there exists a limitation of the maximum control input available to the controller.

Finally, the results given in this paper were motivated by the desire to obtain satisfactory performance with adjustable reset better than that with fixed reset. Consequently, it is concluded that there is little inherent advantages in designing the modified PID controller with adjustable reset. However, since this modified PID control lightens the total amount of control input produced in the controller, it can be good candidates for the next HVAC controllers.

The work reported here is being continued to validate several conclusions obtained by experimental results.

5. Acknowlegdment

This research was partially supported by the National Institute for Environment Studies in Tsukuba. The authors would like to acknowledge staffs of Controls Group for their contribution to this study.

6. Nomenclature

C = overall heat capacity of air-conditioned space (kJ/K)
C_{ad} = overall thermal capacity of humidifier space (kJ/K)
c_p = specific heat of air (kJ/kg K)
a = overall transmittance-area factor (kJ/min K)
a_d = overall transmittance-area factor outside humidifier (kJ/min K)
θ = indoor air temperature (°C)
θ_r = setpoint of indoor air temperature (°C)
θ_s = supply air temperature (in humidifier) (°C)
θ_{sr} = setpoint of supply air temperature (in humidifier) (°C)
θ_c = supply air temperature (in cooling coil) (°C)
θ_0 = outside temperature (°C)
ρ_a = density of air (1.3 kg/m^3)
c_p = specific heat of air (kJ/kg K)
f_s = supply air flowrate (m^3/min)
f_{s0} = reset of supply air flowrate of room (m^3/min)
w_s = $c_p \times \rho_a \times f_s$, heat of supply air flowrate (kJ/min K)
V = room volume (10×10×2.7 m^3)
V_d = room volume of humidifier (m^3)
x = indoor absolute humidity (kg/kg (DA))
x_s = absolute humidity of supply air (kg/kg (DA))
x_{si} = return air absolute humidity at the inlet of air-handling unit (kg/kg (DA))
x_0 = outdoor absolute humidity (kg/kg (DA))
φ = indoor relative humidity (%)
φ_r = setpoint of indoor relative humidity of room (%)
h = rate of moist air produced in humidifier (kg/min)
q_L = thermal load from internal heat generation (kJ/min)

q_B = fan load (59.43 kJ/min)
q_d = load by humidifier ($(190.1 - 1.805\theta_c)h$ kJ/min)
p = evaporation rate of a occupant (0.00133 kg/min)
P = total pressure of mixed air (101.3 kPa)
p_w = partial pressure of water vapor at the inlet of air-handling unit (kPa)
p_{ws} = partial pressure of saturated vapor at temperature θ_c (kPa)
h_0 = reset of rate of moist air produced in humidifier (kg/min)
n = number of occupants in the room (-)
K_P = plant gain of room temperature dynamics
T_P = time constant of room temperature dynamics
L_P = deadtime of room temperature dynamics
K_{Ph} = plant gain of room humidity dynamics
T_{Ph} = time constant of room humidity dynamics
L_{Ph} = deadtime of room humidity dynamics

7. References

Kasahara. M. et al. (2000). Physical Model of an Air-Conditioned Space for Control Analysis, *ASHRAE Transactions*, Vol. 106, Part II: pp. 304-317

Åström. K. & Hägglund. T. (1995). PID controllers: Theory, design and tuning, pp. 164-166 and 196-198, *Instrument Society of America.*

Hartman. T. (2003). Improving VAV Zone Control, *ASHRAE Journal*, pp. 24-32

Taira. U. (2004). Save-Energy Type Temperature-Humidity Simultaneous Control using Model Predictive Control Method, *Journal of Society of Instrument and Control Engineering*, Vol. 43-9, pp. 686-690

Yamakawa. Y. et al. (2010). Compensation of Manual Reset to Offset Thermal Loads Change for PID Controller, *ASHRAE Transactions*, Vol. 116, Part 1, pp. 303-315

Zhang. Z. & Nelson. R.M. (1992). Parametric analysis of a building space conditioned by a VAV system, *ASHRAE Transactions*, Vol. 98, Part 1, pp. 43-48

Matsuba. T. et al. (1998). Stability Limit of Room Air Temperature of a VAV System. *ASHRAE Transactions*, Vol. 104, Part II: pp. 257-265

Yamakawa. Y. et al. (2009). Stability of Temperature Control System in VAV Systems, *ASHRAE Transactions*, Vol. 115, Part 1, pp. 613-621

Ozawa. K. et al. (2003). A Tuning Method for PID Controller Using Optimization Subject to Constraints on Control Input. *ASHRAE Transactions*, Vol. 109, Part 1, pp. 3-11.

ASHRAE. (1989). 1989 ASHRAE Handbook–Fundamentals, *American Society of Heating, Refrigerating and Air-Conditioning Engineers, Inc.*

Wexler. A. & Hyland. R.W. (1983). ASHRAE Transactions 89(2A): 500.

Takahashi. Y., M. J. Rabins & D. M. Auslander. (1969). Control and dynamic systems, *New York, Addison-wesley*

Ziegler. J. G. & Nichols. N. B. (1942). Optimum settings for automatic controllers, *Transactions of the American Society of Mechanical Engineers*, Vol. 64-8: pp. 759

12

Remote-Tuning – Case Study of PI Controller for the First-Order-Plus-Dead-Time Systems

Dennis Brandão, Nunzio Torrisi and Renato F. Fernandes Jr
University of Sao Paulo/Sao Carlos
Federal University of ABC/Sao Paulo
Brazil

1. Introduction

Control loop tuning today is the key factor for quality improvement and optimization of production costs. Simply replacing old systems with modern technology and networked equipment with high-processing capacity does not necessarily imply an improvement in product quality or an increase in plant productivity. However, according to researches conducted in the Industrial sector, most control loops in automatic operation have tuning problems (Harris et al., 1999; Ruel, 2003; Yu, 2006).

Nowadays, systems that aid automatic tuning for control loops can be typically located at two levels in the automation systems hierarchy: they can be installed in the workstations, running together with software control systems and data acquisition or SCADA (Supervisory Control and Data Acquisition) systems; or they can be embedded within a distributed control equipment in the field, for example, in PLCs (Programmable Logic Controllers), smart transmitters, or DCS (Distributed Control System).

When embedded, automated tuning systems are usually operated by an adaptive control where the controller's parameters are continuously adjusted to accommodate changes and process disturbances. Tuning systems already installed in workstations are more advantageous than embedded systems due to their superior processing power and information storage. This characteristic leads to the development of more sophisticated algorithms, and provides additional resources, such as simulation and graphical analysis (Aström & Hägglund, 1995), (Ang et al., 2005).

Remote access systems that use the Internet as the means of communication have become widespread in recent years, both in academic researches and industrial applications. Studies such as (Avoy et al., 2004) show great potential to growth and diversification of remote applications mainly in industrial environments.

Among the advantages of remote access via the Internet in industrial applications, it is important to highlight: enterprises with distributed units can access, share, analyze and process plant floor information in real time and faster; and technical or specialized administrative services can be outsourced with a higher level of interaction between partners, thereby avoiding the need for experts in the staff.

Studies in the Literature report proposals and systems for automatic tuning and dynamic control that use the communication via the Internet in different ways (Yu et al. 2006) (Yang

et al., 2007). An important application registered in those work is related to learning and research centers that provide experiments in robotics, manufacturing control and process control for remote access over the Internet.
However, when considering the use of the Internet directly on the shop floor, observe that the nature of production and automation systems demands certain requirements that must be guaranteed, such as multiple access management, communication and control system security, maximum time interval for process data update, and the integration of different computing platforms and equipments from several technologies.
The aim of this paper is to propose and verify the technical feasibility of a computer architecture in order to achieve remote tuning of control systems on the Internet using open industry communications standard, with requirements satisfactory of performance and security.
The main contributions of this chapter is to propose the use of a Internet link to connect a automation system to a PID tuning tool and evaluate the impact of the communication non-determinism into the final performance of the controller, when compared to a local PID tuning.
The next section will study in detail the main features of the SCADA system communicating remotely with the factory system.

2. Remote SCADA systems

In remote monitoring, SCADA systems are network clients remotely connected to the control system of the shop floor. Typically, control centers, servers and the shop floor are located within the plant, while remote stations, which access data from these servers, are geographically distributed from each other. Remote connections between clients and servers are based mainly on the physical Ethernet, connected remotely via the Internet through the host server. The following figure shows an example of a typical network installation in the industrial environment.

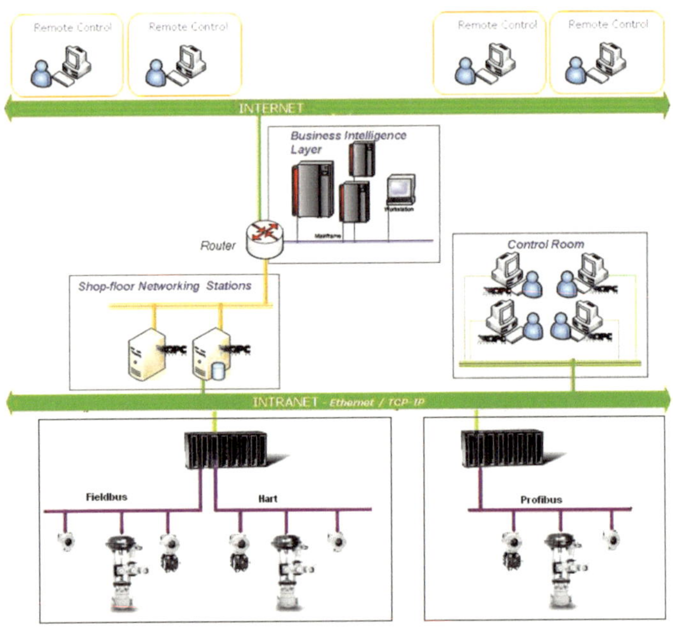

Fig. 1. Example of industrial applications using remote communication.

It is necessary to develop mechanisms for communication networks that can provide services with differentiated quality for real-time applications and multicast, since these applications demand minimum quality in terms of temporal parameters (such as delay and jitter) and effective transmission capacity (such as bandwidth). Several protocols for different network layers attempt to improve quality of service and determinism, for example, RSVP (Reservation Protocol), which only handles the reservation of resources along the route between network nodes. RTP (Real-time Transport Protocol) provides synchronization services, multiplexing, and security for data transfer, and these later two features focus mainly on image processing and voice using the Internet (Hanssen & Jansen, 2003).

Market solutions and academic researches aiming to facilitate the "open" system integration scenario for a shop floor over the Internet, make use of object-oriented technologies such as OPC (OLE for Process Control) through DCOM(Distributed Component Object Model), and DAIS (Data Acquisition from Industrial System) through CORBA (Common Object Request Broker Architecture), that is, all web-based services.

For application layers from the OSI model, there are several technologies to access data from industrial processes, such as: ASP (Active Server Pages) used together with ActiveX objects, and PHP (Hypertext Preprocessor) accessing process database servers available in SQL (Structured Query Language) (Zeilmann et al., 2003). OPC DCOM can be used to access distributed applications, as well as open standard technology such as XML (Extensible Markup Language) or JSON (JavaScript Object Notation), which are currently in widespread use for communication over the Internet.

The OPC Foundation has been developing a new OPC standard based on XML (OPC-XML 1.0 Spec.) since 2003, and included this new standard in a multiple protocol profile specification called OPC UA (Unified Architecture). The OPC UA aims to integrate the various existing OPC specifications (AD, AE, HDA, DX, etc.) into a single database, facilitating the development of applications (OPC Foundation, 2006). In addition, OPC UA offers support to portability and, therefore, can be integrated to any platform. However, this technology is still under approval and there are few commercially launched devices based on this standard.

The new OPC UA specifications show the path to open technologies, like XML, as a major trend in industrial systems interactivity over the Internet (Torrisi & Oliveira, 2007).

3. Current supervision and control researches over the Internet

The World Wide Web has provided opportunities for development and analysis of control systems over the Internet, according to studies by (Yu et al. 2006). Several papers propose the use of the Internet in control systems with different architectures.

Remote access architectures may be implemented at different levels in the manufacturing control hierarchy: at process level, at supervisory level, and at system optimization level.

The works of (Overstreet & Tzes, 1999) and (Yang et al., 2007) include remote control at the process level. In this case, the conventional discrete control structure must be changed to meet the diverging times of the Internet. (Luo & Chen, 2000) analyzed the network delay over the Internet using process control, concluding that the time interval for reading and writing over the Internet increases with the distance, depending on the number of nodes and the occupation of the network.

At the supervisory level, the concern is related to the quality of service. The work of (Kunes & Sauter, 2001) is based on SNMP (Simple Network Management Protocol) in fieldbus

technology systems. This architecture works well for read and write operations and asynchronous notifications, such as alarms. However, most firewalls do not allow UDP (User Datagram Protocol) traffic, and SNMP has low security levels.
Remote Internet protocol solutions, such as OPC-UA and NISV (National Instruments Shared Variables), adopt Client/Server architectures where remote control client applications receive periodic data refresh from Server values and send aperiodic sets of data.
A common practice is grouping the values of items of interest from the Server, with similar change rates, and assigning them to a group in order to allow the remote client to later retrieve all updated values from the group simply requesting the group name identifier.
It is common to schedule periodic updates of data values sent from the Server to the Client using a mechanism called Subscription Polling Mechanism. Although this mechanism speeds up the refresh rate and minimizes the number of update requests to the Server, not all data values might be of interest to the remote control client because some values may not change enough to be relevant for the client application.
In order to minimize the amount of data related to changed values of interest sent from the Server to the Client, the Client can specify a parameter called *DeadBand* for each group, which determines the percentage of range that an item value must change prior to the value being of interest to the Client. Changes in values that are not interested to the Client are not sent to the Client, therefore reducing the amount of data delivered over the network (Torrisi 2011).
(Yang et al., 2004) proposes a remote control at the supervision level for services that do not dependent on the Internet delay, which would be restricted to acyclic services such as SP alteration and tuning parameters of a PID block. The (Yang et al., 2004) studies present a virtual supervisory parameters control. This work shows the control would be invoked only when alterations for parameters such as setpoint (SP) and PID tuning parameters were requested, and then data would be sent to the control. In this context, multiple concurrent accesses are allowed by solving possible conflicts. Also, the security for the whole process is guaranteed since it is possible to provide redundancy and failure diagnostics in remote communication. Another approach at the supervisory level would be remote executing identification and tuning.
(Qin & Wang, 2007) studied the admission control to a web server, which accepts or rejects requests for the system. A Linear Parameter Varying (LPV) method is proposed to identify and control a web server, because the LPV approach tunes the model by specifying the loading conditions of the Internet, allowing the system to adapt to variations in load and operating conditions.
Companies currently offer some programmable logic controllers (PLCs) solutions with embedded web servers, but these solutions have limitations when applied to complex industrial plants (Calvo et al., 2006). For example, the work of (Batur et al., 2000) shows the architecture for remote monitoring and tuning using an SLC 500 Allen Bradley Company. The proposed system uses the measurement variables with the respective sampling times to ensure more determinism in the network. A mechanism for access control is also described, but the disadvantage of the system is that it consists of a proprietary solution, fully based on enterprise software to achieve monitoring and tuning for the controller.
(Yang et al., 2007) presents the architecture for processes control maintenance based on the Internet. The studied characteristics include industrial system performance indices, and failures and successes detection in the degraded control performance. The proposal monitors the system performance index locally, and if any noticeable change occurs in the

index, it will be identified in the system and, then, analysis and tuning will be executed for the stations. In the proposed architecture, the work considered as "heavy", such as the performance index calculation and model identification, is divided and processed locally. Work considered as "light", such as performance test results and process model, is sent to remote analysis. Thus, data analysis would be undertaken by experts who would propose tuning.

Several institutes and companies have conducted researches and provided control and distance learning applications for control systems over the Internet. These works are basically divided in two levels of interaction: the concept of virtual laboratory that brings together a developed physical structure and its subsequent release on the Internet; and distance learning courses that also offer a high level of interactivity, enabling, in some cases, simulation of physical phenomena. These virtual labs allow the user to tune control plants remotely, either through the simulated plant or through a real plant (Ko et al., 2005), (Zeilmann et al., 2003).

4. Common problems for Internet-based supervision and control

The Internet and web services have some obstacles related to their use for industrial control systems, such as delay in communication, data security and latency of web services.

Delay in communication - Over the Internet, a data packet suffers from several types of delays throughout the path from the source to the destination. The main types of delays are: processing delay, queuing delay, transmission delay and propagation delay, for each network node. Processing delay refers to the internal software processing to scan the message and determine where to send it, or check for errors in the message. The queuing delay happens while the message is waiting for queuing during transmission. Transmission delay refers to the time taken to get to the equipment and then be transmitted over the network. Finally, propagation delay refers to the time interval to spread the message on the line. According to (Han et al., 2001), the delay time T_a from the Internet at the time k can be described by:

$$T_a(k) = \sum_{i=0}^{n}\left[\frac{l_i}{C} + v_i^R + \frac{Q}{r_i} + v_i^L(k)\right] = d_N + d_L(k) \tag{1}$$

Where l_i is the distance to the nth link on the network, C is the speed of light coming and the speed of the nth router, Q is the amount of data, r_i is the bandwidth of the nth link and T_a *(k)* is the delay caused by the load of the nth node.

Separating the terms that are dependent and independent of time, there will be a d_N part of time-independent terms and a d_L part of time-dependent terms.

The contribution of each delay component can vary significantly. For example, the propagation time is negligible for the communication between two routers located in the same laboratory; however, it may vary significantly for equipment connected by a satellite link and be the dominant term in the total time delay (Kurose and Ross, 2006).

According to a study by (Luo and Chen, 2000), the performance associated with time delay and data loss shows a large spatial and temporal variation. The average delay of messages increases linearly with the increased traffic, according to (Boggs et al., 1988).

Non-determinism of the network - the Internet network is composed of multiple subnets and multiple routers between the source and destination station. The routers are responsible to select the most appropriate route for the message traffic between those two points. The routing algorithm varies with changes in stability of hardware and software throughout the network. The decision of the best route to be used should be taken for each data packet received. Consequently, there is no guarantee to the determinism of the network (Kurose, Ross, 2006).

However, many techniques have been developed to support real-time traffic, in particular, for the Ethernet. The work of (Wang et al., 2000) proposed management of collisions on the Ethernet. (Loeser & Haertig, 2004) proposing the joint use of intelligent switches and traffic management. (Gao et al., 2005) propose a real-time optimal smoothing scheduling algorithm with the variable network bandwidth and packet loss for data streaming.

The work of (Yang et al., 2004) reports that including the Internet to the levels of industrial control systems would not be practical because the Internet is highly non-deterministic with substantial delays, and the determinism is required on the network.

Data security on the network - It is fundamental to data traffic that distributed control systems meet the requirements for secure communication. In this case, it is necessary to fulfill the following security properties: confidentiality, authentication and message integrity. The confidentiality expects that only the sender and the recipient involved in the connection should understand the message content. Authentication requires that both the source and the destination confirm the identity of the other party involved in the communication. Integrity is required to ensure that the content of the message is not altered during transmission (Kurose and Ross, 2006).

Latency of web services - Despite the advantage of high interoperability, since all SOA (Service Oriented Architecture) entities use common languages for service descriptions, messages and records of services, the use of SOA causes problems of latency and memory space related to the use of web services, and according to (Pham and Gehlen, 2005) these features can be critical depending on the application. For industrial applications where asynchronous and synchronous communication is necessary, jitter effects– in terms of delay variation between successive data packets - may occur due to high internal processing (Torrisi & Oliveira, 2007).

Figure 2 shows the steps for exchanging data using web services. The Application layer represents the boundary between the OPC DCOM client and OPC DCOM Server located locally or remotely.

An application request is made by the remote Internet client through a call to the web server. This web server will receive the request and transfer it to the HTTP-SOAP (Hypertext Transfer Protocol- Simple Object Access Protocol) processor server (this process is shown in Figure 2 as Step 1). In this step, the SOAP/XML request is parsed to recognize commands and parameters, and it could have been binary decoded previously if it were an OPC-UA SOAP/XML request. Then, the corresponding API is invoked (Step 2), to forward the corresponding requested function to the application server (Step 3). The application server requests the message to be handled by the client on the OPC DCOM protocol. After that, the message is passed to an OPC server. Finally, the OPC server will request the data from a field device that will respond, and then the cycle is reversed and the whole process is executed until returning to the Http layer again (Step 6).

According to (Torrisi, 2011), this solution was not developed to meet the requirements and performance standards that are required for the industrial environment.

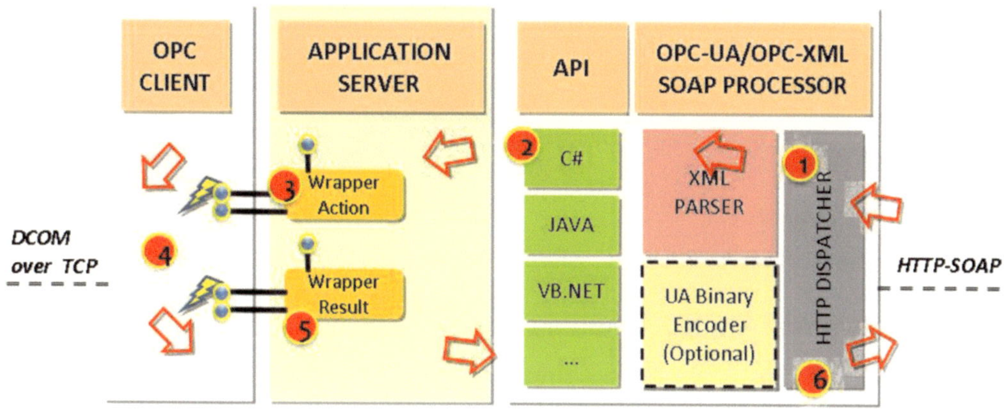

Fig. 2. Data processing path within OPC profiles using web server.

(Torrisi & Oliveira, 2007) proposed a new form of remote communication without the use of web services, called CyberOPC. The CyberOPC solution has a new process data transport protocol with the following characteristics: reduce transport delays for critical time data; ensure security for the communication channel used; ensure integrity and confidentiality for transmitted messages. In order to obtain maximum interoperability with existing shop floor technologies, open standard technologies were used, such as OPC DCOM. CyberOPC communication foresees the use of a gateway station called *CyberOPC gateway* that processes messages sent to the OPC through the public network, and vice versa. Due to the simplicity and short number of CyberOPC commands, the "Parser" containing the rules to recognize these commands is simpler than any XML parser for SOAP messages. Therefore, the OPC commands are executed quickly and, in the case of a periodic request, it is possible to increase the response time using a dedicated cache shared by the OPC client and OPC HTTP Broker.

A quick OPC data cache can be written asynchronously by the OPC client to all periodic data request from the remote Internet client, as shown in Figure 3.

A client application request is received by the gateway (Step 1), which now has the SOAP processor block. Introducing the OPC cache strongly reduces the time taken to call the OPC client. Tests conducted by (Torrisi & Oliveira, 2007) showed a significant reduction for posting time optimization when compared to the gateway-based web services, such as OPC-XML and OPC-UA SOAP/XML. Steps 2, 3, and 4 represent the interaction between the CyberOPC library and the OPC layer.

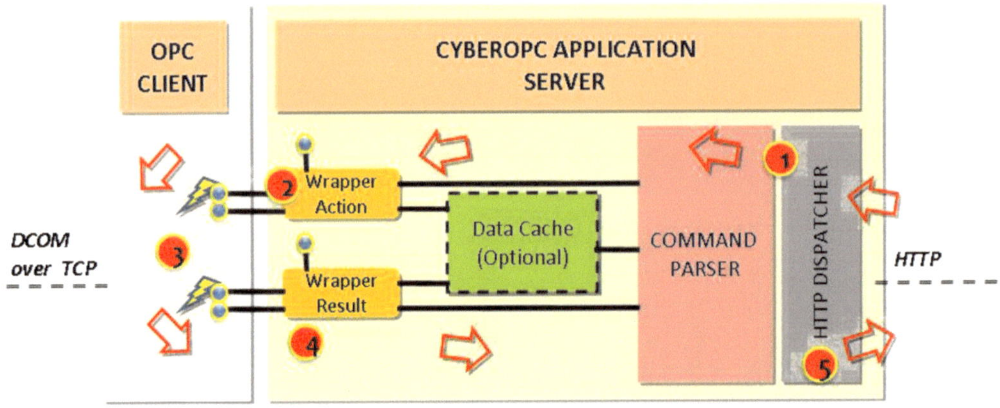

Fig. 3. Data processing path within OPC profiles using CyberOPC web server.

5. Architecture for remote tele-tuning

According to (Zeilmann et al., 2003), software for remote monitoring and acquisition must have a generic framework for data acquisition via the Internet to try to meet the vast majority of industrial automation systems. For such structure to be met, the following characteristics are desirable:

- Remote access to industrial automation system data for clients;
- Network data acquisition performance, specified in the refresh rate for data and maximum data delivery delay;
- Ensuring security in the communication channel to prevent unauthorized access;
- Ensuring integrity, confidentiality and reliability of transmitted messages;
- Open communication interface between software components, as a requirement for system scalability;
- Independence from field devices and protocols in operation;
- Independence from the platform of the remote client and the server, from the industrial automation system.

This section describes a tele-tuning architecture based on the interconnection of modules contained in three different contexts: the industrial plant, the server, and client, as shown in Figure 4. The architecture is based on the client-server application cooperation model, consisting of separated modules that are interconnected in order to provide process and configuration variables from the plant to the remote client. The entire HTTP communication is secured using SSL (Secure Sockets Layer) and, for such reason, HTTPS (HyperText Transfer Protocol Secure) will be cited instead of HTTP.

The Industrial Plant. Nowadays, there are several communication protocols for devices that meet specific applications in industrial environments, for example, process control and manufacturing control. The physical means of communication between these devices also differ from each other, either on the possible topologies, cable types, presence or absence of feeding overlapped communication, adaptation to usage requirements for hazardous areas, among others issues.

The tele-tuning architecture provides a communication channel between the field controller and the device driver for data acquisition, the latter being installed in a computer. Since it is

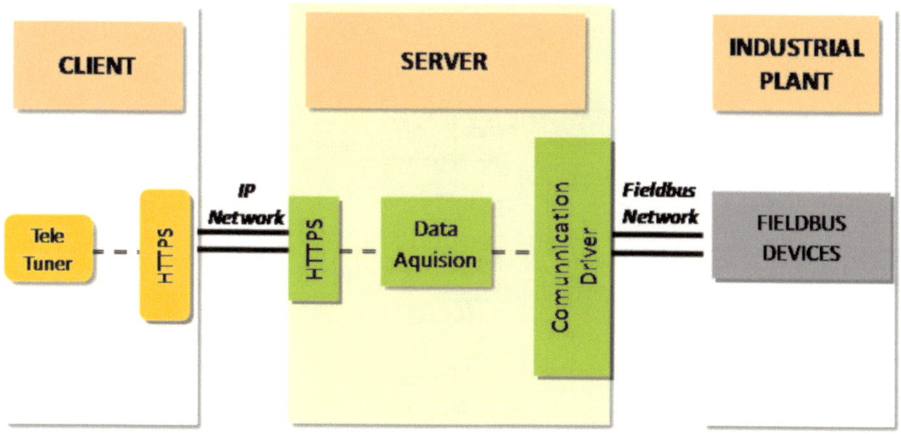

Fig. 4. Tele-Tuning architecture.

assumed the controller or the field devices are generic, the communication protocol between these devices and the related device driver is defined according to the equipment or system in use. In this case, a proprietary or open communication protocol can be used. However, the device driver needs to have an open software interface that can be easily integrated to any software component.

The **network server** is responsible for the interface between the plant and the remote clients. The server consists of several communication modules, as shown in Figure 4: the device driver for communicating to the controller and field devices (OPC Client), the Data Acquisition System, and also, the data web server for remote clients.

The **client** is the remote monitoring and tuning unit, as indicated in Figure 4. The requirement for the client is being an OPC DCOM client that enables communication to several equipment networks. In this architecture, an OPC DCOM client and CyberOPC client were used, which facilitates the implementation of the communication in both local and remote environment. The following section describes more details about the client side.

5.1 Monitoring and tuning system

The control system tuning is typically composed of the following phases: plant data acquisition, system identification, model validation, plant dynamics simulation tuned for verification purposes, control loop tuning, and data effectiveness in the plant (Ljung, 1999).

The proposed tele-tuning called Cybertune, is composed of four main operational modules: data acquisition module, the system identification module, the auto-regressive exogenous (ARX) model to open loop transformation module, and the tuning module. Figure 5 illustrates the relationship between these modules.

The **Data acquisition module** consists of an OPC client or CyberOPC, according to the OPC DCOM specifications (OPC Foundation, 2006) or CyberOPC specifications (Torrisi & Oliveira, 2007). The interface component has the same data access philosophy, consisting of an OPC DCOM library record, groups and items added to the database, and acyclic communication per event, when the client is notified in the occurrence of a new Data event issued by the server.

The **System identification module** is responsible for determining the system transfer function. In this work, ARX model was used due to good results for first and second order linear systems, and it is well-known in the consulted Literature (Aguirre, 2004).

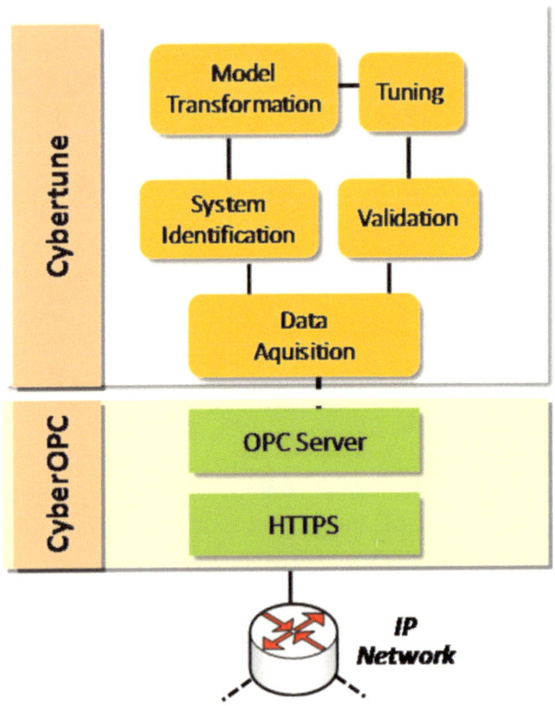

Fig. 5. CyberOPC client Tele-Tuning Schematic.

Because this project aims to validate the architecture for online identification and tuning, in a fair and reliable manner, the main purpose of the identifier is receiving process data and automatically processing the identification, through online identification. It is also important to process the identification offline, where an initial data collection is processed and recorded on the database for later identification and tuning.

The remote online identification presumes that communication delays and sending failures occur. This paper proposes the following solution to prevent these occurrences.

First, all samples collected by CyberOPC are recorded with the timestamp when the gateway acquired the data. Second, since the ARX model requires continuous sampling and CyberOPC sends data in a streaming optimized way (on data change), it is necessary to reconstruct the process signal at a constant sampling rate. To solve this matter, the **pre-identification module** was included. This module is responsible for receiving data from the acquisition module queue and sampling the data to the data identification queue at a constant sampling rate. In order to connect two sampling points, a first-order interpolation is used. Figure 6 show an example for this architecture.

The Cross Test method presented by (Aguirre, 2004) is executed in order to validate the identified model. This method compares the response generated by the identified model and the actual system response, for the same input signal. During the validation, the mean squared error and the percentage rate of the output variation is calculated as a performance measure and method validation measure.

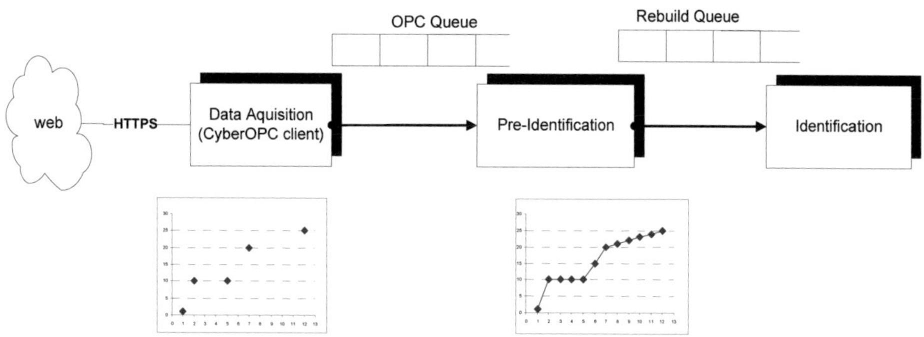

Fig. 6. Processing scheme for data received by OPC and CyberOPC.

In order to obtain the tuning of the model estimated using the model-based methods, it is necessary to obtain the transfer function of the system that in the architecture proposed is obtained by **the model transformation module**. In the identification module was obtained the differences equation (ARX) from the data collected. Thus, it is still necessary to convert the ARX model to the transfer function. The first order plus dead time (FOPDT) transfer function is showed in (2), where K_p it is the static gain, τ the time constant and θ is the system dead time.

$$G(s) = \frac{K_p}{\tau s + 1} e^{-\theta s} \tag{2}$$

In the work (Fernandes & Brandão, 2008) described a mathematical formulation to convert the ARX model in a transfer function (2). Below is described the equations for open loop. Consider a block diagram of a classic feedback controller system as shown in the figure 7.

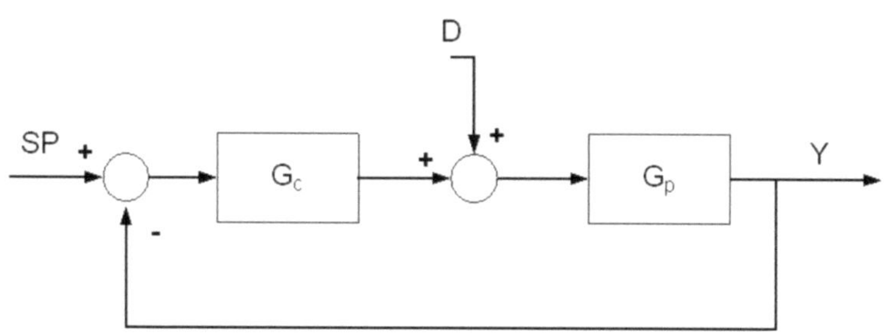

Fig. 7. Classic feedback controller, where G_c is a controller and G_p the plant. The SP is the setpoint , D the load variation, and the Y the system output.

The specification of tuning a controller can be classified as variations due to changes in setpoint (SP) or load variation (D). The control process should act to minimize these disturbances.

$$\frac{Y}{SP} = \frac{G_c G_p}{1 + G_c G_p} \tag{3}$$

$$\frac{Y}{D}=\frac{G_p}{1+G_cG_p}$$

Consider a PID controller ISA standard, as described in (4) where K_c is the proportional gain , T_i integral time constant and T_d the derivative time constant of the controller.

$$G_c=\frac{Y(s)}{U(s)}=K_c\left(1+\frac{1}{T_is}+T_ds\right) \tag{4}$$

The open loop system is accomplished by placing the controller on manual. The real system can be described by equations in open loop showed in (5), where G_p the transfer function of the process, Y the system output, and D disturbance of excitation system.

$$\frac{Y}{D}=G_p \tag{5}$$

For the first-order system shown in (5) and Paddé approach yields the equation (6) in the continuous domain.

$$\frac{Y}{D}(s)=\frac{(-0.5K_p\theta)s+K_p}{(0.5\tau\theta)s^2+(0.5\theta+\tau)s+1} \tag{6}$$

The equation (7) is the conversion from continuous time to discrete using backward difference approximation.

$$\frac{Y}{D}(z)=\frac{(bz_1)z+(bz_2)}{(az_1)z^2+(az_2)z+(az_3)} \tag{7}$$

Where the arguments of equation (7) are showed in (8). T_0 is the system sample rate.

$$\begin{aligned}
az_1&=0.5\tau\theta+0.5\theta T_0+\tau T_0+T_0^2\\
az_2&=-\tau\theta-0.5\theta T_0-\tau T_0\\
az_3&=0.5\tau\theta\\
bz_1&=K_pT_0^2-0.5K_p\theta T_0\\
bz_2&=0.5K_p\theta T_0
\end{aligned} \tag{8}$$

The equation (8) is equivalent to differences equation from ARX model showed in (9).

$$\frac{B}{A}(z)=\frac{B_1z+B_2}{z^2+A_1z+A_2} \tag{9}$$

Then, equating (9) and (7) obtain (10).

$$B_1=\frac{bz_1}{az_1}\quad ;\quad A_2=\frac{az_2}{az_1} \tag{10}$$

Finally, applying (8) and (9) in (10) and making some adjustments are obtained the values of Kp, τ and θ from the FOPDT model as (2).

$$\begin{cases} K_p = \dfrac{(B_1+B_2)\left[(0.5\tau\theta)+(0.5\theta T_0)+(\tau T_0)+(T_0^2)\right]}{T_0^2} \\ \tau = \dfrac{-\left[(0.5A_2\theta T_0)+(A_2T_0^2)+(0.5\theta T_0)\right]}{\left[\theta+T_0+(0.5A_2\theta)+(A_2T_0)\right]} \\ \left[(-0.25A_2T_0)-(0.25T_0)-(0.25A_3T_0)\right]\theta^2 + \\ +\left[(-0.5A_2T_0^2)-(A_3T_0^2)\right]\theta-(A_3T_0^3)=0 \end{cases} \tag{11}$$

For the solution of equation (11) is necessary to solve a second-order polynomial where there are two possible solutions to the system. To determine the solution that best fits the system is compared to identify the model obtained with each solution obtained from the open-loop system. For comparison we used the criterion integral of absolute error (ITAE). Using the same concept we obtain the equations of a closed loop system.

The **Tuning module** is responsible for applying the tuning method to the model obtained by the identification module. Among several tuning methods discussed in the literature, here will be considered the methods that fulfill the requirements related to good performance in first-order systems. This work does not intend to validate tuning methods; therefore only the integral optimization methods of square error and absolute error (ITSE and ITAE) and internal model control (IMC) method will be used in practical experiments for the proposed system, due to their good response to FOPDT systems. However, other developed methods can be feasibly gathered to the tuning knowledge base simply by meeting the requirements of identification systems.

The methods used in this work are based on methods already studied in the literature. First of all, is assumed the controller has original tune based on the classical Ziegler-Nichols (ZN). Then, it is suggested other methods to improve the tuning that are the methods based on performance criterion error integral and by internal model control (IMC). The advantages of these methods include a low overshoot and good settling time (Lipták, 2003), (Seborg et al., 2004), (Zhuang & Atherton, 1993). Below is showed the parameterization of each method used for tuning proposes.

The parameters for ZN method for open loop are showed in (12).

$$K_c = \frac{0.9}{(K_p/\tau)\theta} \quad ; \quad T_i = 2\theta \tag{12}$$

For methods based on integral error criterion ITAE and ITSE, the relationship between the tuning of the controller and the integral criteria is based on the relationship θ/τ (the ratio of dead time and time constant of the system) and expressed in the equation (13), where X is a parameter of the controller (such as Proportional, Integral and Derivative) and m and n constants.

$$X = m\left(\frac{\theta}{\tau}\right)^n \tag{13}$$

The parameters of ITSE and ITAE for PI controller for disturbance due to load change are showed in the table 1.

	P		I	
Method	m	n	m	n
ITAE	0,859	-0,977	0,674	-0,680
ITSE (θ/τ 0.1 - 1.0)	1.053	-0.930	0.736	-0.126
ITSE (θ/τ 1.1 – 2.0)	1.120	-0.625	0.720	0.114

Table 1. Parameters of a PI controller for load change using the methods ITAE and ITSE.

The IMC method which name "internal model control" comes from the fact that the controller contains an internal process model. For a system FOPDT as in (2) the IMC controller equations is showed in (14).

$$G_c(s) = \frac{(1+(\theta/2)\,s)(\tau s+1)}{K_p(\lambda+(\theta/2))s} \tag{14}$$

Where λ is the tuning parameter specified by the user. The choice of design parameter λ is a key decision in more conservative or not controller. In this work is used λ=θ. Arranging (14) for a PI controller as in (4) became the equations (15).

$$K_c = \frac{1}{K_p}\frac{2(\tau/\theta)+1}{2(\lambda/\theta)+1} \;;\; T_i = \theta/2+\tau \tag{15}$$

5.2 Remote tuning use case

Identification tests were performed to validate the proposed architecture simulating FOPDT systems with equation (2) using local and remote identification in a corporate network. The tests use the Cybertune software that allows the monitoring and updating remote data control systems in industrial environments using OPC and CyberOPC technologies.

Tests were conducted using the Fieldbus Plant Simulator (FBSIMU) (Pinotti & Brandão, 2005), which simulates the industrial plant and the fieldbus control logic. (Pinotti & Brandão, 2005) showed that FBSIMU has a good approach to the real system.

Figure 8 illustrates the tests scenarios for local and remote tuning. Local tests involved Cybertune communicating to FBSIMU in the same station using OPC DCOM communication. Remote tests consisted of Cybertune communicating to FBSIMU inside the campus intranet network, using CyberOPC protocol.

Simulation tests were performed using six systems with different characteristics, and considered the relation θ/τ varying from 0.1 to 2.5. The relation is used as a comparison among different types of FOPDT systems, as verified in (Seborg et al., 2004).

To validate the tests, the ITAE performance index and the correlation index (FIT) showed in (16) were used in relation to the real signal and the identified signal. For a higher correlation index, identification is considered good (Ljung, 1999).

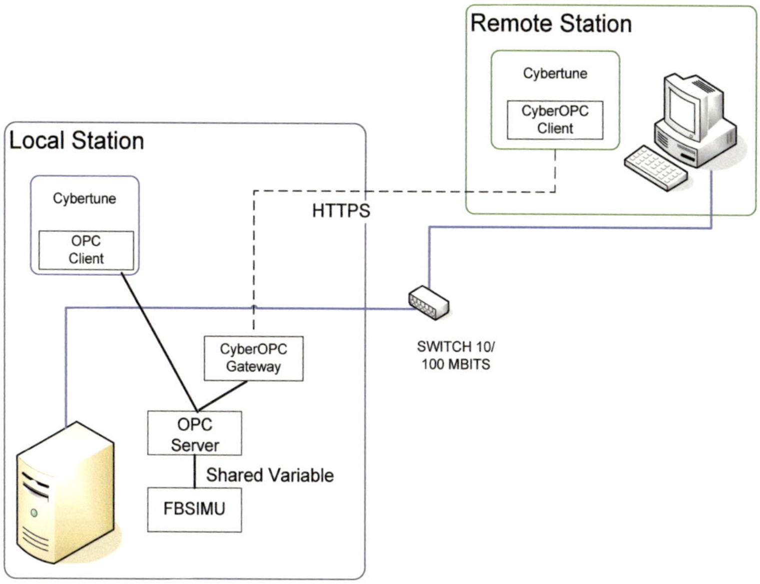

Fig. 8. System architecture for local and remote communication between Cybertune and FBSIMU.

$$FIT = \left(1 - \frac{\sum_{k=1}^{N} Norm\left(y(k) - \hat{y}(k)\right)}{\sum_{k=1}^{N} Norm\left(y(k) - \overline{y}(k)\right)}\right)100 \quad and \quad Norm(V) = \sum_{k=1}^{N} \sqrt{ABS(V)^2} \tag{16}$$

Where, $y(k)$ is the actual process output in the k instant, $\hat{y}(k)$ is the estimated output and $\overline{y}(k)$ is the average of samples throughout the identification.
Consider the FOPDT example from (2) showed in (17):

$$G_p = \frac{2}{100\,s+1} e^{(-50\,s)} \tag{17}$$

Initially, in the model identification phase, to achieve the identification of noisy signals using the parametric models, it is necessary to use a model to identify the highest order to obtain a good approximation of the model. And after the identification of high-order model, the poles and zeros used to describe the noise signal can be canceled (Ljung, 1999). One way of reducing the order of the model is to use only the dominant poles. The equation (9) of the open loop algorithm requires a second-order equation.
Regarding the local identification test, identification is estimated according to approximation using a fourth-order ARX model and sampling rate (To = 1.0 sec). The open-loop transfer function (OPTF) model shown in (18) was obtained. The FIT obtained was 98.50%.

$$OPTF_{LOCAL}(z) = \frac{-0.0170z^3 + 0.0060z^2 + 0.0105z + 0.0023}{z^4 - 1.0030z^3 - 0.4280z^2 - 0.0129z + 0.4448} \tag{18}$$

Using equation (18) was performed a reduction of the model for a second-order equation as in (9) and then applied the equation (11) for the model transformation. The result was the equation (19) which represents the G_p estimated locally.

$$G_{pEST\_LOCAL} = \frac{2.00}{102.05\,s + 1} e^{(-46.62\,s)} \tag{19}$$

The graph in Figure 9 compares the real system and the system identified locally. The final solution has FIT equals to 98.09%.

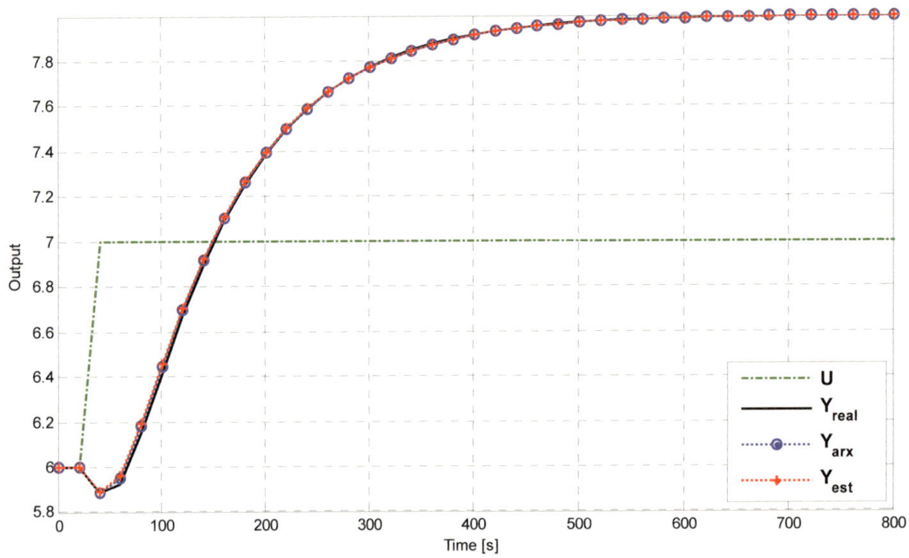

Fig. 9. Graphic comparison of system responses to a step change in an open loop. It shows the original signal (Y_{real}), the 4th order ARX signal (Y_{arx}) and the identified open loop system (Y_{est}).

The remote test with the same system (17) obtained the fourth-order ARX model with sampling rate (To = 2 sec), the estimated model (OPTF) resulted in equation (20) with FIT=95.59%.

$$OPTF_{REMOTE}(z) = \frac{-0.0398z^3 + 0.0065z^2 + 0.0220z + 0.0190}{z^4 - 0.9003z^3 - 0.478z^2 - 0.0808z + 0.4630} \tag{20}$$

After transforming the ARX model into the open loop model, the model approximation is obtained in (21), which represents the G_p from (17) estimated remotely:

$$G_{pEST\_REMOTE} = \frac{2.00}{100.56\,s + 1} e^{(-45.36\,s)} \tag{21}$$

The graph in Figure 10 compares the real system and the system identified remotely. The final solution has FIT equals to 93.63%.

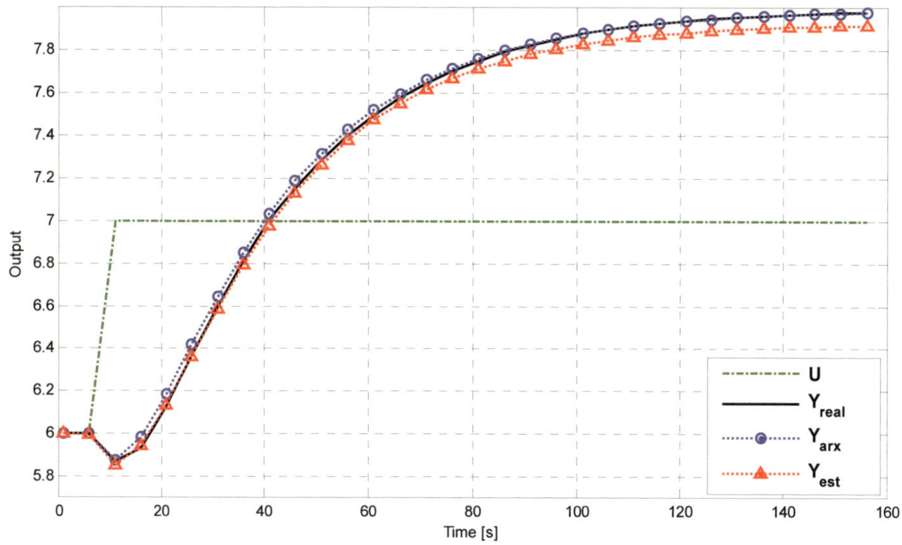

Fig. 10. Graphic comparison of system responses to a step change in an open loop. It shows the original signal (Y_{real}), the 4th order ARX signal (Y_{arx}) and the identified open loop system (Y_{est}).

Table 2 and Figure 11 summarize the results from remote and local ARX identification for each of the six systems. The "Local" identification was realized with the Cybertune running in the same station as showed in the Local Station in the Figure 8. The "Remote" identification was executed with the Cybertune running in a remote station connected to the local station using an internet link of 1 Mbyte.

System	θ/τ	Type Identification	To [s]	FIT [%]	ITAE
1	0.13	Local	1	98.17	1.22E+03
		Remote	2	97.91	1.25E+03
2	0.50	Local	1	98.50	7.92E+02
		Remote	2	95.59	3.42E+02
3	1.14	Local	1	95.24	4.60E+03
		Remote	1	94.82	8.34E+02
4	1.53	Local	1	93.77	3.11E+03
		Remote	2	93.44	2.51E+03
5	1.90	Local	1	98.02	1.80E+03
		Remote	5	96.38	2.02E+03
6	2.33	Local	1	93.77	3.11E+03
		Remote	5	91.60	1.78E+03

Table 2. Local and Remote Identification Results.

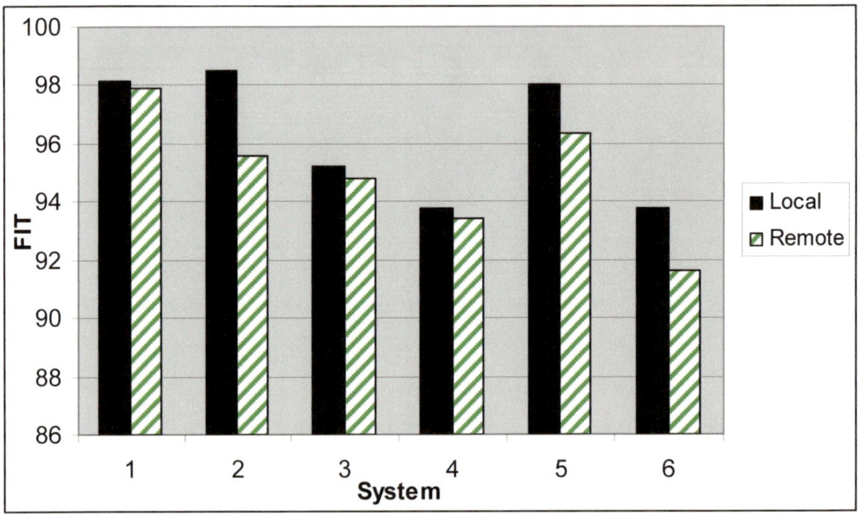

Fig. 11. Comparing local and remote tests for different systems.

The open loop model obtained from the identification phase is used in the tuning phase. This example assumes the controller (Gc) it is PID-ISA as showed in (4).
The tuning parameters of the PI controller for this work, the parameters of a PI controller (K_c and T_i) of current tuning using Ziegler-Nichols (ZN) method and three suggest methods ITSE, ITAE and internal model control (IMC) are summarized in the table 3. The table shows the column "error ITAE" for each method to help in choose the best tuning method.

Method	K_c	T_i	Error ITAE
ZN	0.90	100	1.06E5
ITSE	1.35	298.733	4.05E5
ITAE	0.93	87.1519	8.57E4
IMC	0.8903	146.1284	1.84E5

Table 3. Tuning Results for methods where K_c and T_i are the Parameters of PI controller and ITAE is the error for a step response.

The figure 12 shows the step response for load variation (D) and the system response for the suggest methods ITSE, ITAE e internal model control (IMC).

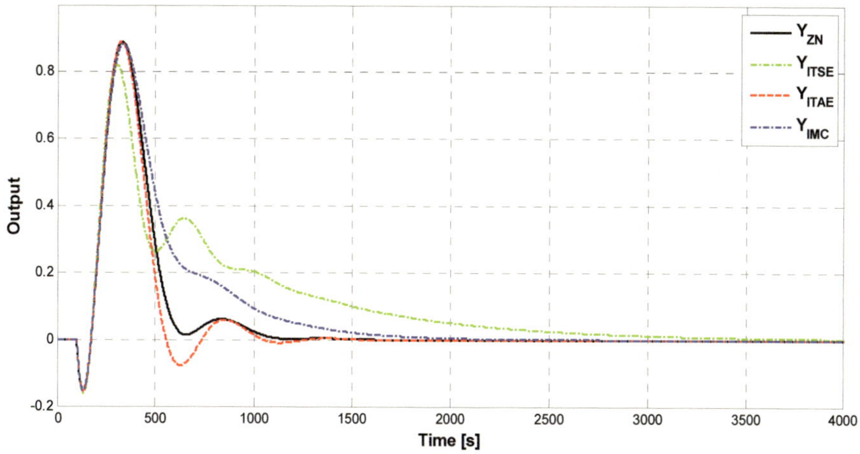

Fig. 12. Results of tuning for the model obtained in the remote station. The step response of the load variation of original tuning ZN (Y_{ZN}) and some common methods based on model (Y_{ITSE}, Y_{ITAE} and Y_{IMC}).

6. Conclusion

The Tele-tuning architecture executes remote tuning for industrial control systems using the Internet, fulfilling acceptable security and performance requirements. In order to validate the architecture, a software application, called CyberTune, using CyberOPC was presented. Validation consisted of a model-based identification and tuning of six FOPDT systems with diferentes θ/τ. This model is a typical class of industrial systems, but the architecture can be extended to other configurations.

The analysis of table 1 demonstrated that remote model identification is really near to local identification and the original system, which validates the architecture for identification and subsequent tuning, implemented with model-based methods.

For remote identification or noise signal, it is necessary to pre-filter the signal and use the highest order of the ARX model to obtain a good approximation of the model as showed in (20).

For remote tuning, the proposed architecture using CyberOPC and reconstruction of the data showed satisfactory results as shown in Fig 12.

Remote monitoring and tuning of control system might be a good solution for process plant companies with multiple sites in remote locations in order to provide the central support for their geographically dispersed control systems. By using this remote monitoring and maintenance system control software suppliers can monitor and maintain their control software products remotely over the Internet.

Nowadays the Ethernet and the Internet are increasing the speed quickly. Industries are beginning to implement networked control systems through this high speed communication. The speed of the next generation Internet might be sufficiently fast to be able to dramatically reduce the transmission delay and data loss. Therefore, it is possible that Internet latency and data loss might become less important issues in future Internet applications. But questions about the security of the Internet shall be continuing existing, because of the public nature of the Internet.

7. Acknowledgment

The authors gratefully acknowledge the Brazilian agency FAPESP for the financial support received, and the academic support and research structure of the Engineering School of Sao Carlos - University of Sao Paulo.

8. References

Aguirre, L. A., 2004. Introdução a Identificação de Sistemas, Técnicas Lineares e Não Lineares aplicadas a sistemas reais. 2ª edição, 2004, Editora UFMG.

Aström, K.J.; Hägglund, T., (1995), Pid Controllers: theory, design and tuning. 2a Edição, 1995, Editora Instrument Society of America (ISA).

Avoy, T. M.; Jounela, S.L.J.; Patton, R.; Perrier, M.; Weber, H.; Georgakis, C., 2004. Milestone report for area 7 industrial applications. In: Control Engineering Practice 12 (2004) 113-119.

Ang, K. H.; Chong, G.; Li, Y. (2005). PID Control System Analysis, Design, and Technology. In: IEEE Transaction on Control Systems Technology, Vol 13, No 4, July 2005.

Batur, C.; Ma, Q.; Larson, K., Kettenbauer, N., 2000. Remote tuning of a PID position controller via internet. In: American Control Conference, Chicago, 2000.

Calvo, I.; Marcos, M.; Orive, D.; Sarachaga, I., 2006. A methodology based on distributed object-oriented technologies for providing remote access to industrial plants. In:Control Engineering Practice 14 (2006), pp. 975-990.

Fernandes, R. F.; Brandão, D. (2008). Online identification method for industrial systems with closed-loop PI controllers, In: INDUSCON, Poços de caldas, August 2008.

Gao, K.; Gao,W.;He,S.; Zhang,Y. (2005). Real-time smoothing for network adaptive video streaming. Journal of Visual Communication and Image Representation Volume 16, Issues 4-5, August-October 2005, Pages 512-526.

Han, K. H.; Kim, S.; Kim, Y. J.; & Kim, J. H. (2001). Internet control architecture for Internet-based personal robot. Autonomous Robots, 10, 135–147.

Hanssen, F. T. Y.; Jansen, P. G. (2003). Real-time communication protocols: an overview. Technical Report TR-CTIT-03-49, Centre for Telematics and Information Technology,Univ. of Twente, The Netherlands, Oct. 2003.

Harris, T.J.; Seppala,C.T.; Desborough, L.D. (1999). A review of performance monitoring and assessment techniques for univariate and multivariate control systems. Journal of Process Control 9 (1999) 1-17.

Hjalmarsson, H., 2005. From experiment design to closed loop control. Automatica, 41 (2005) pp. 393-438.

ITU Recommendation, 1997. X.509 version 3. Information Technology - Open Systems Interconnection - The Directory Authentication Framework, August 1997.

Ko, C.C.; Chen B.M.; Chen, J.; Zhang, J.; Tan, K.C. (2005). A web-based laboratory on Control of two degrees of freedom helicopter. In: International Journal of Engineering, Vol 21, n 23.

Kurose, J. F.; Ross, K. W., 2006. Redes de computadores e a Internet. 3ª Edição, 2006, Editora Addison Wesley.

Kunes, M. & Sauter, T. ,2001. Fieldbus-Internet connectivity: The SNMP approach. In: IEEE Transactions on Industrial Electronics, 48(6), 1248–1256.

Lipták, B. G., 2003. Instrument Engineers' Handbook - Process Control and Optimization, 4nd Edition, 2003, Chilton Book Company.

Li Zheng and Nakagawa, H., OPC (OLE for process control) specification and its developments, SICE 2002 - Proceedings of the 41st SICE Annual Conference, pp. 917-920 vol. 2, 2002.

Loeser, J. E Haertig, H. (2004). Low-latency hard real-time communication over switched Ethernet. In: 16th Euromicro Conference on Real-Time Systems, Catania, Italy, 2004.

Luo, R.C.; Chen, T.M., 2000. Development of a multibehavior-based mobile robot for remote supervisory control through the Internet. IEEE/ASME Transactions on mechatronics, vol. 5, no. 4, December 2000.

Ljung, L., (1999), System Identification - Theory for the User. Second Edition, 1999, Englewood, Prentice Hall.

OPC Foundation, 2006. OPC Standard. In: <www.opcfoundation.org>. Access em: 26 jul. 2007.

Overstreet, J. W.; & Tzes, A. (1999). Internet-based client/server virtual instrument designs for real-time remote-access control engineering laboratory. In Proceedings of the American control conference, Vol. 2, pp. 1472–1476.

Pham, L., Gehlen, G. (2005), Realization and Performance Analysis of a SOAP Server for Móbile Devices, Proceedings of the 11th European Wireless Conference 2005, Vol. 2, Nicosia, Cyprus, April 2005, pp. 791-797.

Qin, W.& Wang, Q., 2007. An LPV approximation for admission control of an internet web server: Identification and control. In: Control Engineering Practice, Feb 2007.

Torrisi, N. M. ; Oliveira, J. F. G. (2007), Remote control of CNC machines using the CyberOPC communication system over public networks. International Journal of Advanced Manufacturing Technology, v. 2007, p. 001, 2007.

Torrisi, N. M. (2011), Monitoring Services for Industrial Applications Based on Reverse AJAX Technologies, IEEE Industrial Electronics Magazine, ISSN: 1932-4529, 2011, Volume 5, Issue 1.

Pinotti Jr, Mario ; Brandão, D. (2005). A flexible fieldbus simulation platform for distributed control systems laboratory courses.. The International Journal Of Engineering Education, Dublin, v. 21, n. 6, p. 1050-1058, 2005.

Ruel, M. (2003). The conductor directs this orchestra -Instrument Society of America. Intech, November 2003, vol 50, no 11, pp. 20-22.

Seborg, D. E., Edgar, T. F., Mellichamp, D.A., 2004 - Process Dynamics and Control, Second Edition, John Wiley & Sons, inc.

Visioli, A.(2001). Optimal tuning of PID controllers for integral and unstable processes. Control Theory and Applications, IEE Proceedings - , vol.148, no.2, pp.180-184, Mar 2001.

Yang, S.H., Zuo,X., Yang,L., 2004. Control System Design for Internet-Enabled Arm Robots. In Innovations in Applied Artificial Intelligence, Volume 3029, Springer Berlin, ISBN 978-3-540-22007-7.

Yang, S. H., Dai, C., Knott, R.P., 2007. Remote Maintenance of Control System performance over the Internet, Control Engineering Practice, Volume 15, Issue 5, May 2007, Pages 533-544.

Yu, C. C., 2006. Autotuning of PID Controllers: A Relay Feedback Approach. 2nd Edition, Springer, 2006.

Wang, Z.P., Xiong, G.Z., Luo, J., Lai, M.Z., Zhou, W. (2000). A hard real-time communication control protocol based on the Ethernet. In Proceedings 7th Australian Conference on Parallel and Real-Time Systems, pages 161-170, Sydney, Australia, Nov. 2000. Springer-Verlag. ISBN 962-430-134-4.

Zeilmann, R.; Gomes Jr.; J.M., Bazanella; A., And Pereira, C. E. (2003). Web-based Control Experiments on a Foundation Fieldbus Plant, 5th IFAC International Conference on Fieldbus Systems and their Applications, Aveiro, Portugal, July 2003, pp. 325-330.

Zhuang, M.; Atherton, D.P. (1993). Automatic Tuning of Optimum PID controllers. In: IEEE Proceeding-D, 140, no 3, pp 216-224.

13

PID Application: RTLS

Jae Ho Hwang and Jae Moung Kim
WITLAB, INHA University
Korea

1. Introduction

Locating systems are categorized as macro or micro locating systems, depending on service coverage. Macro locating systems offer wide coverage of over ten kilometers. These systems are allowed large error margins of over ten meters. They are for use in a global positioning system (GPS) [1] and in mobile locating systems between a mobile station (MS) and a base station (BS) [2]. A micro locating system, by contrast, operates in a small coverage area of less than ten meter, such as indoor environments where it is hard to use a GPS system.
A micro locating system requires a high accuracy of within one meter. Moreover, it requires real-time calculation for more accurate estimation when the target node changes position. Such a real-time locating system (RTLS) is frequently required in robotics applications. However, the conventional direct calculation (DC) method requires a heavy computation load, such as floating point calculation. This not only imposes a high computational load on an embedded system, but also decreases the accuracy [3].
This chapter introduces a PID application in real-time locating system. A ToA algorithm is used to obtain the target node coordinates, but a conventional DC method, which incurs heavy calculation time, is not suitable for embedded systems. This paper proposes the use of a P-control in the PID control algorithm to resolve real-time locating system issues. Performance measures of the accumulated operator number and position error are evaluated. It is shown that the PID method has less calculation and more robust performance than the DC method.

2. Conventional location system

To calculate the target node coordinates, the triangulation method requires more than three fixed reference nodes for which the coordinates are already known. In this method, each reference node calculates the distance between the reference and target nodes to count the propagation time by using time of arrival (ToA) algorithm. The target node estimates two-dimensional coordinates (X_T, Y_T) using this information
The triangulation method is shown in Figure 1 [4]. This method consists of three known reference nodes $((X_1, Y_1), (X_2, Y_2)$, and $(X_3, Y_3))$ and an unknown target node (X_T, Y_T). Each reference node measures its distance (D_{R1}, D_{R2}, and D_{R3}) from the target node. The DC method uses the relation between the coordinates and the distance. Each distance is expressed by the coordinates of the reference and target nodes, as shown in equation 1.

$$\begin{aligned} D_{R_i} &= \sqrt{(X_T - X_i)^2 + (Y_T - Y_i)^2} \\ &= \sqrt{X_T^2 + Y_T^2 - 2(X_i X_T + Y_i Y_T) + X_i^2 + Y_i^2} \end{aligned} \tag{1}$$

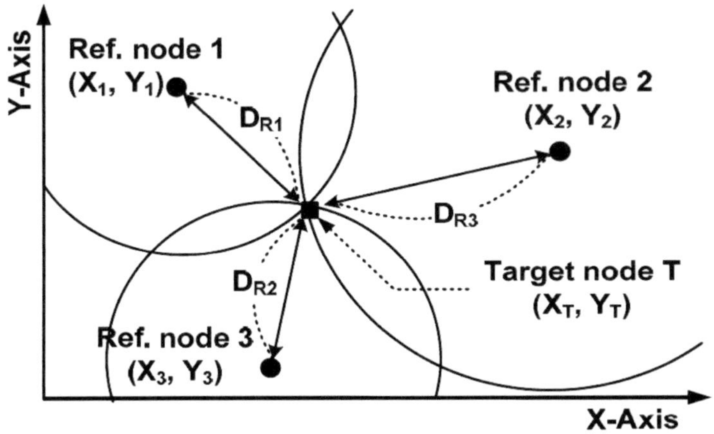

Fig. 1. The triangulation method using the DC method when the number of reference nodes is three

These equations are simplified by squaring and repositioning, as shown in Equation (2).

$$\begin{aligned} X_T^2 + Y_T^2 - 2(X_1 X_T + Y_1 Y_T) &= D_{R_1}^2 - X_1^2 - Y_1^2 \\ X_T^2 + Y_T^2 - 2(X_2 X_T + Y_2 Y_T) &= D_{R_2}^2 - X_2^2 - Y_2^2 \\ X_T^2 + Y_T^2 - 2(X_3 X_T + Y_3 Y_T) &= D_{R_3}^2 - X_3^2 - Y_3^2 \end{aligned} \tag{2}$$

Using these equations, it is complex to calculate X_T and Y_T, so it is assumed that reference node 3 is located at $[0, 0]$. then total coordination are change as $\widehat{X}_T$, $\widehat{Y}_T$, $\widehat{X}_i$, and $\widehat{Y}_i$, which is shifted values as shown in

$$\begin{aligned} \widehat{X}_T &\triangleq X_T - X_3, \ \widehat{X}_1 \triangleq X_1 - X_3, \ \widehat{X}_2 \triangleq X_2 - X_3 \\ \widehat{Y}_T &\triangleq Y_T - Y_3, \ \ \widehat{Y}_1 \triangleq Y_1 - Y_3, \ \ \widehat{Y}_2 \triangleq Y_2 - Y_3 \end{aligned} \tag{3}$$

Then, the third equation $(X_T^2 + Y_T^2 - 2(X_3 X_T + Y_3 Y_T) = D_{R_3}^2 - X_3^2 - Y_3^2)$ is simplified with $X_T^2 + Y_T^2 = D_{R_3}^2$, as shown in Equation (4).

$$\begin{aligned} \widehat{X}_1 \widehat{X}_T + \widehat{Y}_1 \widehat{Y}_T &= \frac{D_{R_3}^2 - D_{R_1}^2 + \widehat{X}_1^2 + \widehat{Y}_1^2}{2} \\ \widehat{X}_2 \widehat{X}_T + \widehat{Y}_2 \widehat{Y}_T &= \frac{D_{R_3}^2 - D_{R_2}^2 + \widehat{X}_2^2 + \widehat{Y}_2^2}{2} \end{aligned} \tag{4}$$

To make the easy calculation, the complex constant value of equation 4 notate simple A and B symbols. those are descried in following equations.

$$A \triangleq \frac{D_{R_3}^2 - D_{R_1}^2 + \widehat{X}_1^2 + \widehat{Y}_1^2}{2}, \quad B \triangleq \frac{D_{R_3}^2 - D_{R_2}^2 + \widehat{X}_2^2 + \widehat{Y}_2^2}{2} \tag{5}$$

Using those notations, the equation 4 will be changed as.

$$\hat{X}_1\hat{X}_T + \hat{Y}_1\hat{Y}_T = A, \quad \hat{X}_2\hat{X}_T + \hat{Y}_2\hat{Y}_T = B \tag{6}$$

Next, the coordinates of the target node are obtained by using elimination of the $\hat{X}_T$, and $\hat{Y}_T$ in both equation.

$$\hat{X}_T = \frac{A\hat{Y}_2 - B\hat{Y}_1}{\hat{X}_1\hat{Y}_2 + \hat{X}_2\hat{Y}_1}, \quad \hat{Y}_T = \frac{A\hat{X}_2 - B\hat{X}_1}{\hat{X}_1\hat{Y}_2 + \hat{X}_2\hat{Y}_1} \tag{7}$$

Then, the real target node coordinates are compensated to the coordinate values of reference 3, which values are shifted as assumed. Finally, the coordinates are obtained, as shown in equation 8.

$$X_T = \frac{A\hat{Y}_2 - B\hat{Y}_1}{\hat{X}_1\hat{Y}_2 + \hat{X}_2\hat{Y}_1} + X_3, \quad Y_T = \frac{A\hat{X}_2 - B\hat{X}_1}{\hat{X}_1\hat{Y}_2 + \hat{X}_2\hat{Y}_1} + Y_3 \tag{8}$$

The DC method is complex and uses many multiple and floating point operations; therefore, it is not suitable to embedded systems. Furthermore, the processing delay caused by use of the DC method results in position errors when the target node is moved, and the method always requires a substantial amount of calculation time whether the target node position changes or not. These features cause a performance reduction when using applications such as a real-time robotics locating system

3. PID algorithm application: RTLS

3.1 ToA algorithm using PID algorithm

The conventional method has the problem of heavy computation load for an embedded system. Therefore, a new coordinate calculation method is necessary for use with RTLSs. In this chaper, the PID calculation method is proposed to meet this need. The PID algorithm is a generic control loop feedback mechanism widely used in control systems. It attempts to correct the error between a measured process variable and a desired set point by calculating and then performing an appropriate action that can adjust the process accordingly. This algorithm involves three separate parameters: the proportional, integral and derivative values. The proportional value determines the reaction to the current error, the integral determines the reaction based on the sum of recent errors and the derivative determines the reaction to the rate at which the error has been changing. The weighted sum of these three actions is used to adjust the process via a control element such as the position of a control valve or the power supply of a heating element, as shown in equation 9 [5] [6].

$$u = K_P e + K_I \int e d\tau + K_D \frac{de}{d\tau} \tag{9}$$

Where K_P, K_I and K_D are tuning parameters and e is the error value. The PID algorithm was applied to the proposed coordinate calculation method. Figure 2 provides a more detailed explanation of the coordinate calculation method.

First, the pseudo target node is placed at an arbitrary position. Subsequently, the distance between the pseudo target node and each reference node is calculated, thereby giving the

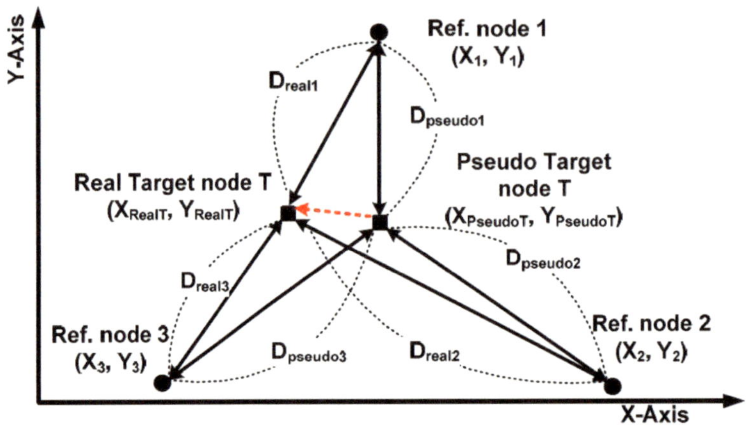

Fig. 2. Triangulation method using the PID calculation algorithm

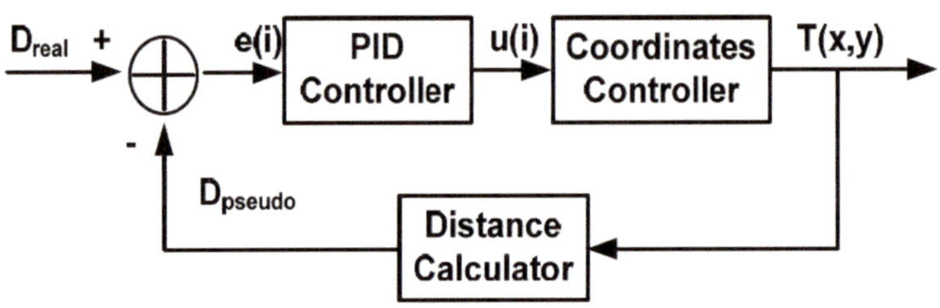

Fig. 3. A PID calculation block to obtain the target node coordinates

pseudo distance ($D_{pseudo1}$, $D_{pseudo2}$, and $D_{pseudo3}$). Using the PID algorithm, the pseudo target node converges to a real target node to control for distance error between the two positions. To utilize the PID algorithm in the coordinate calculation, the PID controller block consists of three blocks to calculate the coordinates, as shown in Figure 3.
First, the PID calculator obtains e by subtracting D_{pseudo}, from D_{real}, as shown in equation 10.

$$e(i) = D_{pseudo}(i) - D_{real}(i), \quad i = \{1, 2, \ldots, N\} \tag{10}$$

where i is the index of the reference nodes. Next, the PID controller block checks the error between the actual distances and the pseudo distances, and the PID controller calculates u using the input e value from each of the reference nodes.

$$u(i) = K_P e(i) + K_I \int e(i)dt + K_D \frac{de}{dt} \tag{11}$$

The coordinate controller adjusts the pseudo target node coordinates to be closer to the real target node. These processes are explained by Figure 4 and in the following equations.
The u values change the coordinate value to compensate for the position of the pseudo target node. It is assumed that the u values have the same vector as the D_{pseudo} value. The

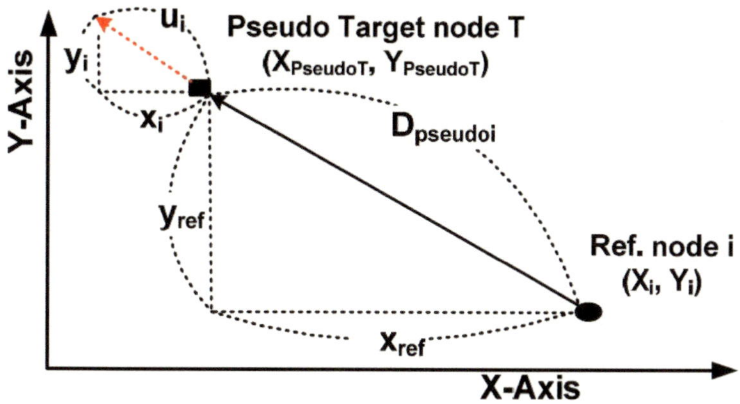

Fig. 4. A coordinate calculation algorithm to obtain the u value from the pseudo target node coordinates

compensation coordinates values ($x_{com}(i)$, $y_{com}(i)$) then have a proportional expression, as shown in equation 12.

$$\begin{aligned} x_{com}(i) : x_{ref}(i) &= u(i) : D_{pseudo}(i) \\ y_{com}(i) : y_{ref}(i) &= u(i) : D_{pseudo}(i) \end{aligned} \tag{12}$$

Here $x_{ref}(i)$ and $y_{ref}(i)$ are calculated as $X_i - X_{pseudoT}$ and $Y_i - Y_{pseudoT}$. The expressions are changed as in the following equation.

$$\begin{aligned} x_{com}(i) &= \frac{u(i)}{D_{pseudo}(i)} \times x_{ref}(i) \\ y_{com}(i) &= \frac{u(i)}{D_{pseudo}(i)} \times y_{ref}(i) \end{aligned} \tag{13}$$

The new coordinates of the pseudo target node are obtained by using the last pseudo target node coordinates($X_{pseudoT}$, $Y_{pseudoT}$), ($x_{com}(i)$ and $y_{com}(i)$).

$$\begin{aligned} X_{pseudoT}^{(n+1)} &= X_{pseudoT}^{(n)} - \sum_{i=1}^{N_{ref}} x_{com}(i) \\ Y_{pseudoT}^{(n+1)} &= Y_{pseudoT}^{(n)} - \sum_{i=1}^{N_{ref}} y_{com}(i) \end{aligned} \tag{14}$$

where N_{ref} is the number of references.
Lastly, D_{pseudo}, which is the feedback value for the next calculation, is obtained using a distance calculator. This is shown in Equation (11). The pseudo target coordinates converge to the real target node after the iteration process.

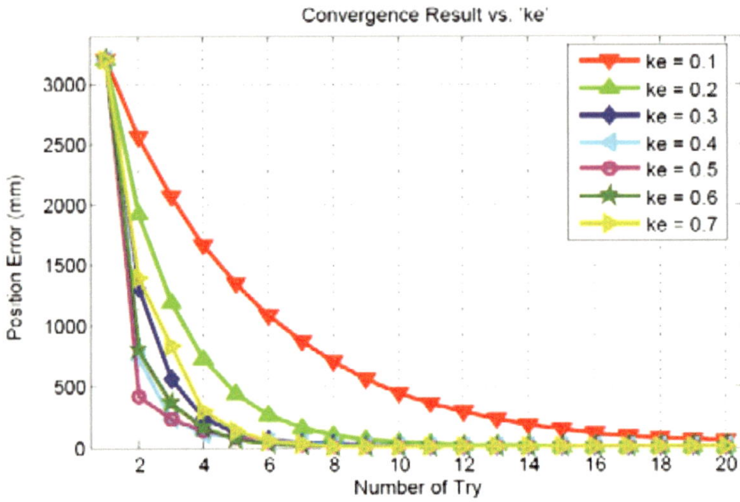

Fig. 5. The position error transition of the pseudo node using the PID method for each value of Kp

$$D_{pseudoT}^{(n+1)}(i) = \sqrt{(X_{pseudoT}^{(n+1)} - X_i)^2 + (Y_{pseudoT}^{(n+1)} - Y_i)^2} \tag{15}$$

The PID calculation method offers the advantage of continuous calculation because, in contrast to the DC method, its calculation time is reduced by using information from past positions. If the target node coordinates are calculated, the processing to calculate the next position is further reduced. Furthermore, the PID calculation method does not require a complex floating point operation. The PID calculation method is useful in embedded systems which use compact micro-processors and specific applications, such as real-time or continuous locating systems. Furthermore, this algorithm reduces the calculation process when the number of reference nodes is increased.

3.2 Simulation and result

In this section, the performances of the DC method and the proposed PID calculation algorithm are compared. The simulation environment is three reference nodes and one target node. The reference nodes are located at [100, 100], [100, 5000] and [5000, 100] (mm), respectively, and the target node is located at [3000, 3000] initially. The ranging result is set to have an error of 0.01 0.1%. The PID algorithm's parameter is set to P-control ($K_P = k_E$, $K_I = 0$, $K_D = 0$).

First, the PID calculation process will be explained. Figure 5 shows the position error of the pseudo target node versus the number of try times for each k_P parameter (k_E). The pseudo target node converges to the target node, where the pseudo target node is located at the coordinates of [2000, 2000] arbitrarily. As a result, the pseudo target node converges to the target node after eight iterations with a high accuracy of less than ten centimeters, except when k_P is 0.1. The parameter of k_P (k_E) has an optimal tuning value when set to 0.5.

Second, the calculation time of the DC method and that of the PID method are compared by counting the number of adders and multipliers when the target node has mobility. The mobility of a target node is generated by a normal distribution random model, and the

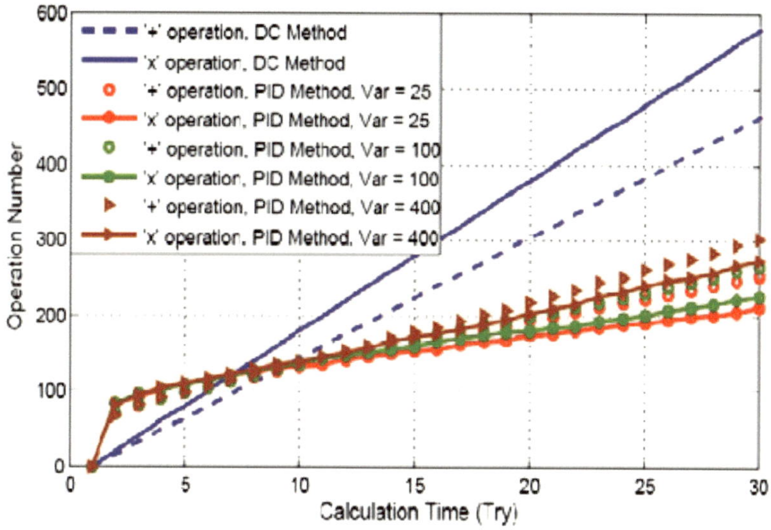

Fig. 6. Accumulated operation number result of the DC method and the PID calculation method when target node has mobility

ranging error is set to 0.01%. The DC method requires 16 adders and 20 multipliers for every calculation time. In contrast, the PID method requires 15 adders and 21 multipliers for each iteration time and 3 adders when the target position does not change. The results show the accumulated calculation time for each method. The calculation time of the conventional method increase linearly. This indicates that the DC method is not related to the mobility of the target node. In contrast, the PID method requires many operators compared to the DC method at the transition region to obtain the target node position. After convergence, the accumulated number of operators for the PID method shows fewer calculation operators compared to the calculation time of the DC method.

Third, the performance of the conventional DC method is compared with that of the proposed PID calculation method in Figure 6. The simulation parameters are the same as in the previous iteration, and the ranging errors are set from 0.01% to 0.1%. As shown in the result, when the ranging errors are small, performance enhancements are small. However, the PID method performs better than the conventional method in cases of a high ranging error because the DC method must use truncated data to operate in a small micro-processor system. In addition, the PID calculation method has better performance when using four reference nodes, but it does not require a complex calculation process.

4. Conclusion

This chapter describes an application of PID algorithm which is a coordinate calculation method in locating systems. The conventional DC method occur a delay of calculation time and gives inaccurate results when used in embedded systems or location system of the moving target. Therefore, a novel method using a PID control algorithm is proposed here. we used a P-control algorithm in this simulation. Therefore we show that this system will require less calculation and shows robust performance when using RTLS applications such as embedded locating systems, home networking systems and robotics positioning systems. If more complex PID control algorithms, such as PI, PD, and PID-control, are used, location

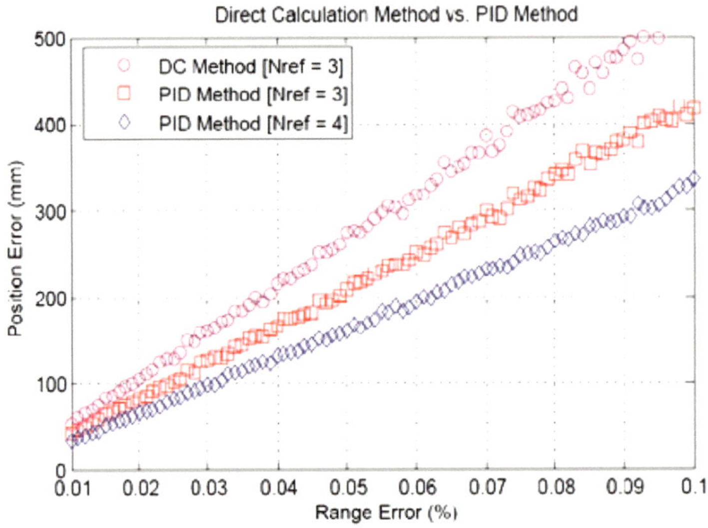

Fig. 7. Position error of the DC method and the PID calculation method when the ranging result has error

accuracy will be enhanced more. furthermore this concept can be also used other location algorithm such as the TDoA method.

5. References

I. Getting. (1993), The Global Positioning System, IEEE Spectrum, Vol.30, Iss.12, Dec. 1993, pp.36-47.

Y. Zhao. (2000), Mobile phone location determination and its impact on intelligent transportation systems, IEEE Transactions on Intelligent Transportation Systems, Vol., Iss.1, Mar. 2000, pp.55-64.

Kegen Yu and Oppermann, I. (2004),UWB positioning for wireless embedded networks, Radio and Wireless Conference, p 459- 462, 19-22 Sept. 2004,

Jourdan D.B. and Dardari, D., Win, M.Z. (2006),Position Error Bound for UWB Localization in Dense Cluttered Environments, IEEE International Conference, p. 3705-3710, June 2006

Panagopoulos H. Astrom, K.J. and Hagglund T. (2002), Design of PID controllers based on constrained optimisation, Control Theory and Applications, IEE Proceedings, p. 32-40, Jan 2002

Fan Yanfie Ren Fengyuan and Lin Chuang. (2003), Design a PID controller for active queue management, Computers and Communication, p 985- 990 vol.2, 30 June-3 July 2003

Jae Ho Hwang, Jae Moung Kim. (2010), ToA Coordinate Calculation Method Using a PID Algorithm, IEICE Transactions on Communications p 1057- 1060 vol.E93-B, April 2010

14

PID Controller Using FPGA Technology

Abdesselem Trimeche, Anis Sakly,
Abdelatif Mtibaa and Mohamed Benrejeb
Ecole Nationale d'Ingénieurs de Monastir-Département de Génie Electrique, Laboratoire d'Electronique et de microélectronique - Université de Monastir, Unité de recherche LARA Automatique de l'Ecole Nationale d'Ingénieurs de Tunis, Tunisia

1. Introduction

Since the Years 60, the law of Moore predicts that the complexity in terms of built-in circuit transistors doubles every two years, remain verified. The programmable FPGA circuits (Field Programmable Gate Array) didn't escape to this law. Since the first FPGA, developed like an evolution natural of the CPLD (Complex Programmable Logic Devices), these circuits didn't stop winning in complexity and integrated henceforth until one billion of transistors for the most recent generations. This increase of the integration level resulted in a similar growth of the power of calculation of these circuits. The FPGAs have been used then to make the fast samples of ASICs (Application Specific Integrated Circuits) and find since some years their place in many domains of applications. However, the order of the processes industrial requires more and more elements of powerful calculations. This type of order is in the same way in perpetual evolution with the development of the numeric circuits of calculation. Thus, the PID controllers represent the majority of the controllers used in the industrial systems control. Of this fact, it will be necessary to digitalize the PID algorithm. The modern digital control systems require more and more strong and fastest calculation components. This type of elements becomes yet indispensable with the utilization of some new control algorithms like the fuzzy control, the adaptive control, the sliding mode control… **[1].** Although the PID controllers are the oldest they represent the most used controllers in the industrial control systems

2. Discrete PID equation

The PID algorithm consists of three basic modes, the Proportional mode, the Integral and the Derivative modes. When utilizing this algorithm it is necessary to decide which modes are to be used (P, I or D) and then specify the parameters (or settings) for each mode used. Generally, three basic algorithms are used P, PI or PID.
The implementation of PID controllers using microprocessors and DSP chips is old and well known [2] [3], whereas very little works can be found in the literature on how to implement PID controllers using FPGAs [4].
Field Programmable Gate Arrays (FPGA) have become an alternative solution for the realization of digital control systems, previously dominated by the general purpose microprocessor systems.

In our work we introduce a simple method for implementing PID controllers. Some other contributions focused on proposing algorithms for tuning the coefficients of PID controllers using FPGAs while the controller itself is still implemented in software.
In [5] the authors describe the architecture of a data acquisition system for a gamma ray imaging camera.
In the past two years, Spartan II and III FPGA families from Xilinx have been successfully utilized in a variety of applications which include inverters [6][7], communications [8][9], imbedded processors [10], and image processing [11].
In our work, data acquisition for the PID controller, which is implemented using Xilinx Spartan-3 Starter Kit Board, is based on 8-bitserial A/D converters extensible from Digilent board AO1. Similar converters are utilized in [12] to implement an adaptable strain gage conditioner using FPGAs.
The application of a PID controller in a feedback control system is shown in fig1, where *ref* is the set point signal, *y* is the feedback signal, *e* is the error signal, and *u* is the control input.

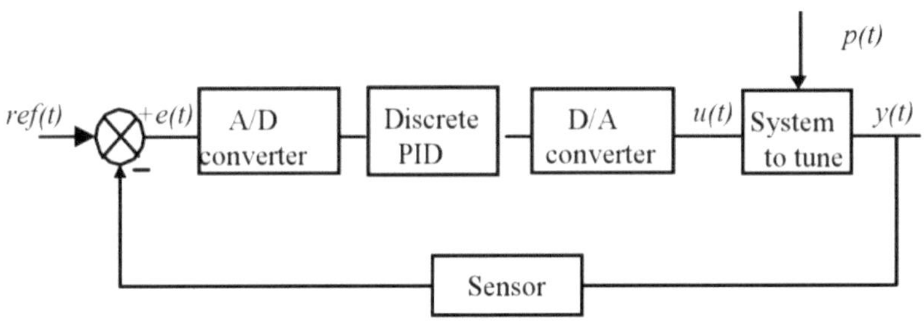

Fig. 1. A PID-based feedback control system.

The simplest form of the PID control algorithm is given by:

$$u(t) = k_p \left(e(t) + \frac{1}{T_i} \int e(t)dt + T_d \frac{de(t)}{dt} \right) \tag{1}$$

According to the study done in [13] the digitized PID equation is brought back to:

$$u_k = u_{k-1} + b_0 \cdot e_k + b_1 \cdot e(k-1) + b_2 \cdot e(k-2) \tag{2}$$

Where the coefficients b_0, b_1, and b_2 are evaluated by the expressions:

$$\begin{aligned} b_0 &= k_p \cdot \left(1 + \frac{T_d}{T} \right); \\ b_1 &= k_p \cdot \left(-1 + \frac{T}{T_i} - 2 \cdot \frac{T_d}{T} \right); \\ b_2 &= k_p \cdot \frac{T_d}{T} \end{aligned} \tag{3}$$

The Kp, Ti and Td, are PID parameters for tuning, and T is the sampling period in seconds.

3. Digital PID architecture

To improve the speed and minimize the cost while offering clearly good performances, the adopted architecture used includes essentially three combinational logic multiplier, one substractor three adders and three registers. The fig 2 gives the adopted architecture.
Indeed, this architecture requires the availability of all calculation operators in each phase.

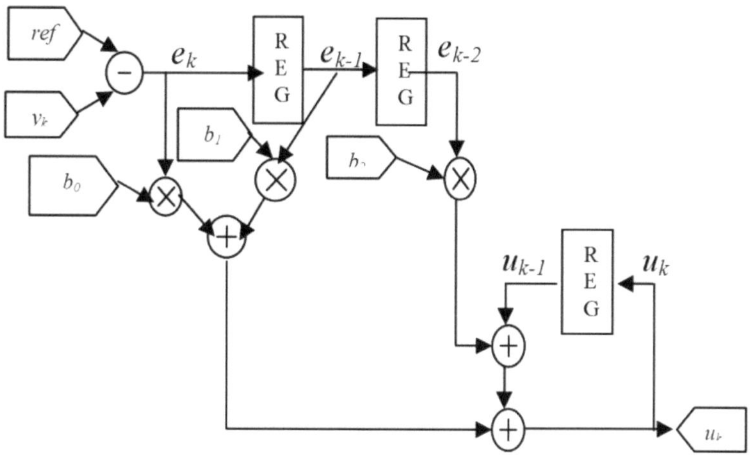

Fig. 2. PID Architecture.

3.1 Conversion blocks presentation

The AIO1 board is a peripheral board designed to work with Digilent's family of system boards. The AIO1 contains analog to-digital and digital-to-analog converters from Analog Devices, two dual op amps, a variety of analog signal I /O connectors, and a solderless breadboard. All analog components use an on-board 5VDC voltage source. All unused I/O signals are passed through the AIO1 board so that it can be used between a system board and other peripheral boards.
The AIO1 uses an 8-bit, 200Ks analog-to-digital converter (the AD7823), and an 8-bit, 1 MHz digital-to-analog converter (the AD7303), both from Analog Devices. The AD8534 op amps (also from Analog Devices) can drive 250mA outputs rail-to-rail with a 3 MHz bandwidth, so many useful devices can be driven directly.
Fig 3 describes all the components of this block.

3.2 Analog input interface

FPGAs are well suited for serial Analog to Digital (A/D) converters. This is mainly because serial interface consumes less communication lines while the FPGA is fast enough to accommodate the high speed serial data. The AD7823 is a high speed, low power, 8-bit A/D converter. The part contains a 4 µs typical successive approximation A/D converter and a high speed serial interface that interfaces easily to FPGAs. The A/D interface adapter (ADIA) is implemented within the FPGA (Figure 5). Inside the FPGA, this adapter facilitates parallel data acquisition. Sampling is initiated at the rising edge of a clock applied at the line sample. The timing diagram of the communication protocol is illustrated in figure 4. The whole conversion and acquisition period is 5.4 µs allowing sampling up to a rate of 185 Kilo Sample per second. This rate is more than sufficient for most PID control applications.

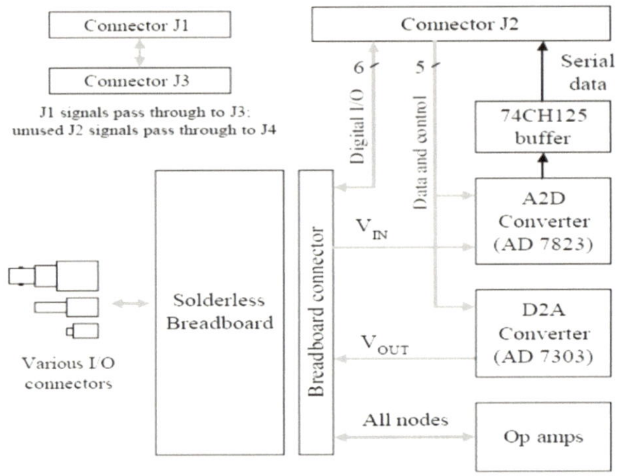

Fig. 3. Diagram of DIGILENT AIO1.

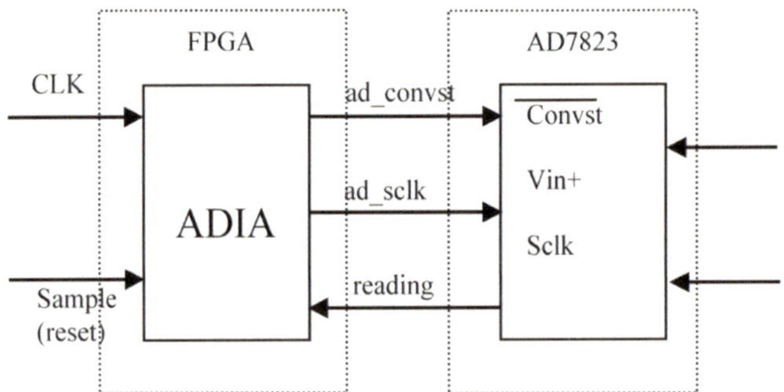

Fig. 4. A/D interface converter.

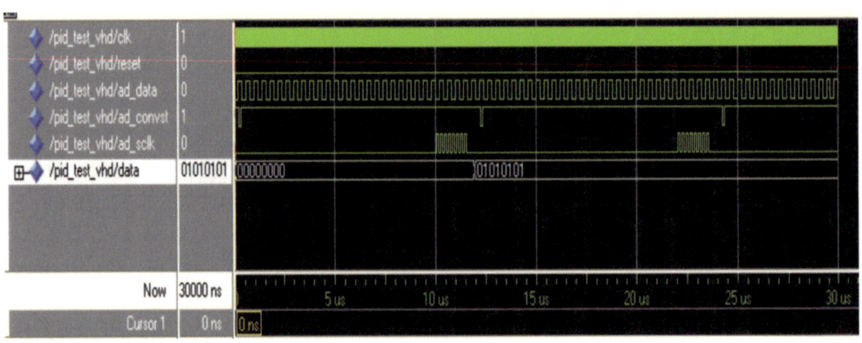

Fig. 5. AD7823 Timing Diagram.

The output coding of the AD7823 is straight binary. The designed code transitions occur at successive integer LSB values (i.e., 1 LSB, 2 LSBs, etc.). The LSB size is = VREF/256.

3.3 Analog output interface

The AD7303 is a dual, 8-bit voltage out Digital to Analog (D/A) converter. This device uses a versatile 3-wire serial interface that operates at a clock up to 30 MHz. The serial input register is 16 bits wide; 8 bits act as data bits for the D/A converter, and the remaining 8 bits make up a control register. It is interfaced to an FPGA as illustrated in Figure 6. The D/A interface adapter (DAIA), which is implemented within the FPGA, facilitates parallel data input for the dual D/A converters. The timing diagram of the communication protocol is illustrated in figure 7. The transmission period of a sample is 680 ns allowing D/A conversion at an excellent rate of 1.47 MHZ.

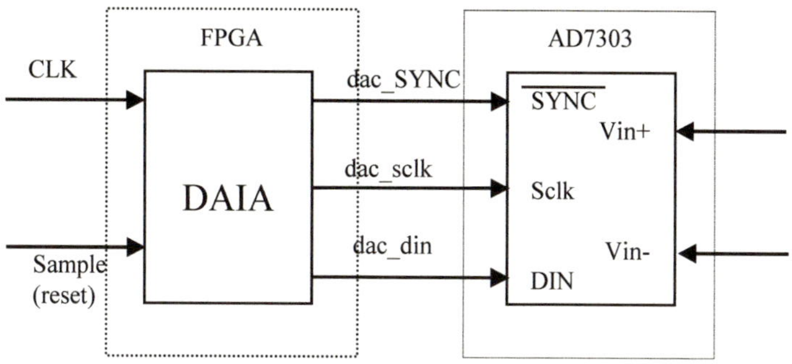

Fig. 6. A/D interface converter.

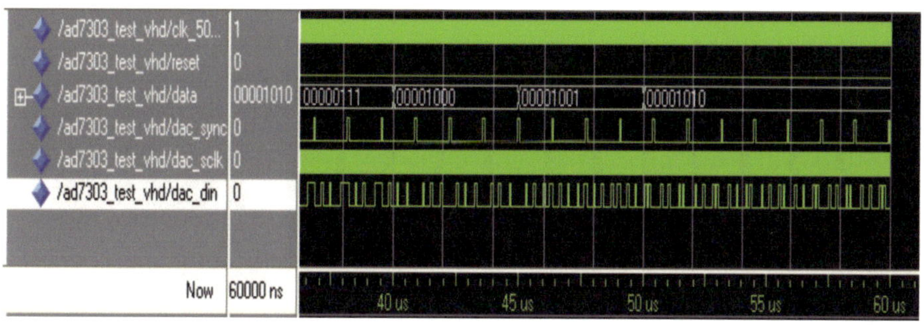

Fig. 7. AD7303 Timing Diagram.

Any DAC output voltage can ideally be expressed as:
VOUT = 2 × VREF × (N/256) where: N is the decimal equivalent of the binary input code. An N range from 0 to 255xVREF is the voltage applied to the external REF pin when the external reference is selected and is VDD/2 if the internal reference is used.

3.4 Implementation results

The proposed based PID controller is implemented using the Xilinx Inc FPGA technology and can be used as a general purpose controller for different applications. The simulation results obtained with the generated VHDL, in this work, the ModelSim® simulator was used. The circuits for the PID controllers have been obtained by logic synthesis and place and route using Xilinx ISE 7.1i, from the VHDL representation generated by the static analyzer. We use a Xilinx Spartan-3 xc3s200-ft256 -4 FPGA. The results presented herein are estimations directly obtained from Xilinx ISE 7.1i.

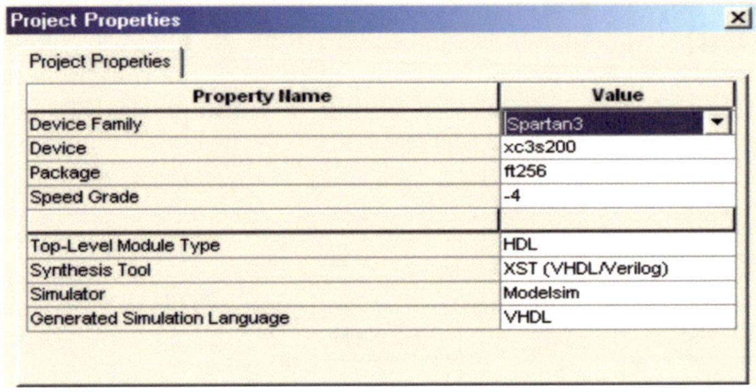

Fig. 8. Design properties.

Table 1 shows the minimum number of multiplications, additions and registers required for the PID controller without conversions block.

Multiplication	3
Addition	3
subtraction	1
register	3
total	10

Table 1. Arithmetic Number for PID controller.

The PID controller block, into a complete control system consisting of analog and digital I/O, is illustrated in figure 9.

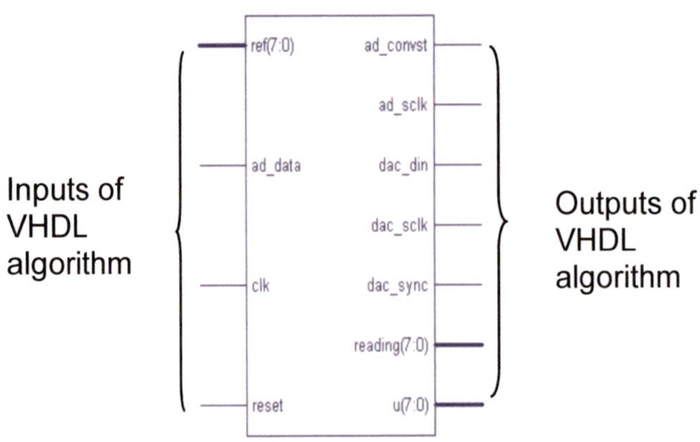

Fig. 9. PID Controller Block.

The simulation results adapted to this block is shown in figure 10.

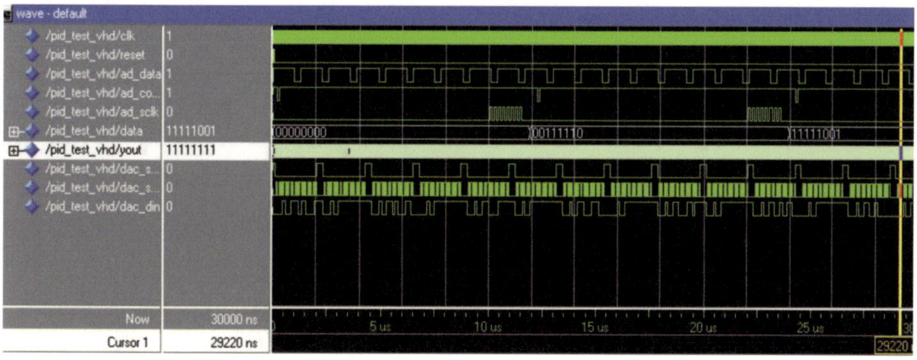

Fig. 10. Simulation diagram of PID controller block.

The synthesis of PID controller block using a Xilinx Spartan-3 xc3s200-ft256 -4 FPGA gives the following results.

Device Utilization Summary

Logic Utilization	Used	Available	Utilization	Note(s)
Number of Slice Flip Flops:	199	3,840	5%	
Number of 4 input LUTs:	539	3,840	14%	
Logic Distribution:				
Number of occupied Slices:	368	1,920	19%	
Number of Slices containing only related logic:	368	368	100%	
Number of Slices containing unrelated logic:	0	368	0%	
Total Number 4 input LUTs:	650	3,840	16%	
Number used as logic:	539			
Number used as a route-thru:	111			
Number of bonded IOBs:	24	173	13%	
Number of GCLKs:	1	8	12%	

Fig. 11. Devices utilizations summary for the PID controller Implementation.

A design which is efficient in terms of power consumption and chip area means that the FPGA chip can be used to accommodate more controllers with adequate speed and low power consumption, resulting in a cost reduction of the controller hardware.

4. Application of PID controller using FPGA technology to command a system of second order

4.1 System presentation

The figure 12 describe the system of second order.

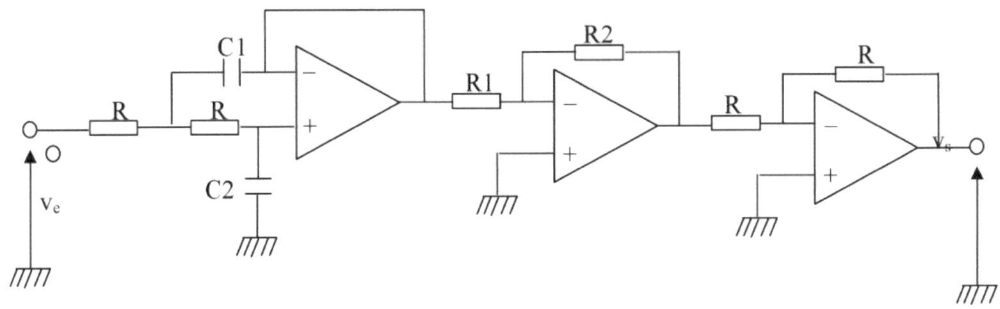

Fig. 12. Installation Diagram of the system of second order.

The values of all the components are:
$R = 12\ k$
$R_1 = 15\ k\Omega$
$R_2 = 10\ k\Omega$
$C_1 = 6{,}8\ nF$
$C_2 = 22\ nF$
The Transfer function of the system is given by the following equation

$$G(p) = \frac{V_s(p)}{V_e(p)} = \frac{K_s}{1+\frac{2m}{\omega_0}p+\left(\frac{1}{\omega_0}\right)^2 p^2} \quad (4)$$

With:
$K_s = \frac{R_2}{R_1} = 0{,}67$: static gain
$\omega_0 = \frac{1}{R\sqrt{C_1 \cdot C_2}} = 6{,}81 \cdot 10^3\ rd/s$
$m = \sqrt{\frac{C_2}{C_1}} = 1{,}80 > 1$: Amortization factor
While permuting the positions of C1 and C2, the amortization factor becomes:

$$m = \sqrt{\frac{C_1}{C_2}} = 0{,}56 < 1$$

4.2 Experimental result

At the time of the order of the system two cases appear; when the amortization is m>1 and m <1.

4.2.1 Experimental results for an amortization m>1

4.2.1.1 System answer results in open buckle

For an order (ref) of the order of 2V applied to the system in BO one gets the answer y (t) presented on the following figure.

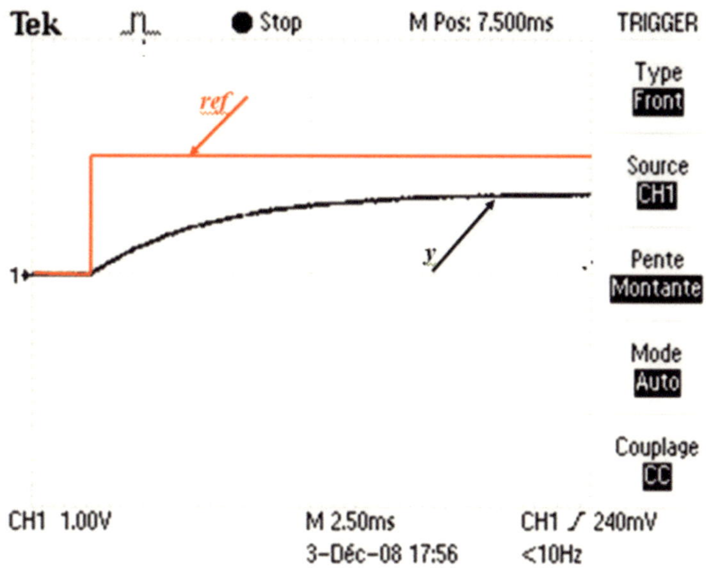

Fig. 13. System Answer in open buckle ($m > 1$).

According to this answer, one verifies that the tension of exit stabilizes without oscillations (m>1) nearly to the value 1,4V, that corresponds more or less to the theoretical value:

$$V_s = K_s \cdot V_e = 0{,}67 * 2 = 1{,}34\text{V}.$$

4.2.1.2 System answer results with P regulator

For an order (ref) of the order of 2V applied to the system ordered by a Proportional regulator (P) with KP = 2, one gets the answer y(t) presented on the following Figure

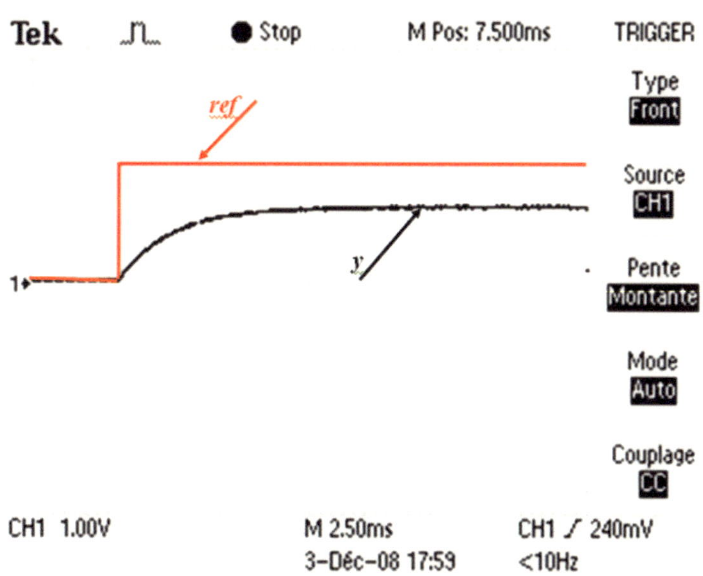

Fig. 14. Answer of the system ordered by proportional regulator P.

One notices that this answer presents a static mistake of the order of 40%. Theoretically this mistake is given by:

$$\frac{1}{1+K_p \cdot K_s} \cdot 100\% = 42\%.$$

4.2.1.3 System answer results with PI regulator

For an order (ref) applied of the order of 2V to the system ordered by a PI regulator with KP=2, KI=0.5, one gets the answer y(t) presented in the following figure:

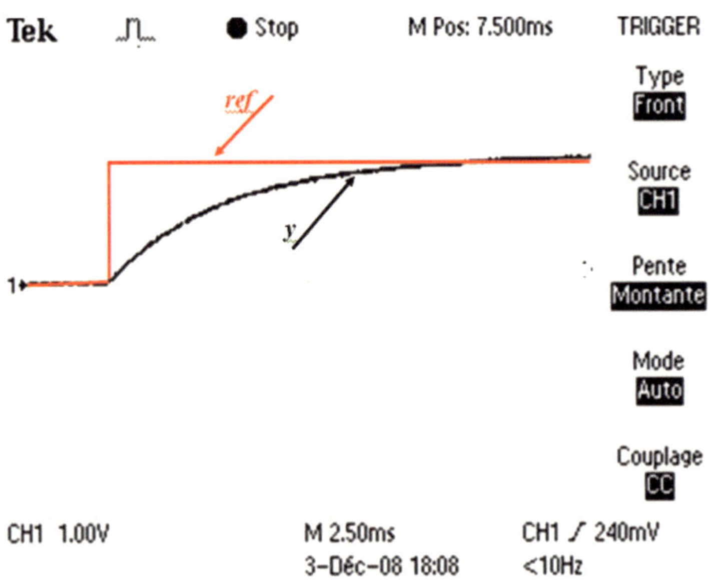

Fig. 15. Answer of the system ordered by PI regulator $(m > 1)$.

One notices the annulment of the static mistake well thanks to the introduction of the I action.

4.2.2 Experimental results for an amortization m<1

4.2.2.1 System answer results in open buckle

For an order (ref) of the order of 2V applied to the system in BO one gets the answer y (t) presented on the following figure

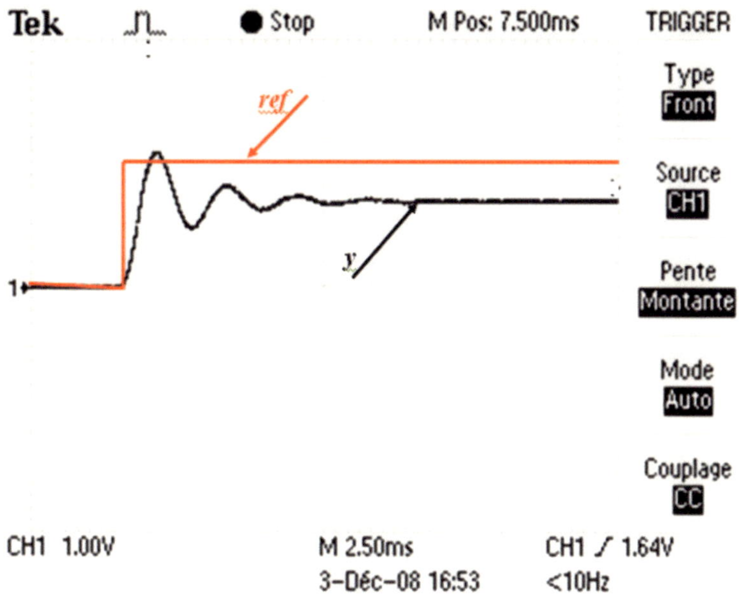

Fig. 16. Answer of the system in open buckle ($m < 1$).

According to this answer, one verifies that the tension of exit stabilizes with oscillations (m <1) nearly to the value 1,4V.

4.2.2.2 System answer results with P regulator

For an order (ref) of the order of 2V applied to the system ordered by a Proportional regulator (P) with KP = 2, one gets the answer y (t) presented on the following Figure

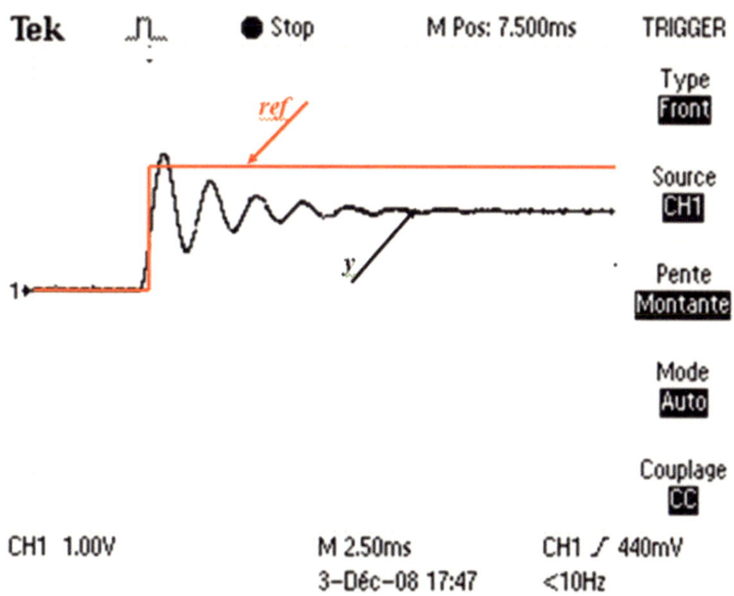

Fig. 17. Answer of the system ordered by proportional regulator P.

One notices that this answer presents a static mistake of the order of 40%. Theoretically this mistake is given by:

$$\frac{1}{1+K_p \cdot K_s} \cdot 100\% = 42\%.$$

4.2.2.3 System answer results with PD regulator

For an order (ref) of the order of 2V applied to the system ordered by a PD regulator with KP = 2,KD=1 one gets the answer y (t) presented on the following Figure

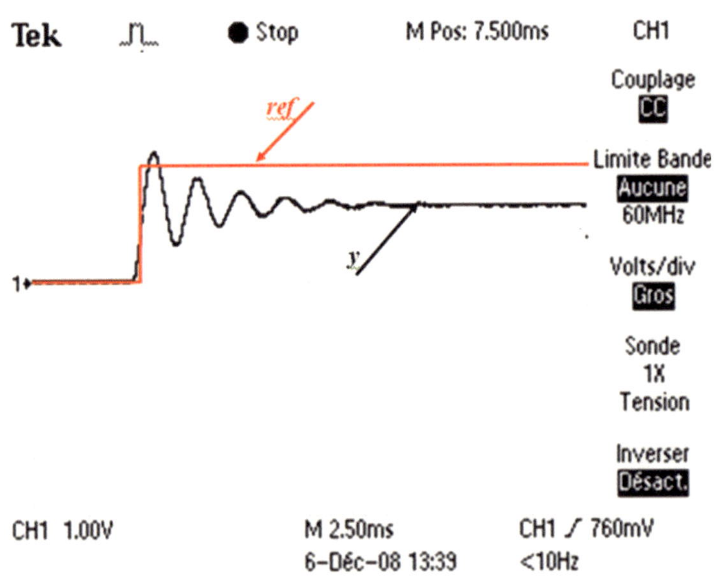

Fig. 18. Answer of the system ordered by proportional regulator P D.

$$(m < 1)$$

One notices that the answer gotten present less oscillations that the one with the regulating P thanks to the Derivative action.

4.2.2.4 System answer results with PI regulator

For an order (ref) of the order of 2V applied to the system ordered by a PI regulator with KP=2, KI=0,5 one gets the answer y (t) presented on the following Figure.

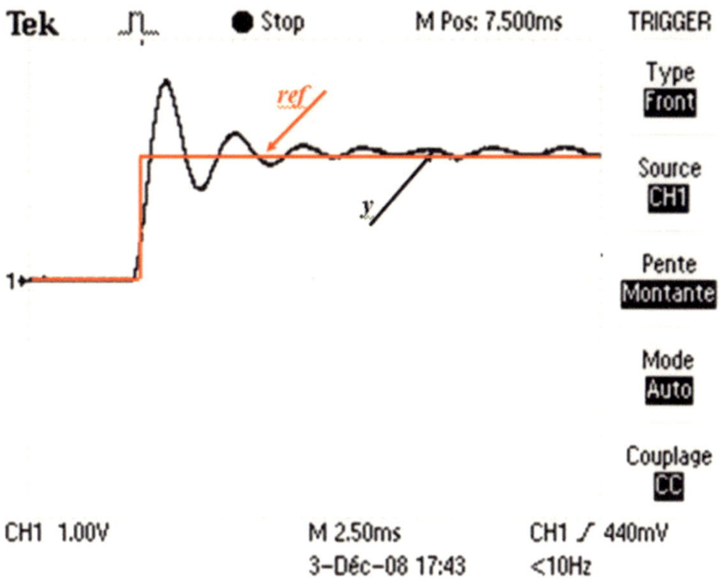

Fig. 19. Answer of the system ordered by proportional regulator P I.

$$(m < 1)$$

one notices the annulment of the static mistake well thanks to the introduction of the I action

4.2.2.5 System answer results with PID regulator

For an order (ref) of the order of 2V applied to the system ordered by a PI D regulator with KP = 2, KD=1 and KI=0, 5 one gets the answer y (t) presented on the following Figure

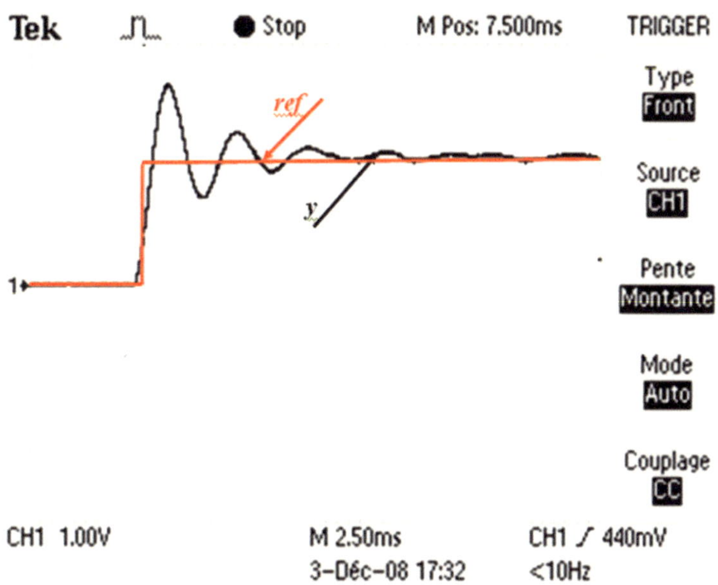

Fig. 20. Answer of the system ordered by proportional regulator P ID.

$$(m < 1)$$

One notices that with the addition of the Derivative action, one has a light reduction of the oscillations in relation to the answer gotten by a regulating PI.

4.3 Tentative evaluation of sampling period

While following the evolution of the order u (t) one could estimate the value of the sampling period experimentally (T) as it's indicated in the following figure

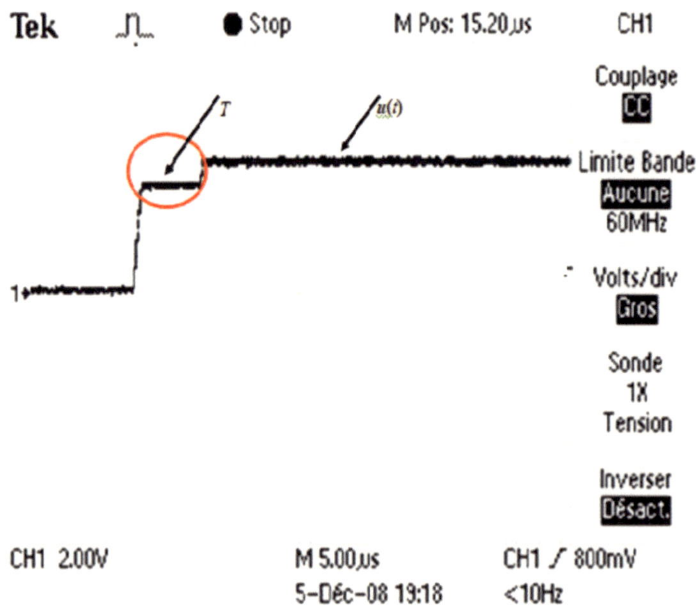

Fig. 21. Tentative evaluation of the sampling period (T).

According to the figure 21, one estimates the value of the sampling period (T) that is the order of 6,7µs.

5. Conclusion

A digital PID controller implemented in FPGA technology is a configurable controller in terms of latency, resolution, and parallelism.

The speed or execution or latency of the controller can be precisely controlled with the amount of reuse of arithmetic elements such as the speed of execution of FPGA based PID controller can be less then 100 ns if desired for high throughput requirements.
Implementing PID controllers on FPGAs features speed, accuracy, power, compactness, and cost improvement over other digital implementation techniques.
In a future fork we plan to investigate implementation of fuzzy logic controllers on FPGAs. Also we plan to explore embedded soft processors, such as MicroBlaze, and study some applications in which design partitioning between software and hardware provides better implementations.

6. References

[1] L. Samet, N. Masmoudi, M.W. Kharrat, L. Kamoun: A Digital Pid Controller for Real Time and Multi Loop Control, 5ème Colloque d'Informatique Industrielle CII'98 8,9 et 10 février 1998,Djerba Tunisie
[2] H. D. Maheshappa, R. D. Samuel, A. Prakashan, "Digital PID controller for speed control of DC motors", IETE Technical Review Journal, V6, N3, PP171-176, India 1989
[3] J. Tang, "PID controller using the TMS320C31 DSK with on-line parameter adjustment for real-time DC motor speed and position control", IEEE International Symposium on Industrial Electronics, V2, PP 786-791, Pusan 2001.
[4] Mohamed Abdelati, the Islamic University of Gaza, Gaza, Palestine:" FPGA-Based PID Controller Implementation".
[5] K. Nurdan, T. Conka-Nurdana, H. J. Beschc, B. Freislebenb, N. A. Pavelc, A. H.Walentac, "FPGA-based data acquisition system for a Compton camera", Proceedings of the 2nd International Symposium on Applications of Particle Detectors in Medicine, Biology and Astrophysics, V510, N1, PP. 122-125, Sep 2003.
[6] R. Jastrzebski, A. Napieralski,O. Pyrhonen, H. Saren, "Implementation and simulation of fast inverter control algorithms with the use of FPGA circuit", 2003 Nanotechnology Conference and Trade Show, pp 238-241, Nanotech 2003.
[7] Lin, F.S.; Chen, J.F.; Liang, T.J.; Lin, R.L.; Kuo, Y.C. "Design and implementation of FPGA-based single stage photovoltaic energy conversion system", Proceedings of IEEE Asia Pacific Conference on Circuits and Systems, pp 745-748, Taiwan, Dec. 2004.
[8] Bouzid Aliane and Aladin Sabanovic, "Design and implementation of digital band pass FIR filter in FPGA", Computers in Education Journal, v14, p 76-81, 2004.
[9] M. Canet, F. Vicedo,V. Almenar, J. Valls, "FPGA implementation of an IF transceiver for OFDM-basedWLAN", IEEEWorkshop on Signal Processing Systems, SiPS: Design and Implementation, PP 227-232, USA 2004.
[10] Xizhi Li, Tiecai Li, "ECOMIPS: An economic MIPS CPU design on FPGA", Proceedings - 4th IEEE International Workshop on System-on-Chip for Real-Time Applications, PP 291-294, Canada 2004.
[11] R. Gao, D. Xu,J. P. Bentley, "Reconfigurable hardware implementation of an improved parallel architecture for MPEG-4 motion estimation in mobile applications",IEEE Transactions on Consumer Electronics, V49, N4, November 2003.
[12] S. Poussier, H. Rabah, S. Weber, "Smart Adaptable Strain Gage Conditioner: Hardware/Software Implementation", IEEE Sensors Journal, V4, N2, April 2004.

[13] L. Samet, "Etude de l'intégration électronique en technologie FPGA d'un algorithme de contrôle de processus: le PID" Thèse Docteur Ingénieur, ENIS-TUNISIE, décembre 1996